Publication No. FHWA-IF-12-027
February 2012

US DEPARTMENT OF TRANSPORTATION
FEDERAL HIGHWAY ADMINISTRATION

MANUAL FOR DESIGN, CONSTRUCTION, AND MAINTENANCE OF ORTHOTROPIC STEEL DECK BRIDGES

SI* (MODERN METRIC) CONVERSION FACTORS

APPROXIMATE CONVERSIONS TO SI UNITS

Symbol	When You Know	Multiply By	To Find	Symbol
		LENGTH		
in	inches	25.4	millimeters	mm
ft	feet	0.305	meters	m
yd	yards	0.914	meters	m
mi	miles	1.61	kilometers	km
		AREA		
in^2	square inches	645.2	square millimeters	mm^2
ft^2	square feet	0.093	square meters	m^2
yd^2	square yard	0.836	square meters	m^2
ac	acres	0.405	hectares	ha
mi^2	square miles	2.59	square kilometers	km^2
		VOLUME		
fl oz	fluid ounces	29.57	milliliters	mL
gal	gallons	3.785	liters	L
ft^3	cubic feet	0.028	cubic meters	m^3
yd^3	cubic yards	0.765	cubic meters	m^3
		NOTE: volumes greater than 1000 L shall be shown in m^3		
		MASS		
oz	ounces	28.35	grams	g
lb	pounds	0.454	kilograms	kg
T	short tons (2000 lb)	0.907	megagrams (or "metric ton")	Mg (or "t")
		TEMPERATURE (exact degrees)		
°F	Fahrenheit	5 (F-32)/9 or (F-32)/1 8	Celsius	°C
		ILLUMINATION		
fc	foot-candles	10.76	lux	lx
fl	foot-Lamberts	3.426	candela/m^2	cd/m^2
		FORCE and PRESSURE or STRESS		
lbf	poundforce	4.45	newtons	N
lbf/in^2	poundforce per square inch	6.89	kilopascals	kPa

APPROXIMATE CONVERSIONS FROM SI UNITS

Symbol	When You Know	Multiply By	To Find	Symbol
		LENGTH		
mm	millimeters	0.039	inches	in
m	meters	3.28	feet	ft
m	meters	1.09	yards	yd
km	kilometers	0.621	miles	mi
		AREA		
mm^2	square millimeters	0.0016	square inches	in^2
m^2	square meters	10.764	square feet	ft^2
m^2	square meters	1.195	square yards	yd^2
ha	hectares	2.47	acres	ac
km^2	square kilometers	0.386	square miles	mi^2
		VOLUME		
mL	milliliters	0.034	fluid ounces	fl oz
L	liters	0.264	gallons	gal
m^3	cubic meters	35.314	cubic feet	ft^3
m^3	cubic meters	1.307	cubic yards	yd^3
		MASS		
g	grams	0.035	ounces	oz
kg	kilograms	2.202	pounds	lb
Mg (or "t")	megagrams (or "metric ton")	1.103	short tons (2000 lb)	T
		TEMPERATURE (exact degrees)		
°C	Celsius	1.8C+32	Fahrenheit	°F
		ILLUMINATION		
lx	lux	0.0929	foot-candles	fc
cd/m^2	candela/m^2	0.2919	foot-Lamberts	fl
		FORCE and PRESSURE or STRESS		
N	newtons	0.225	poundforce	lbf
kPa	kilopascals	0.145	poundforce per square inch	lbf/in^2

*SI is the symbol for the International System of Units. Appropriate rounding should be made to comply with Section 4 of ASTM E380.
(Revised March 2003)

TABLE OF CONTENTS

LIST OF FIGURES

LIST OF TABLES

GLOSSARY

Battle Deck: term first used in the United States to describe an orthotropic steel deck.

Blow-through: excessive, undesirable penetration of the weld application leading to splatter and gaps in the weld root and welded surfaces (see melt-through).

Bulkhead: in orthotropic steel decks, an internal diaphragm placed in closed ribs so as to make a continuous connection of the floorbeam web through the rib. It is intended to dissipate the effects of discontinuous shear forces and distortions local to the rib. Use of the bulkhead detail is not recommended except in specific situations where other alternatives to alleviate stresses are not possible.

Constant Amplitude Fatigue Limit: the constant amplitude stress range under which no crack growth will occur for a particular fatigue detail.

Cut-Out: for orthotropic steel decks, the cut out is a stress-relieving cut made in the floorbeam (diaphragm) web to alleviate the out-of-plane stresses induced by the longitudinal rotations of the rib due to applied loads on the deck and/or to avoid welding to the bottom of the rib where longitudinal stresses are highest.

Crossbeam: alternate name for floorbeam (see Floorbeam).

Deck Plate: the top plate of an orthotropic deck that supports the wearing surface and directly supports the wheel loads. The deck plate is stiffened by longitudinal ribs and transverse floorbeams (diaphragms) on the underside.

Delamination: a separation of the internal layers of a material, shear is no longer transferred though the adjacent layers.

Diaphragm: for orthotropic steel decks, a diaphragm is a transverse component similar to a floorbeam but is typically characterized by not having a bottom flange or being seated atop a sub-floorbeam in the primary bridge framing (see Floorbeam).

Fatigue: the initiation and/or propagation of cracks due to a repeated variation of normal stress with a tensile component.

Fatigue Threshold: see constant amplitude fatigue limit.

Filled Steel Grid Deck: a deck composed of a tightly spaced steel grid, filled with cementitious material to form a riding surface.

Floorbeam: for orthotropic steel decks, a floorbeam is a transverse component which provides support to the ribs and transfers loads to primary girders. Also referred to as a crossbeam, an **intermediate**floorbeam is generally smaller and does not necessarily tie in to a main structural member.

Girder: a main load carrying member that runs longitudinally with the orthotropic deck ribs and the bridge. In orthotropic decks, girders are composite with the deck plate and other components of the orthotropic system.

Intermediate Floorbeam: see Floorbeam

Level 1 Design: in orthotropic decks, design verification by little or no structural analysis, but by selection of details that are verified to have adequate resistance by experimental testing (new or previous).

Level 2 Design: in orthotropic decks, design verification by simplified one-dimensional or two-dimensional analysis of certain panel details where such analysis is sufficiently accurate or for certain details that are similar to previous tested details as described in Level 1.

Level 3 Design: in orthotropic decks, design verification by refined three-dimensional analysis of the panel to quantify the local stresses to the most accurate extent reasonably expected from a qualified design engineer experienced in refined analysis.

Local Structural Stress: the surface stress at a welded detail including all stress raising effects of a structural detail excluding all stress concentrations due to the local weld profile itself.

Melt-through: in orthotropic deck welding, condition where additional weld material penetrates, especially at the back side of the rib to deck weld, and forms additional reinforcing on the opposite side of the weld application.

Orthotropic: derivation of the word comes from two terms. The system of ribs and floorbeams are **orthogonal** and their elastic properties are different or **anisotropic** with respect to the deck: thus **orthogonal-anisotropic** becomes orthotropic.

Orthotropic Steel Deck: A system by which a deck plate is stiffened by longitudinal ribs and transverse floorbeams (diaphragms) directly supporting live loads.

Orthotropic Bridge: A bridge that incorporates orthotropic components (e.g. rib stiffened plates) in its construction. Orthotropic bridges do not necessarily have orthotropic decks; a bridge with an orthotropic deck would be considered an orthotropic bridge.

Redeck (Redecking): The rehabilitation of an existing bridge by removal and replacement of the existing deck with a new deck or deck system.

Refined Analysis: for orthotropic steel decks, evaluation of the local structural stress at fatigue prone details by a detailed three-dimensional shell or solid finite element structural model, including all plate components and connections.

Residual Stresses: Stresses that remain in an unloaded member after the initial cause of the stress is removed. The residual stresses of concern for orthotropic bridges are those resulting from cold-bending, welding, and fabrication (sometimes referred to as locked-in) stresses.

Rib: in orthotropic steel decks, longitudinal members used to stiffen a structural plate that can be open (e.g. angle or plate rib) or closed (e.g. U-shape or trapezoidal ribs).

Rib Span: the span length of a longitudinal rib member between supporting floorbeams.

Seal Plate: in orthotropic steel decks, a plate placed at the end of closed ribs to seal the rib from outside exposure, in particular moisture.

Stress Concentration: Stress at a structural detail that includes the effects of geometric discontinuities and considers the total local stress.

Tooth: in orthotropic steel decks, zone between rib cut-outs on the floorbeam (diaphragm) web

Wearing Surface: placed on the deck plate in order to provide a skid resistant surface with good ride quality, they also provide corrosion protection to the deck plate, level out deck plate irregularities, and last but not the least, potentially contribute to increased fatigue life of the deck plate resulting from reduction in stress levels in the steel plate

GLOSSARY OF ACRONYMS

Acronym	Definition
ADTT	average daily truck traffic
CAFL	constant amplitude fatigue limit (fatigue threshold)
CFRP	carbon fiber reinforced polymer
CJP	Complete Joint Penetration (weld)
CNC	computer numerically controlled
FCAW	Flux Core Arc Weld
FCM	Fracture Critical Member
FB	Floorbeam
FEA	Finite Element Analysis
GMAW	Gas Metal Arc Weld
GVW	Gross Vehicle Weight
LRFD	Load and Resistance Factor Design
LRFR	Load and Resistance Factor Rating
LSS	Local Structural Stress
MT	Magnetic Particle Testing
MPF	Multiple Presence Factor
MTR	Mill Test Report
NDE	Non-Destructive Evaluation
NDT	Non-Destructive Testing
NRL	Notional Rating Load
OSD	Orthotropic Steel Deck
PJP	Partial Joint Penetration (weld)
PQR	Procedure Qualified Record
RD	Rib-to-Deck Plate
RF	Rib-to-Floorbeam (intermediate Floorbeam, or Diaphragm)
RDF	Rib-to-Deck at the Floorbeam
RT	Radiographic Testing
SAW	Submerged Arc Weld
SCF	Stress concentration factor
SHV	Specialized Hauling Vehicle
SMAW	Shielded Metal Arc Weld
TL	Test Level (AASHTO testing criteria for barriers)
UT	Ultrasonic Testing
WPR	Welding Procedure Records

LIST OF SYMBOLS

Symbol	Definition
a	rib width at deck
a+e	center-to-center rib spacing
A	cross-sectional area
A_{eff}	summation of the effective areas of the cross-section based on the reduced effective widths
A_r	area enclosed by the closed rib;
b	rib width at base of rib
b_e	reduced effective width of the plate
b_{od}	effective width of Orthotropic Deck,
B	spacing between Orthotropic Deck girder web plates or transverse floorbeams,
B_1	is an amplification factor to account for additional moment caused by lateral displacements in the panel (P-δ),
C	fatigue detail constant
C_m	the equivalent moment factor
C_r	Capacity (load rating)
d	depth of cross member (e.g. floorbeam or diaphragm)
D_x	plate flexural rigidity in the x-direction
D_y	plate flexural rigidity in the y-direction
DC	Dead load effect due to structural components and utilities
DW	Dead load effect due to wearing surface and utilities
Δ_f	force effect of variable amplitude loads
e	clear space between ribs at deck
E	modulus of elasticity
f	applied stress
F_e	is the elastic critical buckling stress = $\pi^2 \mathbf{E}/(\mathbf{KL/r})^2$
F_y	yield stress of steel
ϕR	factored resistance (LRFD)
γ_i	appropriate load factor (LRFD)
h	height of rib
h'	length of rib along leg
h_{cutout}	height of the rib cut-out
H	effective torsional rigidity of a plate
IM	impact factor
φ	LRFD resistance factor
φ_c	LRFR system factor
φ_s	LRFR condition factor
K	is the effective length factor
K_s	stress concentration factor
L	span length of the Orthotropic Deck girder or Floorbeam
LL	live load effect
m	inverse of slope on the sloping portion of the fatigue S-N curve

LIST OF SYMBOLS (CONTINUED)

M	bending moment
n	cycles per truck passage
N	number of fatigue cycles
p(x,y)	the loading at any point on a plate with coordinates of (x, y)
P	Permanent load other than dead load (load rating)
Q	is the slender element local buckling reduction factor
Q_i	force effect (LRFD)
r	is the radius of gyration of the strut
RF	Rating Factor
S	splice spacing
S_r	fatigue stress range
σ	stress
σ_1	stress at first midpoint node location for extrapolation
σ_2	stress at second midpoint node location for extrapolation
σ_r	stress range
σ_{hs}	stress including concentrations
σ_{lss}	local structural stress
t	plate thickness
t_c	floorbeam web thickness
t_d	steel deck thickness
t_r	rib thickness
u	entire length of the closed rib plate;
v	Poisson's Ratio for steel (v = 0.3)
w	deflection of the middle surface of a plate
ψ	effective width ratio of a box girder flange

FOREWORD

The Manual for Design, Construction, and Maintenance of Orthotropic Steel Deck Bridges has been developed to supplement and modernize the 1963 Design Manual for Orthotropic Steel Plate Deck Bridges written by Roman Wolchuk and published by the American Institute of Steel Construction. Thousands of orthotropic deck bridges have been built around the world. The generous sharing of lessons learned from the performance by the bridge owners, further research in improving the serviceability, and fatigue performance, and the improvement in design, quality of fabrication and inspection, and maintenance of orthotropic bridge decks, have the potential to be one of the most desirable choice by bridge engineers. Good performance has been demonstrated in the laboratory and experienced in the field.

Orthotropic steel decks provide a modular, prefabricated design solution that has proven effective in new construction where speed and extended service life are desired, and in rehabilitation of existing bridges where weight is of primary concern. Orthotropic steel decks have other advantages, such as, low maintenance, suitability for standardization and prefabrication, support of accelerated bridge construction, reduced disruption to traffic during construction, improved work zone safety, and low life-cycle cost. With his extensive research and field experience, Dr. John Fisher of Lehigh University has expressed in many occasions that an orthotropic bridge deck system is most able to provide a 100-year service life.

This Manual, including the new AASHTO LRFD specifications, is the culmination of over four years of diligent effort by FHWA and the HDR Team in working together with the AASHTO Technical Committee T-14 Steel Bridges on a continual basis. The latest research and practice have been synthesized and numerous experts and practitioners, both domestic and international, have been consulted to develop the current state-of-the-knowledge criteria and guidance that will promote cost-effective and durable performance.

The feedback received from participants of FHWA organized workshops on the development of the manual, and the constructive review comments on the final draft of the manual from many engineering professionals are very much appreciated. The readers are encouraged to submit comments for future enhancements of the manual to Myint Lwin at the following address: Federal Highway Administration, 1200 New Jersey Avenue, S.E., Washington, DC 20590.

Myint Lwin

M. Myint Lwin, Director
Office of Bridge Technology

ACKNOWLEDGEMENTS

The authors would like to acknowledge the AASHTO Subcomittee on Bridges and Structures, Technical committee T-14 (Structural Steel Design) under the guidance of Ed Wasserman, as well as Chuck Seim, Roman Wolchuk, Justin Ocel, Dyai Wang, Lois Yeager, Dyab Khazem, Paul Tsokopoulos and Bill Wright for their contributions to the manual. Additionally, we would like to thank Al Mangus and the State of California, Hanna Cheng and the State of New Jersey, and Art Yannotti and the State of New York for answering questions and for their assistance with various details presented in the manual.

The authors also acknowledge the encouragement and guidance provided by Myint Lwin, Director, Raj Ailaney, the Contract Manager, and Brian Kozy, Senior Bridge Engineer of the Office of Bridge Technology throughout the development of the Manual.

1. INTRODUCTION

1.1. INTRODUCTION TO ORTHOTROPIC STEEL DECK BRIDGES

Many of the world's most magnificent modern bridge structures utilize the orthotropic steel plate systems as one of the basic structural building blocks for distribution of traffic loads in decks and for the stiffening of slender plate elements in compression. Examples include the new San Francisco Oakland Bay Bridge, Self Anchored Suspension Span in California and the proposed Strait of Messina Bridge in Italy. Stiffened steel plates have been used for many years in a wide range of steel construction applications. They are particularly prevalent in the ship building industry and for hydraulic applications such as tanks, gates, and locks. The first orthotropic steel deck (OSD) bridge was developed by German engineers in the 1930's and the first such deck was constructed in 1936. In the United States, a similar system was built and often referred to as a "battle deck" because it was considered to be as strong as a battleship.

Generally, the OSD system consists of a flat, thin steel plate, stiffened by a series of closely spaced longitudinal ribs with support by orthogonal transverse floorbeams (Figure 1-1). The deck has considerably different stiffness characteristics in the longitudinal and transverse direction. Hence, the deck is considered to be structurally anisotropic. The name "orthotropic" arises out of a shortened form of the technical structural description of the system. According to **Orthotropic Bridges- Theory and Design** (Troitsky, 1987), "Because [the] ribs and floorbeams are orthogonal and because in both directions their elastic properties are different or anisotropic, the whole system became known as orthogonal-anisotropic, or, briefly, orthotropic." German engineers are credited with creating the word "orthotropic" and a patent was registered in 1948 (Sadlacek, 1987).

The OSD is efficient in that it is integral with the supporting bridge superstructure framing as a top flange common to both the transverse floorbeams (FBs) and longitudinal girders. This results in increased rigidity and material savings in the design of these components. As with other conventional steel-framed construction, loads are generally transferred by FBs transversely to the main load carrying system, such as longitudinal girders. Although, transverse members in the deck are most often referred to as **floorbeams**, there are also other commonly used terms such as crossbeam and/or diaphragm. Diaphragm is most appropriate when the deck is used in a redecking application and the member is rigidly, continuously attached to the existing bridge framing. Intermediate floorbeam may also be used where the FB member is not supported at its ends by main members and it is only serving to provide load distribution in the deck. **The defining characteristic of the OSD bridge is that it results in a nearly all steel superstructure** which has the potential (with minimal maintenance) to provide extended service life and standardized modular design, as compared to more conventional bridge construction.

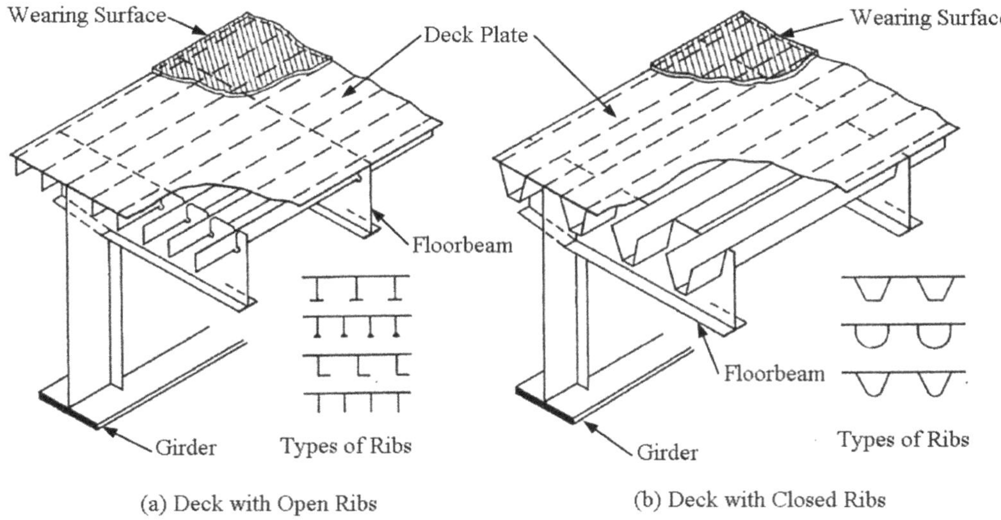

Figure 1-1 Components of the Orthotropic Steel Deck Bridge Girder System showing (a) Open Ribs and (b) Closed Ribs (AISC, 1963. Copyright American Institute of Steel Construction Reprinted with permission. All rights reserved.)

The OSD system has been utilized successfully for thousands of bridges worldwide, particularly in Europe, Asia, and South America. The United States has not yet fully embraced this technology, with only an estimated 100 such bridges in its inventory. The OSD bridge has been most commonly used in the United States for particular design conditions. One condition is for long span structures where the minimization of dead load is paramount, such as the new Tacoma Narrows Bridge. A second example is for box girders which contain slender compressive plate elements requiring stiffening, such as the Alfred Zampa Memorial Bridge (see Figure 1-2) A third example is for redecking of existing major bridges on urban arterials where rapid construction is needed, such as the Bronx-Whitestone Bridge.

OSD construction also has tremendous potential for use in short to medium span "workhorse" girder bridges when located on a high-volume roadway where accelerated construction or extended service life is required. There is a recent trend in the United States towards using bridge systems that are more rapidly constructible to minimize impacts to the traveling public, and to solutions that offer more long term durability and economy with the goal of 100 years of service life (Mistry and Mangus, 2006). The OSD bridge can provide an economical solution to meet these criteria. Furthermore, OSD is able to be constructed quickly because most of the components are prefabricated. Additionally, complete future redecking is rendered unnecessary, which minimizes any major traffic impacts in the future. Furthermore, the OSD provides a smooth continuous riding surface durable against deicing salts with minimal joints to prevent leakage and protect the other bridge components.

Figure 1-2 Erection of Orthotropic Box Girder Segment for the Alfred Zampa Memorial Bridge lifting a Deck Segment into Place

It is recognized that OSD bridges have not been problem-free historically. They present unique challenges in terms of design and construction as compared to conventional bridge construction. Fatigue cracking has been observed more frequently in such decks resulting from the complicated welded details combined with stresses that can be more difficult to quantify and, in particular, early designs which attempted to overly minimize plate thicknesses to reduce weight. In addition, the designs of critical details are not controlled by dead load or ultimate strength, but rather, live load (from individual wheels, in some cases). In any element where cyclic live load stress ranges dominate the design, fatigue will be the controlling limit state.

Early analytical tools were limited in their ability to quantify the stress states at these details and the early experimental fatigue resistance database was limited. Moreover, the fatigue performance of many of these details can be sensitive to fabrication techniques. Design and detailing practices relied heavily on experience gained through trial and error. Unfortunately, many trials were unsuccessful, creating questions among owners as to their long-term effectiveness in United States highway infrastructure. It is unfortunate that many of the reports of cracking have occurred in redecking projects where the interactions between new OSD and existing structure are difficult to account for, and design optimization is not easily achievable if at all possible.

The potential for cracking at the rib-to-deck plate (RD) weld is indicative of this problem. Whereas this one-sided weld was once a source of performance issues, it is now executed with a vast increase in consistency and performance by using a partial joint penetration (PJP) with controlled penetration, and with no melt thru allowed. Cracking is also possible at the rib-to-floorbeam (RF) intersections, where 3-dimensional stresses are generated by the in-plane flexure of the FB response combined with the out-of-plane twisting from the rib rotations. **All of these details have been the subject of extensive research efforts over recent decades, providing better understanding of performance and proper design.** For example, a stress-relieving cut-out in the FB around the rib (Figure 1-3) has performed well when the geometry is appropriately designed.

One clear advantage to the OSD is that it is a highly redundant system and minor cracking is often more of a nuisance to be observed and documented rather than a serious threat to the strength or integrity of the structure.

Figure 1-3 Orthotropic Deck Rib-to-Floorbeam (RF) Connection showing the Stress Relieving Cut-out Detail

Wearing surfaces applied to OSDs have also exhibited performance problems in the past with cracking, rutting, shoving, and/or delamination, which has often resulted in early maintenance and resurfacing. These problems have generally been attributed to inadequate construction control, environmental related degradation of the materials, or flexible design of the steel decking. This is also a symptom of the previous lack of experience. Recent research and development and general design improvements, such as minimum deck plate thickness of 14 mm to 16 mm (9/16 inch to 5/8 inch) have addressed the causes of many of these previous failures. Additionally, **current design concepts have proven successful in many modern OSD bridges in the United States and abroad**

There are two broad categories of surfacing materials currently being used: (a) bituminous surfacing systems including mastic asphalts, latexmodified asphalts, and reinforced asphalt systems; (b) polymer surfacing systems, including epoxy resins, methacrylates and polyurethanes. Although not mandatory, many bituminous surfacing materials used on steel orthotropic deck bridges are 50 mm (2 inches) or greater in thickness, while most polymer surfacing materials are 20 mm (3/4 inch) or less in thickness. The climate generally dictates which type of surface is to be selected since bituminous surfaces are more sensitive to changes in temperature. Both wearing surface types have demonstrated a service life in excess of 30 years. As with all manner of bridges, no matter what type of wearing surface is utilized, regular maintenance and occasional resurfacing will be required during the design life of the deck.

The corrosion resistance of OSDs has historically been very good. The top side is protected by the wearing surface, and the bottom side can be protected with a conventional paint system. Similar to other steel bridge structures, it may require regular maintenance in terms of repainting. Moreover, OSDs are typically made continuous, without joints, for extended lengths, which minimizes potential locations for water penetration. Thus, the coating on the underside of the deck can last for a long period unless it is subjected to direct saltwater spray. The individual ribs are typically sealed with end plates that prevent moisture from entering the interior of the rib. Outside the United States, a common approach has been to use a fully closed box girder cross-section and employ an in-service dehumidification system on the interior to essentially eliminate the possibility of corrosion, and thus there is no need for an interior paint system.

One of the primary reasons for advancement of OSD bridges in the United States is the application of modern techniques for engineering analysis and design. Orthotropic behavior is advanced since the plate distributes lateral loads in two directions and involves integral behavior of the deck with the FBs and girders. Engineers in the United States are generally unfamiliar with these types of structures and existing published manuals, such as (AISC, 1963) and (Troitsky, 1987), although state-of-the-art at the time of their copyright, have become dated. Additionally, early simplified analysis methods do not provide for a complete engineered design. **This Manual aims to gather and disseminate the modern OSD technology based on worldwide practice.** Emphasis is now placed on the fact that the design of these structures is generally controlled by fatigue limit states. Details necessary to make these structures work require advanced fatigue evaluation techniques that must rest on accurate stress range calculations, which is possible with the use of the Finite Element Analysis (FEA) and/or prototype testing. The fatigue testing database has grown considerably over the last few decades, which has provided the necessary data for proper evaluation and detailing for fatigue resistance.

Of equal importance to the engineering design are the means, methods, and quality control of the fabrication and construction. History has demonstrated that refined analysis and design can be rendered meaningless when the construction is not executed properly. Orthotropic details are also advanced in terms of fabrication and must be treated with care. To be successful, they require detailed construction specifications and quality control measures in production. As fabricators in the United States gain more experience in these projects, they will be viewed with less risk, which will promote more bid participation and lower cost. One factor that has exacerbated the risk for fabricators is the unwillingness of engineers to modify construction specifications to reduce risk and increase economy. This has generally been due to lack of data. This Manual provides the necessary information to solve this problem.

General consensus is that for OSD bridges to become cost effective in the United States, standardization is critical. **Standard panel details will promote repetition, economy in design, and fabrication as well as improve quality of the finished product.** Standards would also limit the need to conduct refined engineering analysis or prototype testing for every new project. However, optimum standard designs cannot yet be definitively established until more laboratory testing is conducted and domestically produced systems prove successful in service. This Manual promotes sound detailing concepts based on the current available knowledge, yet leaves room for advancements as more data becomes available and engineers discover improvements to existing designs. It is conceivable that after a period of development (which may last one or two decades) the application of OSD bridge decks will become second nature and standard designs will be available without the need to perform refined analytical procedures or laboratory testing. This Manual is seen as an initial step to achieve this ultimate goal.

OSDs can compete in cost with other suitable deck alternatives, such as the conventional cast-in-place concrete deck and concrete filled grids, when life cycle cost is considered. Because of their higher initial costs, OSDs must necessarily have lower maintenance needs and longer lives in order to be justified. Laboratory studies have demonstrated that an OSD with a 100-year design life may be achievable. Additionally, existing bridges like the Golden Gate Bridge and Fremont Bridge have shown that they are capable of performing with minimal required maintenance. With proper design and detailing, well executed fabrication, and quality construction, the full potential of OSD bridges can be exploited throughout the United States.

The target audience for this Manual includes engineers, owners, contractors, fabricators, and researchers in the bridge industry. The Manual gathers and summarizes worldwide knowledge from countries such as Germany, Denmark, France, Russia, Brazil, China, and Japan, and put it into context with United States design and construction practice. The project types covered in this Manual are long-span bridges, such as cable-stayed and box girder systems, bridge redecking, and decks for the more common "workhorse" girder bridge. It is the last case for which OSDs have not been used frequently in the United States, but with the potential for large volume and standardization of design, the Manual attempts to promote consideration in such cases.

1.2. ORGANIZATION OF THE MANUAL

This Manual covers the relevant issues related to OSD bridge engineering, including analysis, design, detailing, fabrication, testing, inspection, evaluation, and repair. Chapter 2 begins with discussion of some the various applications of OSD bridge construction to provide background with case study examples. Chapter 3 provides basic criteria for the establishment of a cost-effective and serviceable OSD bridge cross-section with detailing geometry that has been used on recent projects worldwide. Chapter 4 provides the relevant information necessary for the engineering analysis of the OSD bridge, including fundamental behavior and application of refined techniques. Chapter 5 outlines the requirements for complete design of OSD bridge superstructures by evaluation of applicable limit states using the Load and Resistance Factor Design (LRFD) methodology. Chapter 6 addresses design details such as materials, corrosion protection, minimum proportions, and connection geometry. Chapter 7 provides basic fabrication, welding, and erection procedures for OSD bridge components, illustrated by photos of shop and field practices. Chapter 8 provides recommended methods for maintaining and evaluating OSD bridges, including inspection, load rating by LRFR, rehabilitation strategies, and fatigue retrofit. Chapter 9 addresses all issues related to wearing surfaces. Chapter 10 covers testing of OSDs. The culmination of all the information provided is demonstrated in the design examples of Chapter 11 which contains one multi-girder bridge example and one cable-stayed bridge example demonstrating engineering design of OSD bridges by refined analytical techniques.

2. BRIDGE APPLICATIONS

2.1. GENERAL BACKGROUND

There appears to be a general lack of experience with orthotropic steel deck (OSD) bridges in the United States with the exception of a few major projects. As such, some time should be devoted to study of previous applications. Success in bridge design has proven to rely on experience and applying lessons learned from past projects. The historical lessons have allowed engineers and researchers to continually make improvements to OSD systems for decades, and since much of the worldwide experience with OSDs lies outside the United States, it is equally important to look internationally to gather the fullest range of examples.

This chapter is intended to highlight a few of the notable successful projects worldwide and let the readers follow through to discover more details. Projects from the Americas, Europe, Middle East, and Asia are discussed herein. Good summaries of OSD bridge projects can be found in the references by Troitsky (1985), Huang et al (2008), Hoorpah (2004), Korniyiv (2004), and Choi et. al. (2008). Although each governing body around the world has adopted subtle differences in terms of design and detailing practice, there is relative consistency as the collective knowledge has been shared in an unprecedented way.

The greatest advantage of OSD construction is realized when it is used in long-span bridges because it is comparatively lightweight, and it can work in composite action with the main longitudinal members. OSDs are also excellent candidates for bridges in seismic zones where they can reduce seismic inertia forces on piers and foundations and are able to undergo ductile deformations without sudden failures. OSDs are also highly desirable in movable bridges where they provide enduring performance with greater ease of movement. Additionally, their ability to perform structurally in a raised position better than other decks renders them superior. ODSs have been used in railway bridges, where the least maintenance, most durable solution is desired. While they are not effective at resisting direct train loads (especially where heavy locomotives are involved) they can perform well when ballasted for load distribution. In fact, many of the fatigue concerns discussed throughout this manual can be reduced or eliminated when ballast is utilized. In summary, OSD design can provide cost-effective solution in the following conditions:

- Long-span bridges
- Movable bridges
- Bridges in seismic zones
- Bridges where rapid construction is required
- Bridges where extended service life is required
- Cold weather conditions where cast-in-place concrete is difficult

Since many of the existing applications of orthotropic construction are long-span, signature bridges, they are all very special designs in terms of the bridge cross-section and superstructure configuration. The designs are often controlled by particular design criteria, constructability, or other considerations that in retrospect may not be obvious to the outside investigator. Prudence must be exercised when attempting to reuse any design detail that was used in a past project. For instance, internal rib bulkheads have been used on recent bridge redecking projects to control certain secondary stresses due to unique interactions with the existing superstructure. Thus, it should not be assumed to be the ideal solution for new construction (see Chapter 4 for more on this subject).

In the following sections, a number of notable orthotropic bridge projects are discussed based on the general categories: Plate Girder Bridges, Box Girder Bridges, Suspended Span Bridges, Bridge Redecking, and Movable Bridges. Although there are many different bridge types that utilize orthotropic construction, it is emphasized that basic design of the OSD is quite similar in all cases. The commentaries provided are not intended to make judgments on previous designs based on present knowledge. They are instead intended to be discussions of typical examples of bridge types, and what they achieved at the time of their construction.

2.2. PLATE GIRDER BRIDGES

OSD bridges of this type were originally built in the 1930s under the name "battledeck floor" structures. A deck plate was welded to the tops of longitudinal I, T, or WF sections. The sections acted as ribs and were spaced up to 0.838 m (33 inches) on center, depending on the deck plate thickness. Transverse delivery of floor loads to the main girders was provided by floorbeams. Local load distribution, if deemed necessary in addition to that of the deck plate itself, was often provided by additional transverse plate stiffening welded to ribs and deck plate. Remnants of this type of construction still exist in prototypes such as the North Saginaw Road Bridge in Michigan (1920). Similar examples were produced in Germany with bent plate flanges over the Autobahn.

OSDs were used in the early 1950s for longer span plate girder bridges. Ribs always run longitudinally with transverse framing acting to distribute loads to several longitudinal girders and as support for the deck plate. Examples of these are the Kurpfalz Bridge in Mannheim and the Eddersheim Bridge Germany of 1950 and 1953, respectively.

The year 1956 saw a major achievement in the construction of this bridge type with a three-span continuous structure in Belgrade (the Save River Bridge). Two variable depth girders spanning 75 - 261- 75 m (246 - 856 - 246 ft) over two interior piers, supported a deck with flat plate ribs. The 25 mm by 267 mm (1 inch by 10.5 inch) ribs spanned continuously between FBs spaced at 1.585 m (5.2 ft) centers. The roadway width of this bridge was 11.98 m (39.3 ft).

The Golden Horn Bridge, erected in 1974 in Istanbul, consists of eight continuous spans of two plate girders, for a total of 819.3 m (2688 ft) and with the major span of 132.9 m (436 ft). The 24.7 m (81 ft) wide roadway deck is supported by an OSD consisting of closed ribs, 0.222 m (83/4 inches) deep, spaced at 0.622 m (24.5 inches) on center and spanning 4.50 m (14.75 ft) to FBs. Two plate girders, spaced at 24 m (78.75 ft), support the entire structure. Pedestrian walkways cantilever outboard of the plate girders.

In the United States, the use of OSDs did not take hold as it did in Europe. Instead, urban congestion demanded better use of space with smoother roadway alignment at bridge approaches and crossings. Curved bridges, which seemed to better meet this need and were built with the use of composite concrete decks on steel plate girders, would require complex details such as rib kinks to make OSD work. Three examples of orthotropic plate girder bridges used in the United

States are the Crietz Road over I-496 Bridge in Lansing, MI, (1969), BART Bridge No. A-096 in Berkeley, CA, (1972), and the 680-580 Test Bridge in Dublin, CA, (1965). The 680-580 Bridge was built as an experimental bridge to verify the performance of two deck sections with different steel details and wearing surfaces (Figure 2-1). The bridge is still in service to-date, but the wearing surface has been replaced on the thin section.

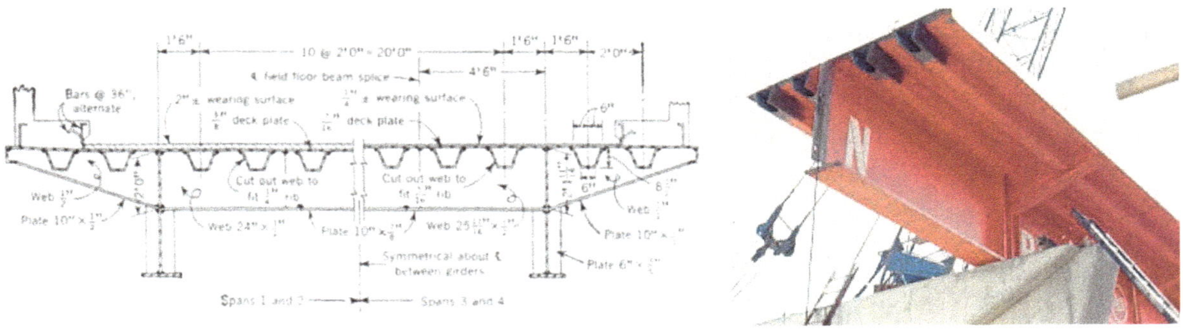

Figure 2-1 Detail Cross Section Drawing and Photo of the 680-580 Bridge in Dublin, CA under Construction

In general, multi-girder bridges are selected in short to medium span applications because of their redundancy and for the ease of staging future resurfacing. **Use of OSD in a plate girder bridge eliminates the need for complete redecking in the future since the deck is designed to last as long as the superstructure.** This allows for the use of wide girder spacing, and even a two girder cross-section (if the fracture critical concerns can be addressed by increased material toughness and/or testing). Thus, the overall cost of such bridges can be competitive, even though the deck may be more expensive than comparable cast in place concrete. These types of bridges also have good potential for use in urban applications requiring rapid construction or extended service life. This bridge type is the subject of Design Example 1 in Chapter 11.

2.3. BOX GIRDER BRIDGES

It is often useful and necessary to provide bottom lateral bracing systems for plate girder designs, to create a tube-shape cross-section that provides torsional closure. In this manner deflections on one side of the road subjected to asymmetrical loads are reduced. Lateral systems are often needed to reduce wind effects on bottom flanges of outer girders. The box girder is typically found to provide a less complicated and more elegant solution to address these problems.

Often, two box girders are connected by FBs on the inboard side and cantilever FBs on the outboard. This is done to extend the width of the roadway and to balance the load in the boxes. An example is the Dusseldorf-Neuss Bridge (1951). The box widths are 7.315m by 0.152 m (24 ft by 6 inches). The Luxemburg Bridge spanning the Alzette River Valley in Luxemburg and the Saint-Christopher Bridge in Lorient, France, are other examples of early double box spans.

The center FBs joining the two boxes of the Dusseldorf-Neuss serve as supports for stringers carrying street car tracks. Stringer beams were used instead of an orthotropic deck to carry trolley tracks.

The Concordia Bridge, built in 1965 in Montreal, Canada, is an example of a single box with three internal cells. That is, the two exterior boxes were closed off in the middle with a continuous bottom plate. This bridge, a major achievement, is 690.4 m (2,265 ft) long between abutments and four internal piers. The roadway is 25 m (82 ft) wide, and the three cell box is 17.4 m (57 ft) wide at the bottom.

The Weser Bridge (1954) and the Speyer Bridge (1957), both in Germany, were early examples of single box cross-sections. The Speyer Bridge boxes are 7.99 m (26.2 ft) wide, and those of the Weser 5.91 m (19.4 ft) 4 wide. The Weser and the Speyer are single boxes with cantilever arms, but the roadway is contained primarily over the box.

Additional early examples of self-supporting single box girders are the Europa Bridge near Innsbruck, Austria (1963), and the Coronado Bridge in San Diego, CA, (1969).

The Poplar Street Bridge in St. Louis, MI, (1967), the San Mateo-Hayward Bridge in the lower San Francisco Bay (1967), the Rio de Janeiro-Niterio Bridge in Brazil (1974), and the Yukon River Bridge in Fairbanks, AK, (1975) are examples of double boxes built on the new continent. These major structures span a total of 355 m (1,165 ft), 1689.2 m (5,542 ft), 847.9 m (2,782 ft), and 701 m (2,300 ft), respectively.

The San Mateo-Hayward Bridge is a prime example of a successful application of OSD construction in a continuous box girder (Figure 2-2). The bridge carries six lanes of heavy traffic across San Francisco Bay and the original epoxy asphalt wearing surface is still performing well after 45 years of service.

Figure 2-2 Rendering of the San Mateo-Hayward Bridge showing the Variable Depth Box Girders and Orthotropic Steel Deck

OSD construction is found most often in box girder sections due to the requirement to stiffen wide, slender plate components. **Steel box girder crosssections with an OSD result in a very light superstructure,** often allowing for preassembly and launching or float-in construction of large sections. Box girders also provide the benefit of facilitating routine inspection and maintenance from within and providing protection of the orthotropic components. The box below the deck level essentially acts as an inspection walkway that can be accessed without disruption to traffic.

2.4. SUSPENDED SPAN BRIDGES

The lightness of orthotropic bridge cross-sections makes them excellent candidates for use in suspended spans for minimizing dead load in the global superstructure system. This includes cable-stayed, suspension, and arch bridges. Cross-sections that have been used include plate girder, multi-cell box girder, single cell box girder, and many combinations of stiffening trusses and FBs.

Cable-stayed bridges first emerged with the use of OSDs for relatively short spans, by today's standards. Erected in 1957, the Dusseldorf-North Bridge spanned 260 m (853 ft) between towers and 475.8 m (1,561 ft) between end piers. Box girders were used to anchor the stays and support the deck. In 1959 the Severin Bridge in Cologne, Germany achieved a 452.3 m (1,484 ft) span between piers using a single tower and asymmetrical spans of 301.8 and 150.6 m (990 and 494 ft). Two continuous box girders were used to anchor the stays.

The Viaduc de Millau Bridge was completed in Millau, France in 2004 (2,460 m [8071 ft] long and 32 m [105 ft] wide). It is a world record holder for suspended spans and is also the highest from grade. The bridge crosses the entire River Tam valley. This was achieved with the use of seven cable staying towers, supporting an aerodynamic three-cell box girder, which is 91 ft wide.

OSDs are particularly advantageous for cable-stayed bridges for their large compressive strength and for the elimination of the need to accommodate future redecking, which is typically a controlling design criterion for a composite deck system with post-tensioned concrete deck.

Over the last 20 years, many cable-stayed and main-cable suspension bridges have been erected with OSDs as part of the superstructure. In the 1960s, suspension bridge decks were still most commonly composed of concrete filled steel grids. But, by the 1980's designers of suspension bridges began to shift to OSDs. The reason was lightness and the need to span even greater lengths. The list of OSDs used in main cable suspension bridges grows yearly worldwide. They are too numerous to mention here. In the United States alone, the Alfred Zampa Memorial Bridge was erected with a suspended box girder in the North San Francisco Bay Area in 2004; in April 2007, the New Tacoma Narrows Bridge was built with an OSD supported on a stiffening truss system. The suspended spans of the new San Francisco-Oakland Bay Bridge are being erected, with the likelihood that the bridge will be completed in 2012. It is composed of two near-record large box girders that serve to anchor the main cables.

In Asia, suspension bridge construction is proceeding at an astounding pace. OSDs are the option of choice. The current main span world record holder (with 1993m [6538 ft] between towers), the Akashi-Kaikyo Bridge, in Japan, was erected in 1998. It is composed of a suspended truss with a jointed OSD (with expansion joints). The Tsing Ma Bridge in Hong Kong is presently the 7th longest span (1,377 m [4,517 ft]). Opened in April 1997, it carries two three-lane roadways on top of a hybrid open box girder tube, with two railways inside the box.

Europe also has its prototypes. The Forth Roadway Bridge (1964), Severn Bridge (1966), and Humber Bridges (1981) in the United Kingdom are prominent examples, as is the more recent Storebaelt Bridge in Denmark (1998). All were built with OSDs. It is clear that the defining features of these very long span structures are a cable system, the lightness of OSDs used to support the roadway, and the steel-suspended structures.

A prime example of a successful application of orthotropic construction in a suspended-span bridge is the Fremont Bridge over the Willamette River in Portland, OR, which opened in 1973 (see Figure 2-3). This bridge has proven to be a trouble-free design under continuous heavy interstate traffic in an urban setting (Abrahams and Hirota, 2004). The original epoxy asphalt wearing surface performed well, but did require replacement after approximately 20 years due to normal wear. Recent inspection revealed that of the 155,000 square ft of OSD area, there is essentially no visible fatigue damage (HDR 2008).

Figure 2-3 Photo of the Fremont Bridge over the Willamette River in Portland, OR. The Upper Deck is an Orthotropic Steel Deck for the Full Length of the Main Span Unit.

2.5. BRIDGE REDECKING

OSDs have been introduced in the United States as an option to replace aging concrete filled steel grid or reinforced concrete decks on suspension bridges. The George Washington Bridge (New York, 1977), Benjamin Franklin (Philadelphia, 1984), and Golden Gate Bridge (San Francisco, 1986), are early examples of deck replacements for suspension bridges. As of today, orthotropic deck replacements are on the drawing board for the Throgs Neck and Verrazano Narrows Bridges, in New York; both suspension bridges.

OSDs have also been used to replace reinforced concrete decks on approach viaducts of major bridges, such as the Throgs Neck Bridge and the Triborough Bridge. There is a great potential for development of this bridge type, especially in urban areas where traffic congestion during repairs is best handled by the faster construction permitted by OSDs.

Internationally, there have been many examples of successful application of redecking bridges with OSDs. The reference by Huang and Mangus (2008) provides a comprehensive list of relevant projects. One notable example is the Lions Gate Bridge in Canada, for which the approach viaduct was first redecked with OSD in 1975, and then the main span unit was subsequently redecked in 2002 (see Figure 2-4). In both projects, prefabricated deck units were installed quickly during nighttime bridge closures. The design employs the cost effective solution of using U-shaped ribs without expensive cut-outs in the FB intersections. The wearing surface on the viaduct deck has also proven successful, exceeding the predicted design life (Buckland, 2004).

Figure 2-4 Aerial Photo of the Lions Gate Bridge showing the Redecking with Orthotropic Deck Segments being Lifted into Place

2.6. MOVABLE BRIDGES

As orthotropic deck technology advanced, it became clear that OSDs are ideally suited for movable bridge construction for two main reasons. First, they are light and require less power from prime movers to lift and lower the leaf(s); similarly they require smaller ballast counterweights. Second, they deliver the entire floor load (when the bridge is lifted) to the girders directly through the deck plate with much less difficulty than their counterparts, the open grid and concrete filled grid decks. The internal forces in the trunnions are also reduced. OSDs have been used successfully in many different types of movable bridges, such as lift bridges, bascule, articulating ramps, floating bridges, and swing spans. A comprehensive review of projects can be found in the work by Mangus (2001).

There are approximately 140 major bascule bridges worldwide, including 48 in the United States, 21 in Holland, and nine in the United Kingdom. Europe has approximately 50 bascule bridges designed with OSDs, all of them built in the last 20 years. The most notable are the Erasmus Bridge in Rotterdam, Netherlands, and the Gateway to Europe in Cadiz, Spain. In the United States, some decks on existing bascule bridges have been replaced with OSDs. The trend is expected to continue.

3. TYPICAL BRIDGE SECTIONS

This chapter provides basic criteria for the establishment of a cost effective and serviceable orthotropic steel deck (OSD) bridge cross-section, which includes both the overall cross-section (global) and the panel section details (local). Criteria for the detailed layout of the panel geometry, such as rib and floorbeam (FB) proportions, spacing, and span, are discussed. The criteria in this chapter are provided for preliminary design layout.

The orthotropic panel should be considered a structural component "module" that can be assembled into a bridge cross-section in any number of geometries that the designer may conceive. In fact, orthotropic panel construction allows the designer to consider sections that are not possible otherwise, such as wide, single cell box girder sections. Figure 3-1 shows an example of such as section, with modular panels making the enclosure. The possibilities are quite broad depending on the ingenuity of the designer. However, the recommended details of the modular panel itself are somewhat established based on the cumulative worldwide experience through successful laboratory testing and in-service performance.

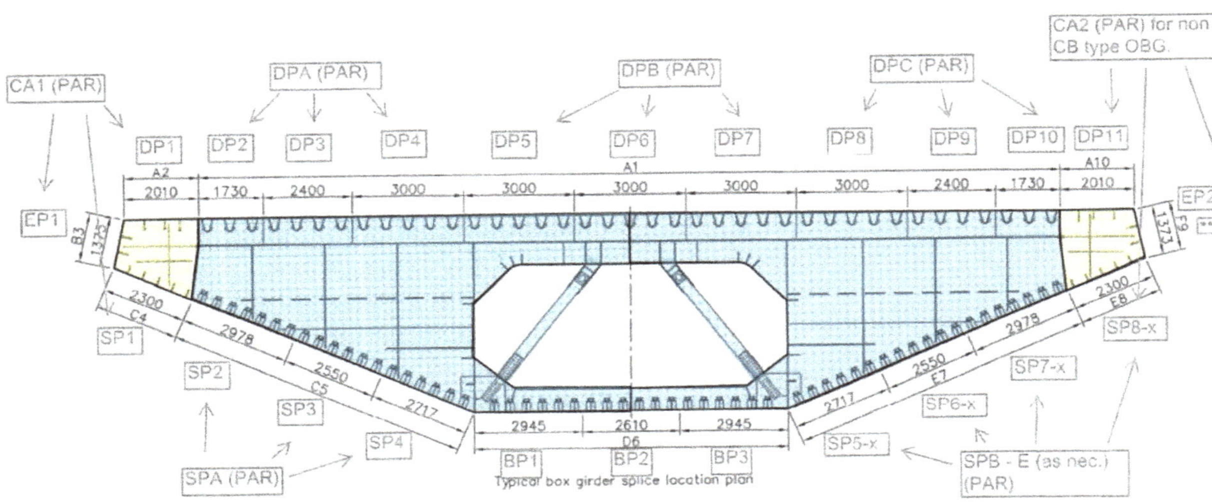

Figure 3-1 Example of OSD Box Girder Bridge Section for Modular Construction. Note the Repetition of Each Sub-panel of Cross-Section.

3.1. GENERAL LAYOUT

The important issues to consider in layout of the bridge and panel sections are constructability, serviceability, and inspectability (aerodynamics aside). Panel joints and splices should be proportioned to facilitate fabrication, handling, assembly, shipping, erection, and to keep the deck plate splices out of the primary wheel paths. Orthotropic fabrication requires careful planning due to unique challenges related to fit-up, weld execution, and distortion control. These issues are covered in more detail in Chapters 6 and 7. The general rules for steel bridge constructability apply also to OSD bridges. However, there is often motivation to maximize shop connections and minimize field welding. Economical solutions often involve bigger field sections since they are lighter than comparable concrete solutions. The designer should consult potential fabricators for their input if they wish to explore layouts that may be outside the limits of conventional steel bridge construction.

The OSD allows for the development of modular multi-girder bridge sections that are rapidly constructible These may include smaller closed box girders or open T-shaped girders (which is the focus of Design Example 1 in Chapter 11). These types of modular girders are very stable during erection and are pre-decked, which minimizes construction time. Since conventional crossframes can be eliminated from these bridge cross-sections, very little elevated assembly work is performed by ironworkers. Also, once the girders have been erected, a safe working platform is established, providing easy access for workers without the need to tie off and for the staging of materials. Furthermore, the construction can be executed more quickly than a conventional bridge with cast in place concrete deck, which minimizes traffic congestion during construction. The conventional cast in place concrete deck, which is one of the most time consuming stages of the construction due to curing requirements, is replaced with a simple wearing surface overlay.

Since OSD bridges have a relatively large amount of steel surface area with corrosion potential, serviceability must be considered in the development of the bridge cross-section. For this reason, closed sections that minimize exposed area are preferred. This will minimize the initial cost of the coating system, as well as future maintenance costs. Designs of many of the modern long-span bridges worldwide have employed a single-cell box girder section with a dehumidification system for the interior air space (Sorensen, 2004). No interior corrosion is possible if the relative humidity is kept below 60 percent. Such an approach only requires a simple prime coat on the interior surfaces, which can reduce the painted area by up to 80 percent and provide essentially unlimited resistance to corrosion (Gimsing, 1998).

In selecting the general cross-section, consideration of inspectability is also very important. The OSD bridge has the potential for fatigue damage due to the sheer volume of welded steel details, and should be routinely inspected to identify any cracking that may develop; particularly in the early years of service. This is discussed in more detail in Chapter 8. Any closed box girder should be detailed carefully to permit inspector access. The single cell box girder cross-section provides the best solution to facilitate this inspection. Although the interior of a box girder is considered a "confined space" that may require special inspector training, this bridge type can easily be routinely inspected without disruption of traffic. More importantly, all welded details can easily be inspected at close proximity by the inspector, if necessary.

For examples of OSD bridge cross-sections that have been constructed throughout the world, the reader should consult Troitsky (1985), Mangus and Sun (2000), or Huang et al (2008).

3.2. ORTHOTROPIC PANEL DETAILS

OSD panels are generally classified as either open-rib systems or closed-rib systems (Figure 3-2). In either system, the ribs are arranged in the longitudinal direction of the bridge for distribution of wheel loads to FBs, and to provide increased flexural rigidity to the primary girder(s). When ribs are oriented in the transverse direction, the situation for the durability of the surfacing worsens due to the "washboard effect" that is created by passing wheel loads (Sadlacek, 1992). The ribs can be made discontinuous to fit between the FBs; however, current practice is typically to make them continuous through cut-out "windows" in the web plates of the FBs (such a discontinuous connection would not be allowed by current United States welding standards). OSDs subjected to direct wheel loads are typically stiffened with closed ribs, while other structural elements such as box girder flanges may be stiffened with either closed or open ribs.

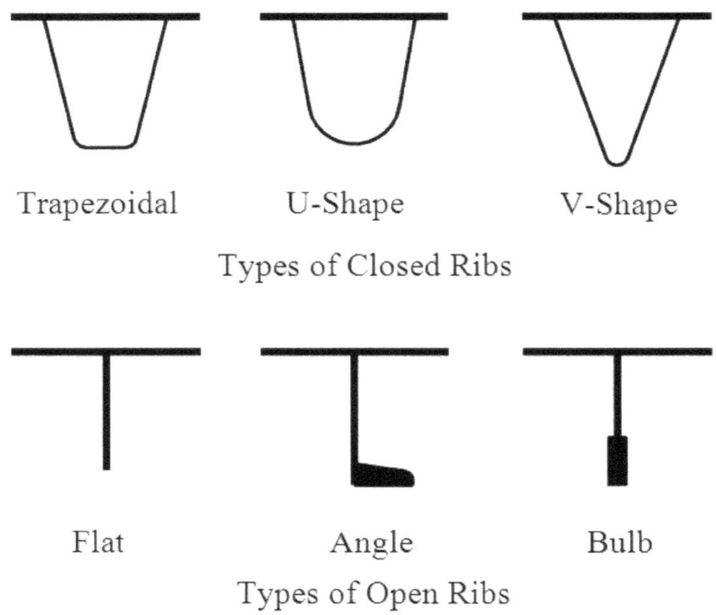

Figure 3-2 Common Rib Types for Orthotropic Decks including Closed and Open Ribs

The FBs are usually made from steel plates welded together or a rolled section in the shape of an inverted T-section, and the top flange is formed by the deck plate. FBs are usually spaced from 3.05 to 6.1 m (10 to 20 ft), depending on the rib system employed. Obviously, increasing the FB spacing requires fewer of the costly rib-to-floorbeam (RF) intersections. Longer rib spans have been utilized in a few instances (Wolchuk, 2004), **but** the performance of such designs under heavy traffic has not been well proven. It is difficult to control the fatigue stresses at the RF intersections for long rib spans. An additional consideration for FB spacing is the transverse spacing of the main girders (i.e., FB span). That is, a relatively larger FB spacing should accompany a larger FB span.

3.2.1. Open Rib Systems

Open ribs can be made from flat bars (most common), bulb shapes, inverted T-sections, or angles (Figure 3-2). They usually vary in size from 9 mm by 203 mm to 25 mm by 305 mm (3/8 inch by 8 inches to 1 inch by 12 inches) along the cross-section of the bridge, and are spaced approximately 305 mm to 406 mm (12 inches to 16 inches) on center. Span lengths of open rib systems are generally in the range of 1.52 m to 3.05 m (5 to 10 ft).

Experience indicates that it is simple to fabricate the open ribs and vary the rib dimensions as required for the various parts of the OSD. The field splicing of the open ribs is also relatively simple, and the bottom of the open rib deck permits easy access for inspection and maintenance.

There are two main disadvantages of the open rib deck system. First, it is torsionally soft relative to closed shapes. This means that it is not very efficient in distributing transverse loads such as wheels from one rib to adjacent ribs. This results in more ribs, closer FB spacing, and hence, more steel per square foot than a comparable closed rib system. Secondly, the total amount of welding required to fabricate the system is approximately double the amount required for a comparable closed rib system (i.e., a pair of vertical rib elements requires four lines of weld to the deck plate as opposed to two lines for a closed rib).

While modern OSD designs typically do not use the open rib for deck plates subjected to direct traffic loading (see article 3.2.2), open ribs are still often used to stiffen box girder webs and bottom flanges. Open ribs can be preferred in curved bridges due to ease of bending. They are also sometimes used in the vicinity of other structural components, such as barriers, to provide space for connections. Thus, economical designs may include some combination of open and closed ribs in the bridge cross-section.

3.2.2. Closed Rib Systems

Most common among the many types of closed ribs are the trapezoidal, U-shaped, and V-shaped ribs (Figure 3-2). The most commonly used section is the trapezoidal rib. It has been found to be the most useful by engineers and the worldwide steel industry.

The closed-rib system is the preferred system relative to open-ribs for a number of reasons. First, it has much higher flexural and torsional rigidity. The high torsional rigidity contributes to better distribution of concentrated transverse loads and, consequently, to a reduction in stresses in the deck plating. Fewer welds, less distortion, and reduced steel weight are further advantages.

A complication of the closed rib system is in the execution of the one side partial penetration weld for the rib connection to the deck plate (RD connection). This fatigue sensitive weld requires care for fabricators to execute with consistent quality. Also, due to its geometry and inherent torsional strength, closed rib decks are subject to local secondary deformations and stresses that make them vulnerable to fatigue at the intersection with the FB (RF connection). Furthermore, field splices of the ribs are also more complicated, and this system requires tolerance control in fabrication and erection to ensure proper fit at the splices.

Although the trend to greater rib span lengths is present with closed rib systems, this is limited by larger cut-outs in the webs of the FB that are detrimental to FB shear resistance. In addition, deflections associated with large local transverse curvatures may lead to premature failure of the wearing surface.

3.2.3. Rib Proportioning

It is typically during preliminary design (i.e. prior to the execution of any analysis or testing) that the rib spacing, span, and stiffness, as well as the deck plate thickness, must be selected. General rules for limits on these proportions are provided in the Eurocode (ECS 1992). Generally, ribs for deck plates should be spaced center-to-center at no more than 760 mm (30 inches) with rib walls separated by 380 mm (15 inches), which provides a uniform support spacing of 380 mm (15 inches) for the deck plate in the transverse direction. This limits local bending of the deck plate and differential displacements between ribs from wheel loads to increase longevity in the wearing surface and reduce stresses in the rib-to-deck (RD) weld. Ribs for panels not subjected to traffic loading can be spaced further apart.

As stated previously, the deck rib span should be made as long as possible while still limiting stresses at the FB to safe levels. The Eurocode (ECS, 1992) provides a useful guideline for the minimum rib stiffness as a function of rib span (Figure 3-3). Note that two limits are provided, one for ribs adjacent to the girder web and one for typical ribs away from the girder. Increased rib stiffness adjacent to the girder web is necessary to prevent cracking of the wearing surface at the girder.

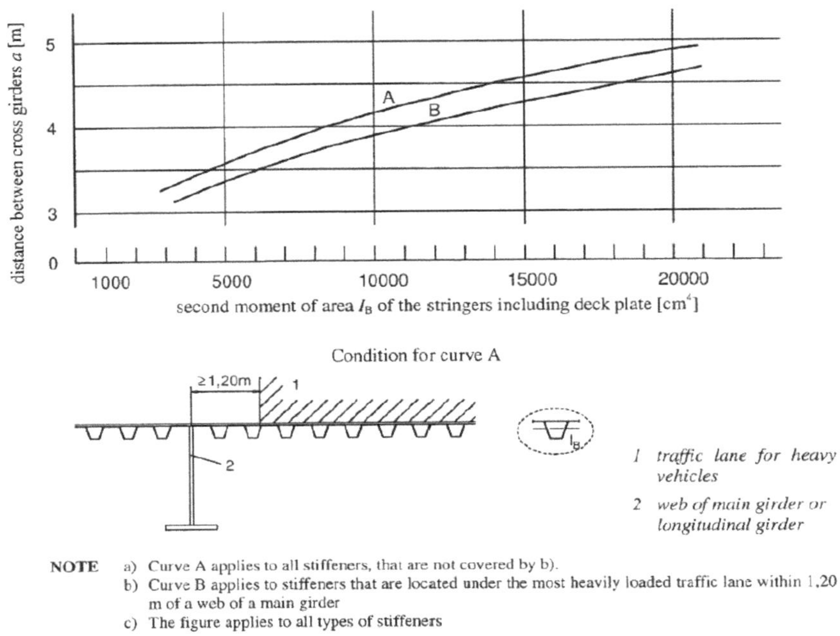

Figure 3-3 Guidance on the Minimum Rib Stiffness as a Function of Rib Span as provided in the Eurocode (ECS, 1992)

As a design aid for preliminary sizing and stress computations, section properties for trapezoidal closed ribs are presented in Appendix A. There, rib depths vary from 203 mm to 356 mm (8 inches to 14 inches), wall thicknesses vary from 6 mm to 11 mm (1/4 inch to 7/16 inch), and the upper rib width varies from 292 mm to 387 mm (11 1/2 inches to 15 1/4 inches) with a fixed bottom flange width of 165 mm (6 1/2 inches). For further guidance on selection of reasonable OSD rib proportions, Figure 3-4 summarizes detailing proportions used on a number of recent (as of 2011) projects worldwide that used the trapezoidal rib section. Note that there is little variation in the proportions as these details have proven to be cost effective and have performed well. The final detailing dimensions, including the FB size and cut-out geometry (if used) must be determined by testing and/or analysis as outlined in the remaining chapters of this Manual.

Bridge Name	Bridge Type	Location	Year Opened	Superstructure Type	h (mm)	a (mm)	b (mm)	t_r (mm)	a+e (mm)	L (mm)	t_d (mm)	d (mm)	t_c (mm)	FB Cutout?
The New Little Baelt Bridge	Suspension bridge	Denmark	1970	Box girder	244	144	287	6	598	3000	12	3000	8	yes
The Faroe Bridges	Cable stay / continuous beam	Denmark	1985	Box girder	294	148	287	6	620	4000	12	3250	10	yes
Pont De Normandie	Cable Stayed	France	1994	Box girder	243	193	293	7/8	605	3930	12/14	3000	16	yes
Hoga Kusten Bridge	Suspension bridge	Sweden	1996	Box girder	294	150	287	6	600	4000	12	4000	10	yes
Great Belt East Bridge	Suspension bridge	Denmark	1998	Box girder	294	150	287	6	600	4000	12	800	12	yes
Great Belt East Approach	Continuous beam	Denmark	1998	Box girder	294	150	287	6	600	4022	12	900	14	yes
Sutong Bridge	Cable Stayed	China	2008	Box girder	292	164	284	8	600	4000	14	4000	20	yes
Stonecutters Bridge	Cable Stayed	China	2009	Box Girder	339	150	298	9	600	3800	18	varies	12	yes
The Megyeri Bridge	Cable Stayed	Hungary	2008	Box girder/ I girder	292	184	284	8	600	4000	14	1696	12	yes
Millau Viaduct	Cable Stayed	France	2004	Box girder	300	200	300	7	600			600	20	yes
Incheon Second Bridge	Cable Stayed	Korea	2009	Box girder	260	188.5	304.1	8	600	3750	14	3000	11	yes
Irtysh River Bridge	Suspension bridge	Kazakhstan	2002		262	207.7	324.1	8	628.1	4000	14			
San Mateo Hayward	Continuous Beam	Caltrans	1967	Box girder	203/305	-	-	16		3167	16/19	838		yes
Fremont	Tied Arch	Oregon	1973	Girder/FB	305	152	305	8	600	3430	13	1270	11	no
Golden Gate Redecking	Suspension	Caltrans	1985	Truss/FB	279	152	356	9	673	7620	16	305	13	no
Williamsburg	Suspension	New York City	1998	Girder/FB	279	165	356	9.5	724	3050	16	-	8	yes
Bronx Whitestone	Suspension	New York City	2005	Girder/FB	343	127	330	8	660	3010	16	-	19	yes
New Tacoma Narrows	Suspension	Washington	2007	Truss/FB	305			8		6100	16	1690	9	yes
Alfred Zampa	Suspension	Caltrans	2003	Box girder	305	166	356	8	660/726	6200	16	3000	26	yes
SFOBB	Suspension	Caltrans	2012	Box girder	345	-	300	12	600	5000	14	1370		yes

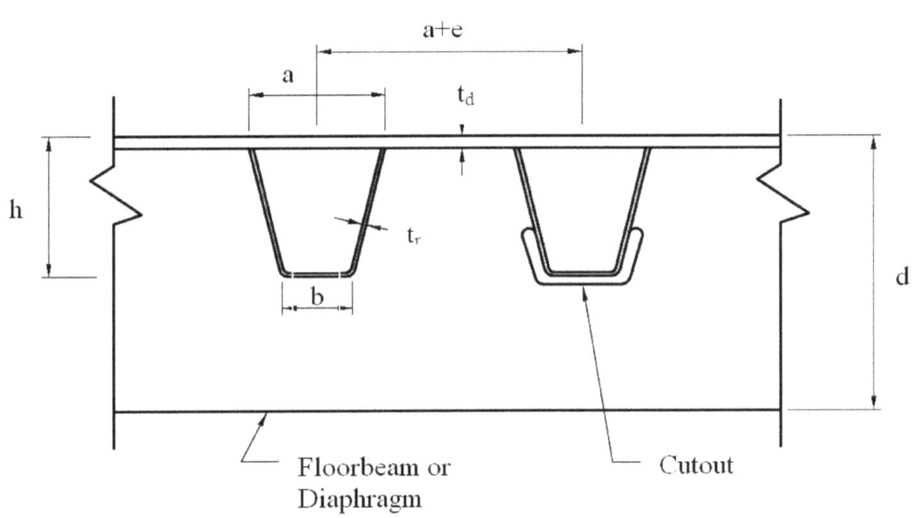

Figure 3-4 Orthotropic Deck Proportions on Recent Worldwide Projects including the Basic Rib and Floorbeam Dimensions as Shown in the Sketch

4. STRUCTURAL BEHAVIOR AND ANALYSIS

This chapter provides the relevant information necessary for the engineering analysis of the orthotropic steel deck (OSD) bridge. The majority of the chapter is focused on the subject of fatigue behavior and evaluation of fatigue stresses using the finite element analysis (FEA) method. A cursory review of plate bending theory, Huber's equations, and the Pelikan Esslinger Method for simplified analytical treatment of the OSD is provided for understanding the background and development of the theoretical behavior of OSDs based on classical (direct) analysis methods. This knowledge is valuable to the engineer for perspective on the design concept and potentially for future developments of more robust, simplified analysis methods. A synopsis of the historical evolution of OSD analysis can be found in the article by Kurrer (2011).

It is well documented that OSDs have tremendous reserve strength for local lateral loads, such as truck wheels due to the phenomenon of membrane stiffening. Thus, fatigue limit states driven by local distortional mechanisms at critical details typically control the design. It bears repeating that in addition to global and local stresses, the conditions affecting fatigue in OSDs are generally from different mechanisms and must all be combined for analysis. Computational based structural analysis has been demonstrated to greatly facilitate this. The chapter ends with discussion on the stability of the orthotropic panel for evaluation of strength limit states. The behavior and analysis of wearing surfaces are covered exclusively in Chapter 9.

4.1. EVOLVING PRACTICES

The understanding of behavior in the OSD has been continually evolving since the original designs of the 1930s. In post-war Europe, structural steel was scarce when these structures were first being built, and minimization of material was the essential common practice. This approach to minimize material seemed to satisfy design requirements at a time when fatigue issues were not fully understood. Additionally, reliance was placed on asphalt wearing surfaces to spread the wheel live loads. The result was that deck plates spanning 305 mm (12 inches) between ribs could be as thin as 9 mm (3/8 inch). As such, deck plates began failing by fatigue cracking completely through the plate and would require periodic repairs by gouging and rewelding. Of note is that the current European trends that developed to retrofit these thin deck plates is to first repair the steel and then to provide shear studs and a thin layer of reinforced concrete up to 101 mm (4 inches).

In the 1950s, the practice of using OSDs advanced, as did structural steel which was typically 227.7 MPa (33 ksi grade) at that time. More recently, Grade 345 (Grade 50) steel began being used starting in the early 1990s, yet rib sections have hardly changed, while deck plate thickness is on the increase. A major reason for this discrepancy is that one rib displacement relative to another impacts the performance of the wearing surface. Another reason is that design of OSDs is governed by fatigue, mainly where the ribs interact with the FBs.

Historically, the area which has undergone the most dramatic evolution may be at the rib-to-floorbeam (RF) connection. This issue evolved recognizing the rotation of the rib at the FB support and the resulting stresses. Evolution of the joint determined that in some cases a stress-relieving cut-out was needed in the Floorbeam (FB) web to preclude high localized out-of-plane stresses at the bottom of the rib, and to remove the weld from a point of maximum stress. Initially, engineering intuition as to how stresses flowed around the opening in the FB guided the design practice. That is, the need for a cut-out was correctly recognized, but a limited understanding about stress flow in the connection in combination with a scarce fatigue database (at the time incorrectly treating only perpendicular stresses at the toe of weld) steered engineers to conceive of the details such as those illustrated in Figure 4-1.

Though illustrated in Figure 4-1, early OSDs did not have an interior bulkhead. It was later introduced, similar to the cut-out, for intuitive reasons such as the presumed in-plane stress flow in the web. It was thought that adding the bulkhead minimized excessive stress concentrations introduced by a large hole in the web. Moreover, it was thought to minimize the distortion of the rib stem that would be created with the FB web stiffening the exterior side of the rib and no internal plate to provide resistance on the inside of the rib. The bulkhead was seen as a solution to this concern, and was proposed in response to cracking detected in the Westgate Bridge in Melbourne, Australia in the early 1990s. The cut-out was placed at the bottom of the rib, far from the deck plate, and was made shallow to achieve in-plane stress flow without excessive stress concentration around the hole.

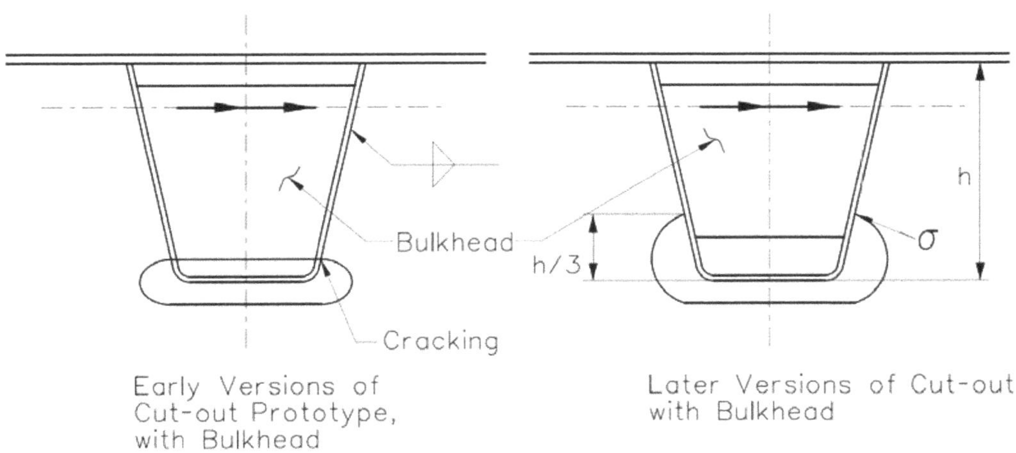

Figure 4-1 Early Versions of Stress Relief Cut-out Shapes for the Trapezoidal Shaped Rib showing the Use of and Stress Flow through Bulkheads

For these initial designs, the length of the cut-out was determined mostly by trial and error and was intended to minimize the out-of-plane effects on rib and FB engendered by rib rotation about the transverse horizontal axis.

Decks with this earlyshallow cut-out version outs did not fair wel Rib stems were failing below the top edge of the cut-out. Although analytical tools, such as the finite difference method, were available as far back as the 1970s to determine the cause of these failures, they were still a new technology, and engineering intuition again provided the next step. Often mentioned by OSD practitioners is the "Ostapenko Effect," which surmised that the enlargement of the rib bottom face due to Poisson's effect was the cause of stem cracking. This effect has been shown to be of little substance, yet led to the criterion that the height of the cut-out above the bottom of the rib needed to be one-third the height of the rib. This concept was written into the AASHTO LRFD

Bridge Design Specifications (AASHTO, 2010). These cut-out versions maintained that the termination of the cut-out be perpendicular to the stem, such that the principal stress in the diaphragm would be perpendicular to the rib-to-floorbeam (RF) weld. There is a known fatigue resistance database for this RF detail. Also, contrary to customary practice, the AASHTO LRFD Specifications prescribed fillet welds for the RF connection. The non-prescribed practice was to provide a wrap-around at the weld termination.

To evaluate all of the conjecture with respect to the behavior of the cut-out, in 1998 a major full-scale test was conducted at Lehigh University to assess the performance of an OSD prototype designed for the Williamsburg Bridge deck replacement in New York (Connor, 2002). This test and the FEA conducted in association with it indicated that the prevailing ideology was poorly conceived. The findings were as follows:

- The bulkhead did not behave like a link for continuity in the stress flow. But, because of its disconnection to the deck plate, discontinuous horizontal shear caused it to act more like a beam in double curvature. The resulting stress fields are graphically illustrated in Figure 4-2.
- The weakest ligaments in the continuum were the tensile portions of the diaphragm and bulkhead that showed root cracking in the bulkhead and toe cracking in the diaphragm. This has been called "type b" weld termination cracking by some researchers (Figure 4-3). Complete Joint Penetration (CJP) welds where root cracking takes place, instead of fillet welds, would have made the prototype last longer.
- The predominate stress patterns in the FB and bulkhead were in-plane, not out-of-plane stresses, as it was assumed. The out-of-plane components were found to provide approximately 15 to 20 percent of the combined stresses, depending on the thickness of the FB.

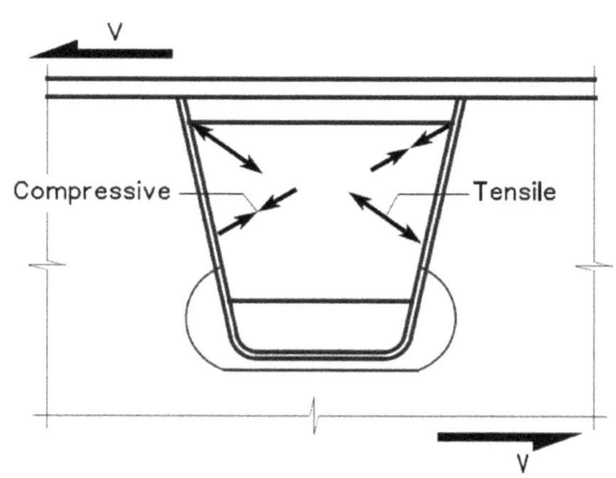

Figure 4-2 Horizontal Shear Stress Field through the Bulkhead showing the Tension and Compression Field Acting Diagonally

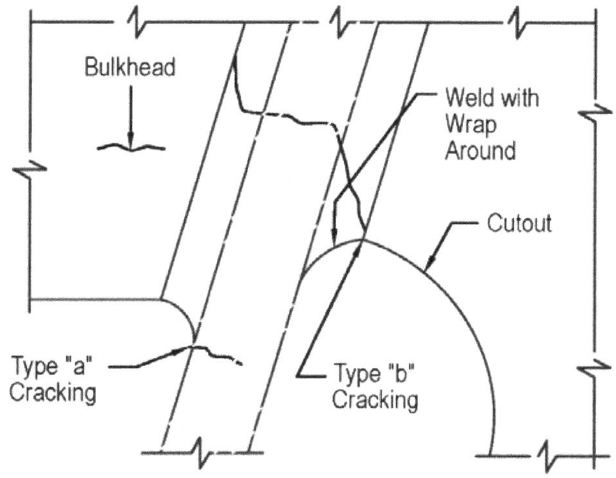

Figure 4-3 Common Cracking that May Occur at the Cut-out when Bulkhead is Used.

Equipped with this knowledge, engineers studying the alternatives for redecking of the Bronx-Whitestone Bridge conducted analytical parametric studies using FEA techniques. These studies were not necessarily exhaustive by research standards, but led to revelations that could not have been conceived by the use of simplified analysis techniques. The studies indicated that:

- Where (following simplified analysis) the longitudinal stress (σ_{lr}) at the RF weld would be compressive, regardless of wheel position relative to the center line of the rib, the real response of the rib could be tensile or compressive, depending on transverse wheel position applied in the bays adjacent to the diaphragm (see Figure 4-4). This is because wheel-eccentric loading produces torsion, displacing the rib laterally at mid-span, but with restraint at the diaphragm.
- Tensile and compressive stresses could alternate on each face of each rib stem, depending on wheel transverse position.
- Where early editions AASHTO LRFD displayed lack of awareness of the local vertical stress in the ribs stem (σ_{vr}), following simplified analysis, FEA indicated that the perpendicular edge of the cut-out to the rib represented an abrupt transition to these stresses, thus requiring a smooth transition as shown in Figure 4-4. AASHTO's implication that a two-inch radius is just as bad as an abrupt transition was incorrect and laboratory testing provided better data. This was reported and discussed in detail at the Orthotropic Steel Bridge Conference of 2004 in Sacramento, CA.
- Although the bulkhead reduced in-plane diaphragm displacements, an advantage for the deck plate at the rib-to-deck-at-floorbeam (RDF), the internal abrupt transition of the bulkhead presented additional stress concentrations, fabrication problems, and extra cost.
- It was realized that, although a thickening of the web would increase the out-of-plane bending rigidity of the diaphragm plate (and increase stress), it would reduce the in-plane stresses by a greater amount. It would also reduce stresses in other stress concentration areas (such as at the RDF), where resistance is low. Thus, the optimum diaphragm thickness depends on the entire geometry configuration. The current trend is toward a thicker diaphragm web to reduce the RDF stresses, when an internal bulkhead is not used.

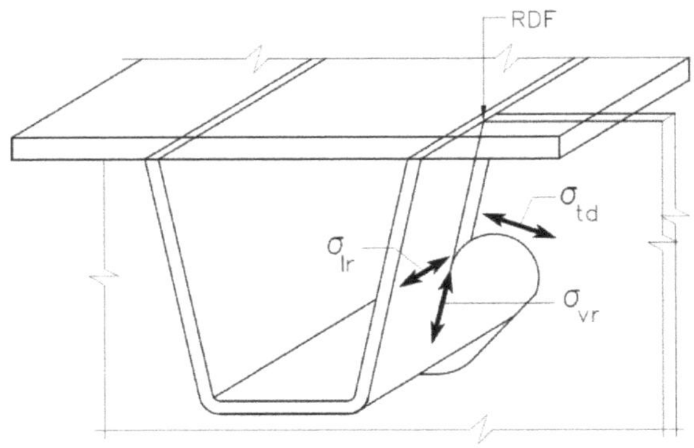

Figure 4-4 Rib-to-Floorbeam (RF) Connection Showing the Improved Smooth Cut-out Termination and Relevant Stresses to be Considered at the Joint

Test results published in August 2007 from the University of California - San Diego indicate that behavior as predicted by FEA analyses relative to rib stresses (σ_{vr}) could also occur in new major structures fabricated with an abrupt transition cut-out or bulkhead (Sim, 2007). These cracks were not at the toe in the FB cut-out termination but in the toe in the rib stem (Figure 4-3). This has been called type "a" weld termination cracking by some researchers. Note that type "a" cracking could emanate from either face of the rib, depending on which weld toe is lower in the bulkhead/cut-out arrangement.

In Europe and Canada there were parallel developments in a design without a cut-out in which a round-bellied rib, which passes through the FB, was welded all around it from end-to-end of the rib. Practitioners who developed this design determined that, as a rule of thumb, when the depth of the FB and the rib depth have a ratio $d/h > 2$, the design would succeed. Obviously, this ratio is predicated on specific rib rigidities and FB spacing. It is clear from Figure 4-5 that the concern is to minimize out-of-plane bending stresses in the FB web, and that, should the ratio d/h be less than 2, this must be remedied by a more rigid and heavier rib that would rotate less. Also, an intermediate diaphragm that would spread the load to more ribs can be used. The advantages and disadvantages of such a design without cut-out are described below:

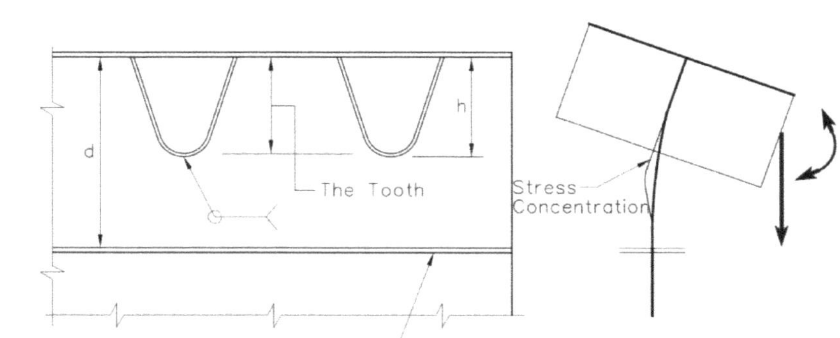

Figure 4-5 Sketch of Rib-to-Floorbeam (RF) Connection Elevation and Section Showing Exaggerated Out-of-plane Deformation of the Rib and Floorbeam Web Bending when no Cut-out is Used

Advantages

- Fabrication would be less costly. The history of the cut-out design shows that smooth terminations are often required with associated grinding. Estimates of savings range from 25 to 40 percent less costly when considering the cost of the FB connection to the rib. Total reduction in deck costs have been reported as 15 percent. The validity of these figures should be taken cautiously, as they represent the opinions of a small sample of fabricators.
- From the point of view of stresses at the RDF, this design is much better than the cut-out design without the bulkhead because it reduces the in-plane distortion of the tooth (see Figure 4-5) as it engages the deck plate. This reduces the tooth's displacements (or leading edge upward vertical displacement, trailing edge downward), which in turn reduces bending of the deck plate spanning over the trough of the rib, thus helping the longevity of the RDF detail. Another advantage is that the wheel load cannot cause as large a vertical displacement of one tooth relative to its neighbor, as would a design with the cut-out. This also helps reduce deck plate bending at the RDF.

Thus, the advantage is not only in reducing internal effect at the RDF and eliminating the stresses at the termination of the cut-out, but also of eliminating a significant number of connection details where a bulkhead might be needed otherwise.

Disadvantages:

- To reduce longitudinal stresses at the bottom of the rib to a level somewhat below Category C (fatigue categories are discussed later in this chapter), the rib needs to be made considerably stiffer than as in the case of the cut-out.
- FB web out-of-plane stresses are higher.

Notwithstanding the good performance of these designs over the past 30 years in many bridges, the data come from field experience with many unknowns. That is, there have been some notable failures due to improper joining, and there is not any significant amount of data from testing with known loads. While such performance may be indicative of good design, it is not of long enough duration to guarantee good performance past 100 years. Furthermore, the system may be satisfactory only for limited spans when applying the rib proportions used in past practice. Regardless, the system has very good potential to be successful.

Future Use ofthe Internal Rib Bulkhead

There is a widespread belief in the industry that bulkheads do not provide sufficient benefits and present problems, such as:

- They are costly to fabricate and present abrupt terminations, which should be alleviated by grinding if analysis indicates as such.
- Failures of the welds attaching the bulkhead are not able to be inspected and cracks could eventually turn into the stem.
- Alignment is not assured with the FB, since the rib and deck plate are attached prior to FB placement, possibly introducing additional secondary displacements and stresses.

Therefore, it is recommended that the bulkhead be avoided when possible. As such, new designs should start without it. In cases of introducing a new OSD on an existing structure that offers little head room and requires an exterior cut-out, and has a relatively flexible FB, the bulkhead may help reduce stresses at the RDF sufficiently to make an OSD design viable.

4.2. BEHAVIORAL MECHANISMS

The conventional method of bridge analysis is to assume that the complete structural system is comprised of several basically independent sub-elements such as the deck, stringers, FBs, and primary superstructure. This approach is based on the assumption that each element acts relatively independently and transfers load from itself to the next element without consideration of the real interaction among the various sub-elements. For most applications, this has shown to yield conservative designs and can be implemented with a simplified design methodology. In the OSD bridge, the deck plating, ribs, FBs, and main girders are all integrated into one structural unit. The deck panel must perform several functions simultaneously, including distribution of wheel loads and acting as the top flange of both the FBs and main girders. The fact that the OSD performs all these functions results in a very efficient utilization of material but the interactions cannot be ignored.

In the mid 1900s, prior to the widespread availability of modern FEA software, it was proposed to decompose the OSD structure into a series of pseudo sub-systems that are more easily understood and analyzed individually by simplified (non-computerized) methods. This led to the identification of a series of simple mechanistic "systems" that contribute to stresses in the panel. Once the stresses are calculated from analysis on each of these systems, they can be combined by the principle of linear superposition for verification of the applicable design limit states. The same basic approach to describing fundamental behavior is applied herein, with some extensions. Many of the behavioral systems identified by the pioneers in the field are still valid today. In addition, research conducted in recent decades has brought to light many local distortion mechanisms that were not originally apparent. The relevant behavioral mechanisms in the OSD when subjected to wheel loads are summarized in table 4-1 and discussed in detail in the following sections.

Table 4-1 Orthotropic Steel Deck Deformation Mechanisms

System	Action	Figure	Result
1	Local Deck Plate Deformation	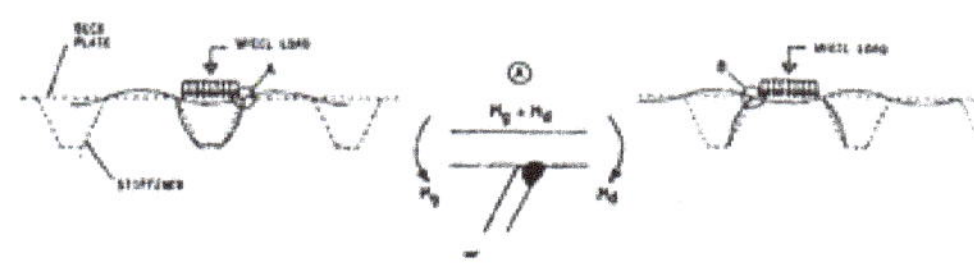	Transverse flexural stress in deck and rib plates and RD
2	Panel Deformation		Transverse deck stress from rib differential displacements
3	Rib Longitudinal Flexure	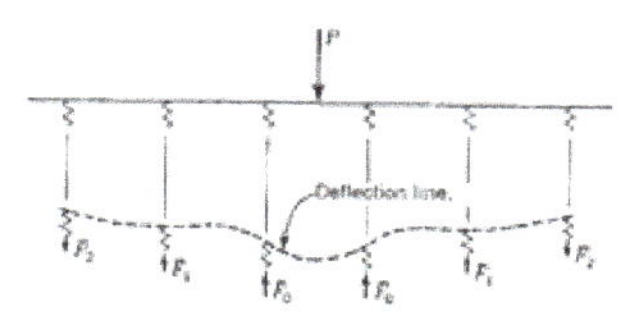	Longitudinal flexure and shear in rib acting as a continuous beam on flexible FB supports
4	Floorbeam In-plane Flexure	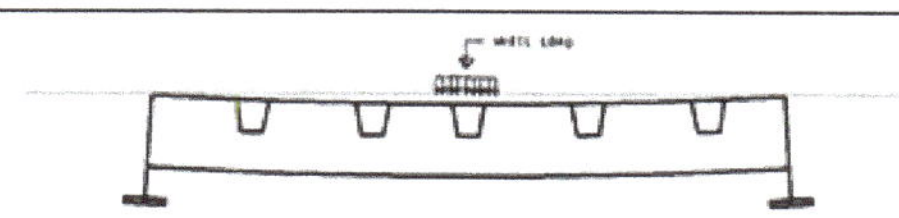	Flexure and shear in FB acting as beam spanning between rigid girders
5	Floorbeam Distortion	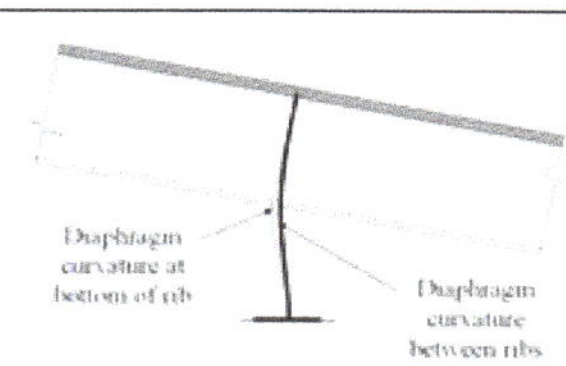	Out-of-plane flexure of FB web at rib due to rib rotation
6	Rib Distortion		Local flexure of rib wall due to FB cut-out
7	Global	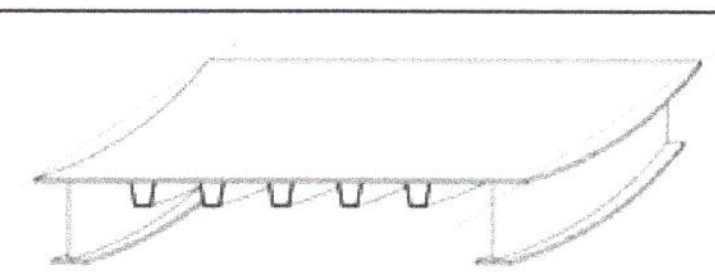	Axial, flexural, and shear stresses from supporting girder deformations

4.2.1. Local Deck Plate Deformation – System 1

Load distribution begins with transfer of local wheel loads from the deck plate to the supporting rib walls. The response is influenced by the spacing of the rib walls and relative thickness (flexural stiffness) of the deck plate and ribs. It is noted that the stresses generated from this mechanism are localized and thus sensitive to the size of the wheel patch load and any load dispersion that may occur through the wearing surface. Very often, it is the front axle of the design truck with only a single tire that can maximize response from this mechanism. It is also noted that this system is one driving factor in the fatigue of the RD, but is generally not a concern for strength based limit states. Simplified analysis of System 1 stresses can be accomplished by employing a finite strip (frame) model of the OSD cross-section in the transverse direction with assumed rigid supports for the ribs, or elastic flexible supports based on the rib flexural stiffness and span length.

4.2.2.Panel Deformation – System 2

By far, the most complicated system to analyze is System 2. Analysis of this system requires an understanding of the two-way load distributing behavior of the OSD panel when subjected to out-of-plane loading, which is a complex problem. Early engineering solutions that used OSDs had an ideological underpinning to the theory of elasticity of plates (plates loaded normally to the plane of the plate). This solution was founded in Huber's Equation:

$$\mathbf{D}_x \frac{\partial^4 \mathbf{w}}{\partial \mathbf{x}^4} + 2\mathbf{H} \frac{\partial^4 \mathbf{w}}{\partial \mathbf{x}^2 \partial \mathbf{y}^2} + \mathbf{D}_y \frac{\partial^4 \mathbf{w}}{\partial \mathbf{y}^4} = \mathbf{p}(\mathbf{x}, \mathbf{y}) \qquad (4\text{-}1)$$

This equation represents the static equilibrium of a plate of uniform thickness with orthogonal and torsional properties,

where:

D_x	=	plate flexural rigidity in the x-direction
D_y	=	plate flexural rigidity in the y-direction
H	=	effective torsional rigidity of the plate
p(x,y)	=	the loading at any point on the plate with coordinates of (x, y)

To solve for the stresses, presuming D_x, D_y, and H can be estimated, it is necessary to find w(x,y), the vertical displacement in the z direction that satisfies the homogenous and particular solution of the loading p(x,y). The moments per unit length are then given by:

$$\mathbf{M}_x = -\mathbf{D}_x\left(\frac{\partial^2 \mathbf{w}}{\partial \mathbf{x}^2} + \nu_y \frac{\partial^2 \mathbf{w}}{\partial \mathbf{y}^2}\right); \qquad (4\text{-}2)$$

$$\mathbf{M}_y = -\mathbf{D}_y\left(\frac{\partial^2 \mathbf{w}}{\partial \mathbf{y}^2} + \nu_x \frac{\partial^2 \mathbf{w}}{\partial \mathbf{x}^2}\right) \text{; and} \qquad (4\text{-}3)$$

$$\mathbf{M}_{xy} = 2\mathbf{D}_{xy} \frac{\partial^2 \mathbf{w}}{\partial \mathbf{x} \partial \mathbf{y}} \qquad (4\text{-}4)$$

where:

D_{xy}	=	$\frac{1}{4}(2\mathbf{H} - \nu_x \mathbf{D}_x - \nu_y \mathbf{D}_y)$,
ν	=	Poisson ratio in each respective direction.

Stresses are then derived from cross-section properties per unit length. Huber's equation is derived with the assumption that axial forces in the plane of the plate are not present.

The differential equation lends itself to solutions by Fourier Series. Rigorous solutions of this equation for OSDs are truly daunting tasks when FBs represent either discontinuities in the continuum or uncertain boundaries with which compatibility is necessary, but whose displacement is not easily understood. Recently, Higgins (2003, 2004) successfully applied this solution technique for calculation of moments and deflections in filled grid deck systems, which is supported by FEA and experimental results. However, OSDs with closed ribs have added complications due to increased torsional rigidity of the ribs. Single closed ribs must be smeared into a uniform property of an idealized plate. This is an idealization whose errors are not known. The most widely accepted solution to this equation was first proposed by Pelikan and Esslinger (1957).

The pioneers who developed these analytical techniques were practical individuals who knew how to make reasonable assumptions in an era when little was known about fatigue and the aim was to determine conformance to allowable stresses for service evaluation, in which even large local error is not important, because of the high ultimate strength of the deck. As expressed in the Design Manual for Orthotropic Steel Plate Deck Bridges (AISC, 1963), the simplifications attributed to Esslinger and Pelikan are as follows:

Open Ribs

- The deck plate is treated as a beam, i.e., the plate is given rigidity in the short direction, from rib to rib. Deflection and flexure (at 178.7 MPa [25.9 ksi]) and shear criteria governed, giving a 9 mm (3/8 inch) thickness over a 305 mm (12 inch) rib spacing, for a 53.4 kN (12 kip) wheel load.
- The wheel load is distributed to adjacent ribs as in a beam on elastic foundations.
- Effective width of deck plate (used to calculate the rib/deck plate composite properties over major rib carrying load) is a function of its share of the wheel load and of the "effective" rib span. The effective width is usually larger than the actual rib spacing. The effective rib span is always 0.7 times the actual span.
- Ribs "near" the FB support are treated as resting on rigid foundations, and ribs "near" FB mid-span are treated as resting on flexible foundations. Ribs near mid-span have larger positive moments and smaller negative moments than those near FB support. The Design Manual for Orthotropic Steel Plate Deck Bridges (AISC, 1963) gives moment relief formulae, based on sinusoidal deflection of the FB.
- In short, concepts of orthotropy are abandoned in favor of partial compatibility between beams. Global transverse rigidity is ignored; influence lines for beams are invoked.

Closed Ribs

- The torsional rigidity of the deck plate is governed by **G**, **K**, and μ as defined by the following equation:

$$\mathbf{H} = \frac{1}{2}\left(\frac{\mu \mathbf{GK}}{\mathbf{a} + \mathbf{e}}\right) \quad (4\text{-}5)$$

- Where G is the shear modulus for steel, **K** is a factor representing the physical properties and geometries of the rib such:

$$\mathbf{K} = \frac{4\mathbf{A}_{\mathbf{r}}^{2}}{(\mathbf{u}/\mathbf{t}_{\mathbf{r}}) + (\mathbf{a}/\mathbf{t}_{\mathbf{p}})} \quad (4\text{-}6)$$

where

$\mathbf{A_r}$	=	area enclosed by the closed rib;
u	=	entire length of the closed rib plate;
a	=	rib width where it is joined to the deck plate;
$\mathbf{t_r}$	=	thickness of the rib plate;
$\mathbf{t_p}$	=	thickness of the deck plate;
e	=	spacing between ribs stems of adjacent ribs – i.e. a + e = rib spacing;
μ	=	a number less than 1 which accounts for the reduction of the torsional rigidity due to the flexibility of the deck plate. The Design Manual for Orthotropic Steel Plate Deck Bridges (AISC, 1963) provides lengthy formulae for evaluating this factor for four closed-rib geometries; and
H	=	distributed torsional rigidity per unit width of deck.

- The transverse rigidity of the deck plate and the ribs are ignored.
- Esslinger/Pelikan solved the Huber differential equation and developed charts for longitudinal moments for various loads and spans.
- Adjustments are made to moments based on FB rigidity, the same way as is done for open ribs.

Unfortunately, torsional moments at the deck FB support were not sought in the original work. Also, the introduction of a stiffening intermediate FB that is not supported on the girders, but merely spreads the load to more ribs, is a complication that was not dealt with.

The primary value of this method is that it provided a direct solution technique for the OSD. The solution also reveals that the response of the orthotropic panel under System 2 is influenced primarily by the flexural and torsional stiffness of the ribs. Closed ribs have increased torsional stiffness over comparable open rib sections and thus provide increased load sharing and reduced differential deflections to minimize rib, deck, and wearing surface stresses. This solution is now mostly obsolete, as alternate solutions and the introduction of FEA have demonstrated its shortcomings. This method, however can provide insight as a secondary check for rib moments and shears, as well as provide a basis of comparison for overall geometry.

It is noted that deck plate and wearing surface stresses are caused primarily by the combination of: 1) flexure of the deck plate between the rib walls due to the wheel loading (System 1) and 2) flexure of the deck plate due to differential deflection of the adjacent ribs (System 2).

4.2.3. Rib Longitudinal Flexure – System 3

After loads are distributed transversely among ribs by System 2, the individual ribs then transfer load in the longitudinal direction to the FBs. In this mechanism, the rib can be considered as a continuous beam on discrete flexible supports. The System 2 will provide the rib moments and shears for the ideal case when the FBs are rigid. The System 3 provides the rib moments and shears that result from the FB flexibility. The flexible FBs cause an increase in positive rib moments and a decrease in negative rib moments, as well as a decrease in FB positive moment. To complicate matters, the ribs have continuity across FBs in the longitudinal direction, and the

FBs have flexibility that interacts with the ribs, which is difficult to quantify by simplified analysis. Indeed, De Corte and Van Bogaert (2006) recently found that effect of shear deformations in the FB can have a large influence on the bending moments that occur in the longitudinal ribs. This is especially true when the FB contains cut-outs at the ribs, which reduces the overall shear stiffness.

The Pelikan-Esslinger Method partially addressed this problem and provided a direct technique for assessment of changes in the rib and FB moments based on their relative flexibility. These techniques are demonstrated by numerical examples in the Design Manual for Orthotropic Steel Plate Deck Bridges (AISC, 1963) and Troitsky (1987). However, their work neglects the effects of shear deformation and is bound by the simplifying assumptions as described previously. An effective approach to assessment of this mechanism is to employ a simple 2-D "grid" model of a single rib with intermediate FBs or by analysis of a 1-D continuous beam with applied support settlement. Solutions to these simple problems can readily be found in the literature.

4.2.4. Floorbeam In-Plane Flexure – System 4

The next step in the load path is in transfer of load from the ribs to girders through the FB. The stresses in the FB are a combination of in-plane stress (flexure and shear) and out-of-plane stress (twisting) from rib rotations. The latter is discussed in more detail later. Much of the research on this system has been focused on determining the optimum cut-out geometry based on in-plane testing and analysis (Kolstein, 2007). Cut-outs with carefully selected geometry have demonstrated experimentally that the out-of-plane stresses can be kept below 17.25 MPa (2.5 ksi), or 25 percent of the in-plane stress (Williamsburg and Bronx-Whitestone). Thus, the focus on the in-plane behavior may be justified in some cases.

For simplified 2-D analysis of the FB, an equivalent Vierendeel model approach was proposed by Haibach and Plasil (1983) and has been adopted in the current Eurocode (ECS, 1992) (see Figure 4-6). In this model, the upper chord is equivalent to the actual deck plate. The chord is then pinned to vertical posts consisting of two parts. The upper part reflects the web area between the lower point of the cut-out and the deck plate (tooth). The lower part reflects the non-disturbed web areas between the position of the lower chord of the model and the lower point of the cut-out. The lower chord consists of the lower flange of the FB and the non-disturbed web area. This model provides the horizontal shear force in each tooth, which is then resolved into shear and flexural stress by simple mechanics calculations.

Although the Vierendeel model is a simple and direct analysis method for the in-plane behavior of the FB, it provides limited accuracy at the free edge of the cut-out. De Corte et al. (2007) proposed a two-step approach involving refined FEA modeling of a single tooth for determination of a geometric shape factor (concentration factor) that can be applied to the nominal stresses obtained from the Vierendeel model. This additional step is necessary to accurately quantify the cut-out stress. The other limitation of the Vierendeel model is that it provides no assessment of stress at the cut-out termination, which is arguably the most critical location for fatigue. Since refined FEA of the entire FB is no more difficult than FEA of a single tooth, it is recommended that the entire FB be modeled for more accurate in-plane analysis.

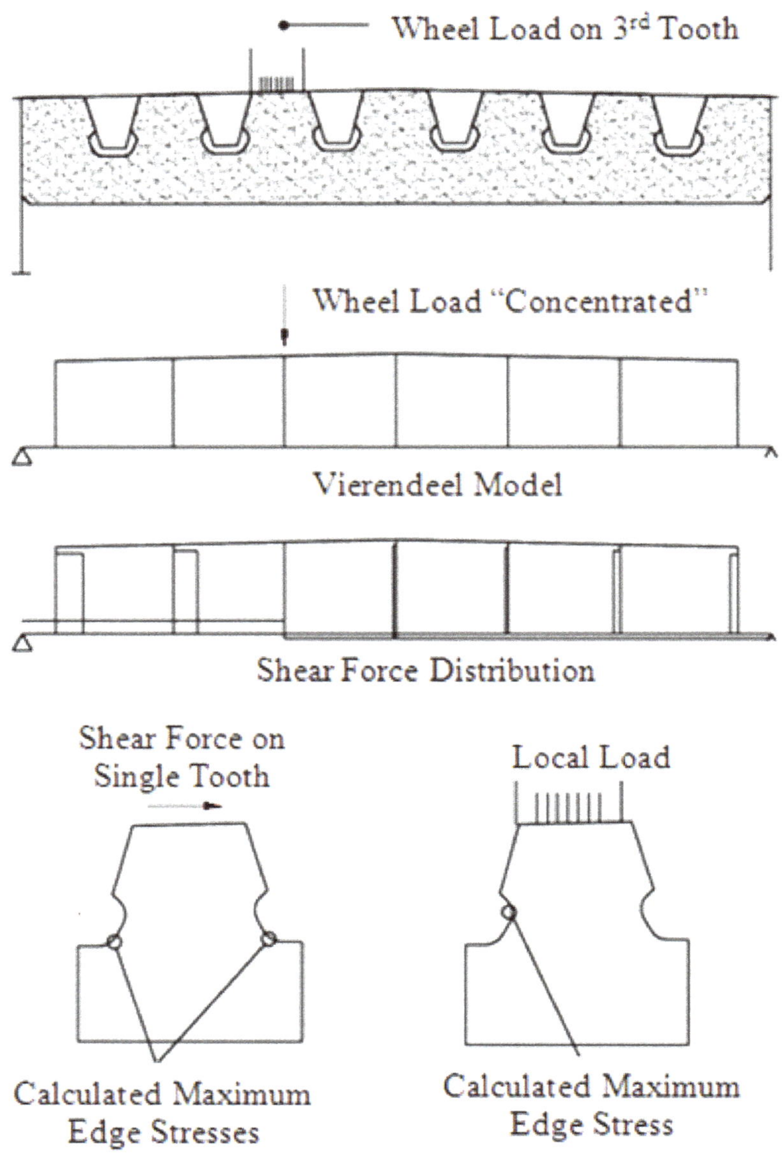

Figure 4-6 Vierendeel Model for Simplified Analysis of an Orthotropic Deck Floorbeam Showing the Qualitative Distribution of the Shear Force through the Tooth (adapted from De Corte, 2007)

4.2.5. Floorbeam Distortion – System 5

OSDs respond to the imposed load with effects that are primarily in two orthogonal directions and involve localized distortions at the FB. Immediately under the wheel, stresses in three orthogonal directions invariably materialize. The behavior of the deck insofar as it delivers loads to several ribs and to the FBs was described above.

Over the past 15 years, the engineering community has come to realize that the most important aspect of OSD design is not how wheel loads are shared by adjacent ribs or what rib moments can be expected over the FB at midspan or near support. Rather, the focus now is on what effects occur at the intersections of the ribs and FB and how these effects impact the local stresses in the plane of the FB, and on their survivability.

During the last decade, laboratory testing in the United States and the use of FEA shed light on the behavioral effects along the FB and how the rigidity of the FB and deck plate interact in ways that are often in opposition. For example, an increase in FB thickness may improve RDF stress ranges, but could exacerbate them at the cut-out or at the welded all around detail. Or, introducing a cut-out may not need a heavy rib to keep stresses low where the rib is welded to the FB (RF), but it will weaken the performance at the RDF.

The local mechanisms that impact all rib/FB details along the FB or diaphragm are:

- Out of plane distortion from rib rotation
- In plane distortion from horizontal shear
- In plane distortion from vertical displacement of the tooth.

These deformations by diagrams are discussed and illustrated below, including the impact on points of stress concentration.

Out of plane distortion from rib rotation:

Figure 4-7 shows concentrations resulting from rib rotation at support. This behavior was one of the early industry concerns, but it has been discounted as the most important since the studies that have been previously mentioned in this chapter were completed.

In plane distortion from horizontal shear:

Horizontal shear effects exist internally in any flexural member where there are shear forces transverse to the axis of the member. From basic mechanics, this is represented by the quantity VQ/I. In OSDs, where discontinuities exist by virtue of the rib passing through the FB web opening, these effects take on a peculiar form that was illustrated in part in Figure 4-2 when a bulkhead is present.

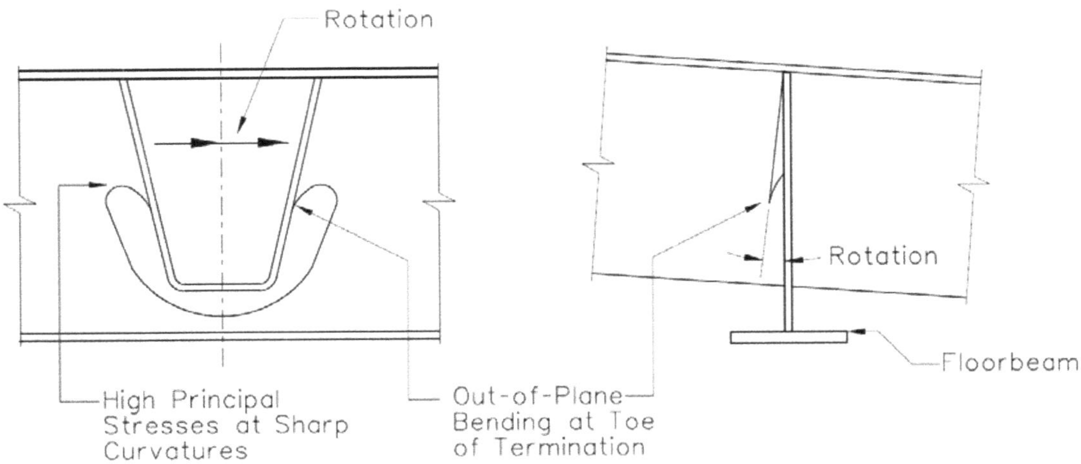

Figure 4-7 Sketch of Rib-to-Floorbeam (RF) Connection Elevation and Section Showing Exaggerated Out-of-plane Deformation of the Rib and Floorbeam Web Bending when a Cut-out is Used

Figure 4-8 and Figure 4-9 show the VQ/I effect at the FB in the condition found in a simply supported FB, with and without a cut-out, respectively.

The VQ/I effect causes flexing of the tooth, which creates stress concentrations near the base and in the deck plate. The concentrations indicated are more severe in the detail of the cut-out (see Figure 4-9) because the tooth is much weaker in-plane than the detail in Figure 4-8. It is noted that Figure 4-8 and Figure 4-9 are exaggerations intended to illustrate the VQ/I mechanisms, which in Europe is referred to, instead, as the Vierendeel effect.

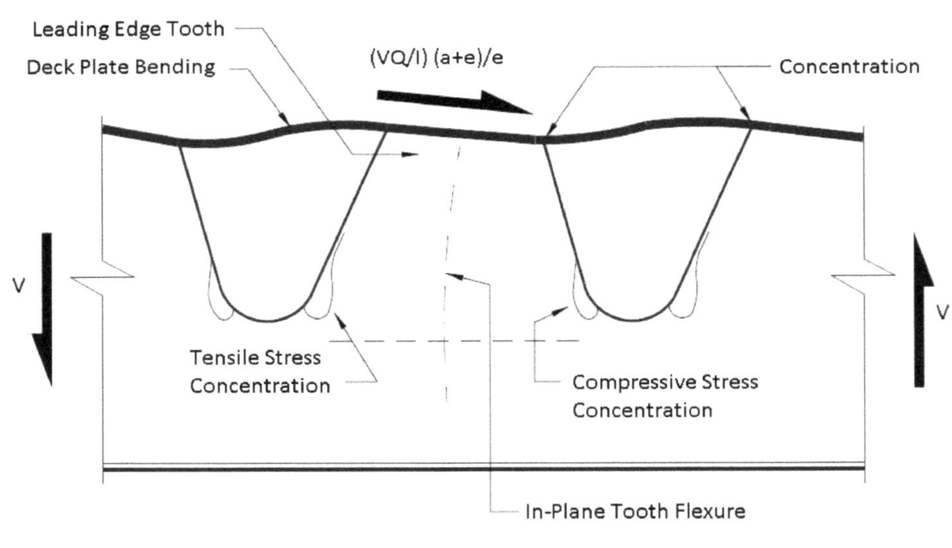

Figure 4-8 Deformation of the Deck, Floorbeam, and Rib Resulting from the VQ/I Effects on the Floorbeam Tooth, No Cut-out. Stress Risers are Shown near the Bottom of the Rounded Rib

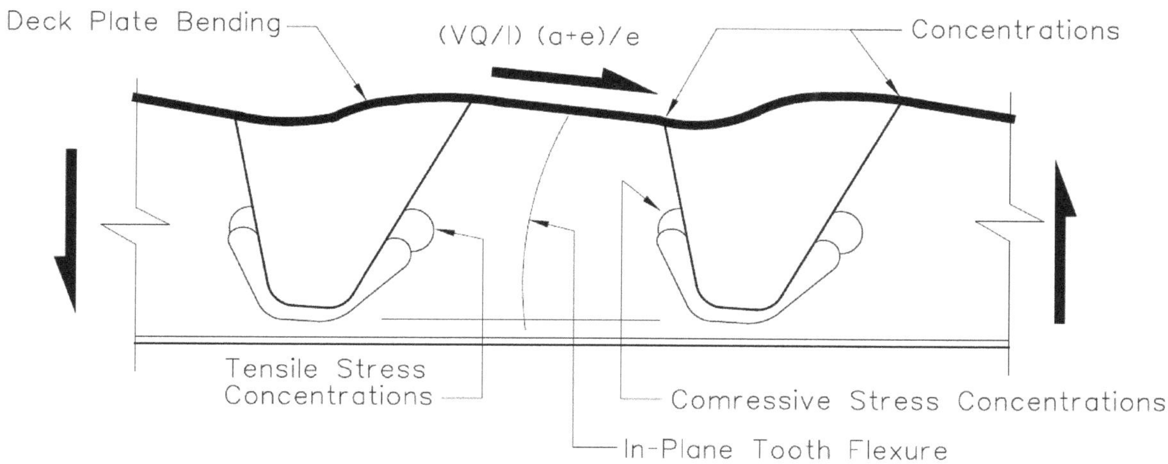

Figure 4-9 Deformation of the Deck, Floorbeam, and Rib Resulting from the VQ/I Effects on the Floorbeam Tooth, with Cut-out. Stress Risers are Shown near the Top of the Rounded Cut-out Transition

It is noted that the higher the cut-out is made, the worse the stress at the deck plate becomes, while the stresses at the rib stem diminish. Hence, optimal cut-out geometry needs to be developed with consideration of many factors.

Finite element analysis (FEA) shows that stresses at the bottom of the deck plate are more concentrated at the leading edge of the tooth (where they are compressive) than at the trailing edge where they are tensile. Also, laboratory tests of full-scale models show greater damage at the RDF of the leading edge.

The bulkhead detail is not shown under this behavior as the stresses were already shown in Figure 4-2. It is common knowledge that the bulkhead helps counteract the damaging effects of this behavior at the RDF, but it shifts concentrations to the terminations of the bulkhead where abrupt discontinuities exist (see Figure 4-3).

In plane distortion from vertical displacemt of the tooth

Figure 4-10 illustrates the vertical displacement of the tooth to both flexure and compression caused by the wheel. This, in turn, impacts on the stresses in the deck plate where it meets the rib (the RDF detail). These phenomena were observed during parametric studies for design projects and were also reported by European researchers (De Corte and Van Bogaert, 2007).

The size of the cut-out and the thickness of the FB impact the stresses at the RDF resulting from this distortional mechanism as well. To limit vertical displacements of the deck plate, the total remaining tooth dimension D_t should be as large as possible with the smallest cut out as feasible using a minimum radius after the termination, consistent with required fatigue resistances at rib and on the FB surfaces. A thick FB tooth reduces this effect, while it may increase out-of-plane effects by smaller amounts.

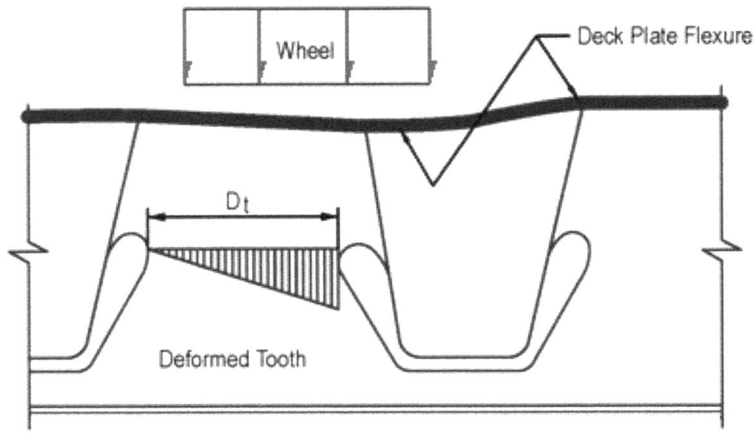

Figure 4-10 Application of a Wheel Load to the rib with a Cut-out and the resulting Vertical Tooth Displacement

4.2.6.Rib Distortion – System 6

This very important phenomenon was completely ignored until recently. In a closed-rib system, the rotation of the rib, when the wheel is at midspan and eccentric about the axis of the rib, causes the rib to twist about its center of rotation with consequent lateral displacement at midspan. The FB, however, represents a fixed boundary (in the plane of rotation).

When there is a cut-out with or without a bulkhead, the boundary is partially fixed and has discontinuities that impose out-of-plane deformations in the rib stems, which engender high stresses relative to the available fatigue resistances.

Figure 4-11 shows that these stresses are both longitudinal and "vertical" at the intersection of two hypothetical planes: 1) of the FB and 2) of the ribs stem. They are shown in the diagram with the heretofore conventional cut-out (on the order of h/3 in depth with an abrupt transition). This detail is presently not deemed optimal in many cases, as it does not provide sufficient resistance to fatigue stresses engendered by the distortion.

By observation of the curvatures in Figure 4-11, it is evident that a shallower cut-out would create more severe effects at the cut-out termination. Also, while one stem displays tension due to distortion on the outside face of the rib, the opposite stem displays compression on the outside. Thus the stresses at the inside faces of the stems are reversed. Therefore, as wheels pass on opposite sides of the rib center line, reversal of stresses occurs in both stems at these concentrations.

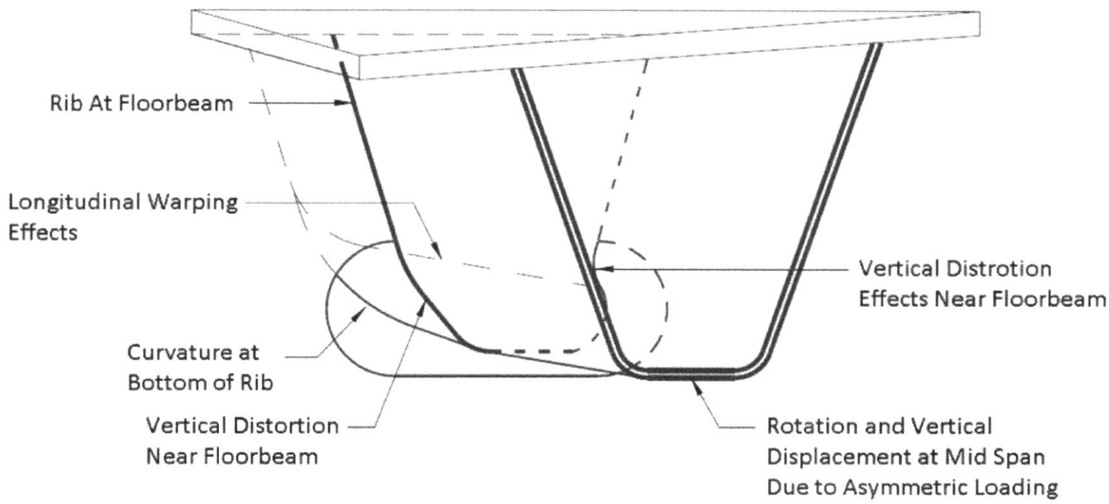

Figure 4-11 Rib Distortion Effects showing the Various Points of Distortion at the Rib-to-Floorbeam (RF) Connection

To make the cut-out detail work well in combination with a bulkhead, the FB and bulkhead would both require smooth terminations to address the vertical distortion effects. While this satisfies theoretical considerations, difficult fabrication problems are envisioned on the bulkhead side to perform necessary grinding.

4.2.7. Global – System 7

This mechanism involves displacement of the primary girder plus orthotropic panel system as it spans between points of global support. This can be evaluated using conventional methods of structural analysis. For a suspended span structure, these stresses may include local demands from the girder spanning between cable anchorage points, as well as global demands from cable sag. Review of the various possibilities for superstructures is beyond the scope of this manual. To calculate the System 7 stresses, modeling very often uses simplified "spine" elements to represent the entire bridge cross-section, with application of the "effective width" of the OSD. This is covered in more detail below.

4.3. EFFECTIVE WIDTH

The orthotropic plate, when made an integral component of the bridge superstructure system, is subjected to stresses from both local and global response, as discussed previously. Engineers often conduct global analysis of bridges by modeling the entire superstructure cross-section with a simplified two-dimensional "spine" element and then determining the resulting plate component bending and axial stresses by subsequent calculation. OSDs often involve wide proportions relative to the girder web or FB spacing, which calls into question the traditional assumption of plane sections remaining plane from elementary beam theory. In this case, due to the action of in-plane shear strain in the deck plate, the longitudinal stresses in the parts of the plate remote from the web lags behind those nearer the web. For design purposes, it is often convenient when calculating stresses to replace the actual width of the flange with an "effective width," which will produce the same maximum stress resulting from the shear lag phenomenon by elementary beam theory.

The effective width need not be determined when using refined analysis (as described in Chapter 5), but the concept of effective width is intended to enable the design engineer to easily calculate the maximum stress on the OSD plate from actions on the primary girder or FB to ensure that, for applicable limit states, it is within allowable limits. It is noted that some definitions of effective width are made based on elastic first order analysis of stress, while others are based on inelastic behavior considerations and post-buckling behavior. The limit state under investigation must be consistent with the assumptions made in the definition of effective width. An example of this is the case of concrete slab effective width in a composite steel girder; for Service and Fatigue limit states the slab is considered fully effective, while for Strength limit states the slab has a reduced effectiveness.

The first approach proposed for calculation of stiffened flange plate effective widths is found in Wolchuk and Mayrbaurl (1980) and is based on research by Moffatt and Dowling (1975 and 1976). This work is based on linear elastic stress distribution observed from FEA. The following discussion focuses on the effective width of the box flange between the webs within the positive moment regions and in the negative moment regions in the vicinity of interior supports. Using the symbology provided in the reference, the effective width of a box flange is expressed as $b_{eff} = \psi B$, where B is the total width between the webs and ψ is the effective width ratio (see Figure 4-12).

Curves (1) and (2) in Figure 4-12 apply to the maximum positive moment region of simply-supported girders and continuous girders. The distance $L = L_1$ is taken as the simple-span length or the distance between the points of inflection in determining the value of ψ for these regions. Curve (1) applies to unstiffened box flanges, while Curve (2) applies to stiffened box flanges with a ratio of the stiffener area to the box flange area $A_s/Bt = 1$. The values of ψ are to be determined for intermediate values of A_s/Bt by interpolation. One can observe that even for the extreme case of $A_s/Bt = 1$, ψ is approximately equal to 0.9 at $L/B = 5$. This value is corroborated by the work of Goldberg and Levy (1957), in which a minimum effective box flange width of 0.89 was found for the case of $L/B = 5.65$.

Curves (5) and (6) in Figure 4-12 apply to the cross-section over interior supports in continuous-span girders. In this case, $L = L_2$ is taken as the distance between points of inflection on each side of the support. If the distances between the support and the points of inflection on each side, C_1 and C_2, are unequal, ψ is determined as the average of the values of ψ for $L_2 = 2C_1$ and $L_2 = 2C_2$. One can observe that at $L/B = L_2/B = 5$, curves (5) and (6) indicate a range of ψ values of only 0.55 to 0.62. This reduction in effective width is attributable to the large vertical shear force near the support, and not only due to the fact that the flexure is negative.

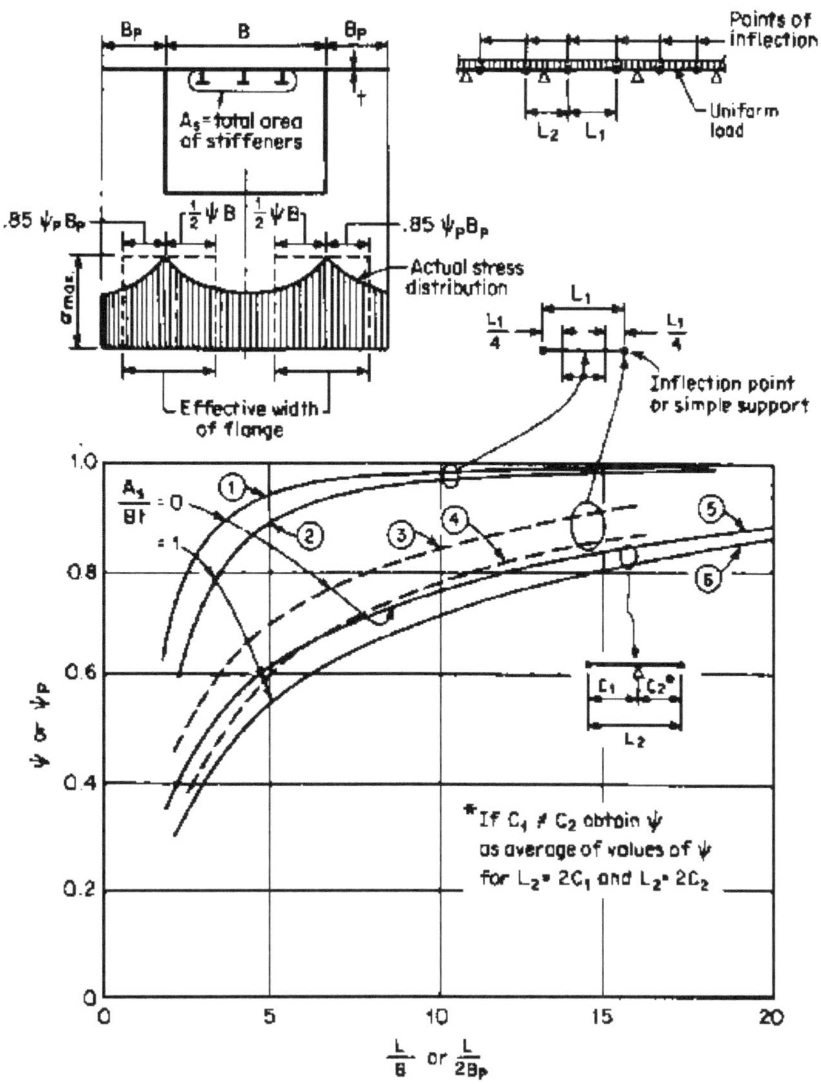

Figure 4-12 Flange Effective Widths for Orthotropic Deck Panels and Design guide Chart (Wolchuk and Mayrbaurl, 1980)

Curves (3) and (4) in Figure 4-12 apply at the inflection points or simple-support locations. In Wolchuk and Mayrbaurl (1980), the value of ψ is taken as a constant based on the value from Curves (1) and (2) within the middle $L_1/2$ of the span, and is varied linearly between this value and the inflection point or simple support values, and linearly between the inflection point values and the interior support values. Obviously, this level of refinement in the assumed effective width may not be necessary in general.

For simplified analysis, the effective width of the deck, including the deck plate and ribs, acting as the top flange of a longitudinal superstructure component or a transverse beam may be taken as:

- $L/B \geq 5$: fully effective
- $L/B < 5$: $b_{od} = 1/5\ L$

where:

L	=	span length of the OSD girder or transverse FB,
B	=	spacing between OSD girder web plates or transverse FBs,
b_{od}	=	effective width of OSD,

for Strength limit states for positive and negative flexure. Tests have shown (Dowling et.al., 1977) that for most practical cases shear lag can be ignored in calculating the ultimate compressive strength of stiffened or unstiffened girder flanges (Lamas and Dowling, 1980) (Burgan and Dowling, 1985) (Jetteur, 1984) (Hindi, 1991). Thus a flange may normally be considered to be loaded uniformly across its width. Only in the case of flanges with particularly large aspect ratios [L/B<5], or particularly slender edge panels or stiffeners (Burgan and Dowling, 1985) (Hindi, 1991) is it necessary to consider the flange effectiveness in greater detail.

Furthermore, consideration of inelastic behavior can increase the effective width as compared to elastic analysis. At ultimate loading, the region of the flange plate above the web can yield and spread the plasticity (and distribute stress) outward if the plate maintains local stability. Results from studies by Chen et al. (2005) on composite steel girders (including several tub-girder bridges) indicate that the full slab width may be considered effective in both positive and negative moment regions. Thus, OSD plates acting as flanges are considered fully effective for Strength limit state evaluations from positive and negative flexure when the L/B ratio is at least 5. For the case of L/B less than 5, only a width of 1/5th of the effective span should be considered effective.

For Service and Fatigue limit states in regions of high shear the effective deck width can be determined by refined analysis or other accepted approximate methods. Additionally, consideration of effective width of the deck plate can be avoided by application of refined analysis methods.

If the designer prefers to do so, the procedures in the Design Manual for Orthotropic Steel Plate Deck Bridges (AISC, 1963) may be used as an acceptable means of simplified analysis. However, it has been demonstrated that using this procedure can result in rib effective widths exceeding the rib spacing which may be unconservative.

4.4. LOAD PATHS AND LOAD DISTRIBUTION

The important issues on the subject of load distribution in OSD bridges are related to: 1) the local dispersion of wheel loads through the wearing surface and 2) how wheel loads are transferred from the deck plate to the supporting superstructure girders. Accepted practice has been to assume that wheel pressure loads are dispersed through the wearing surface at an angle of 45 degrees in all directions when a thick bituminous material is used. However, recent research has indicated that this approach is not always correct, and any load dispersion is more dependent upon the tire contact hardness (Battista et al, 2008). Although stresses in some of the OSD components are sensitive to the dimensions of the wheel patch area, any load dispersion may be lost at higher temperatures when the wearing surface material softens. Also, it is conceivable that in the future the thick wearing surface may be replaced by a thin surface with little dispersion capability. For these reasons, no dispersion will typically be assumed for design purposes, which is not considered overly conservative.

The second important issue on the subject of load distribution is related to how wheel loads are transferred from deck to ribs to FBs to girders. The sequence of this load transfer has been described in detail previously in the section on behavior mechanisms. Most relevant to the analyst is the fraction of a wheel load that is carried by an individual rib in the OSD system. This is used to facilitate simplified one-dimensional analysis of the rib. For the panel mechanism (System 2), analytical and experimental results have indicated that wheel loads are typically shared transversely between three ribs for most typical panel geometries, i.e. the rib directly under the wheel and the neighboring ribs on either side. Thus, there is minimal accumulation of load in an individual rib from two side-by-side trucks. The single truck event controls the rib response.

Field monitoring tests have also revealed that load distribution between ribs is highly dependent upon temperature and loading velocity. This is due to the viscoelastic material response and thermal properties of the wearing surface material. For accurate assessment of the load distribution accounting for these factors, prototype testing or refined FEA modeling is required.

Based on typical proportions of modern closed rib panel designs ($H/D_y > 0.05$ and rib spacing > 0.610 m [24 inches]), the Pelikan Esslinger Method indicates that using a wheel load distribution factor of 0.5 for rib design will yield conservative estimates of moment when the FB spacing is no less than 3.05 m (10 ft). Wheel load distribution factor should be taken as 1.0 for rib shear. Note that a ratio of $H/D_y > 0.07$ is recommended to provide sufficient resiliency for the wearing surface.

4.5. FATIGUE PERFORMANCE OF STEEL CONNECTION DETAILS

4.5.1. Rib-to-Deck Plate (RD) Weld

The rib-to-deck plate (RD) weld is likely one of the most studied welded joints in OSDs. Interest in this connection is obvious as there can be 50 times the bridge length of the RD connection in a typical OSD, depending on the number of ribs in the panel section. Hence, economic fabrication and long-term performance of this detail are essential.

Fatigue cracking at this detail has been observed to initiate at several locations, depending on various parameters pertaining to the connection, as will be discussed. Much of the cracking in the field has been in Europe (Kolstein, 2007), but there is at least one known case in the United States. Because of the importance of this connection and the observed cracking, there has been much laboratory testing of the joint. In fact, Kolstein reports that there have been 245 tests distributed over nine independent research programs from 1974 to 2000. Unfortunately, many of the test programs, primarily the earlier data, does not realistically replicate the actual in-service stress range cycle.

Field measurements at the RD indicated that the response of this apparently simple connection is rather multifaceted. The effects of multiple vehicles in different lanes, vehicles changing lanes, and wearing surface stiffness all can influence the stress ranges applied to the detail. And although the previous issues with respect to loading are important, the most studied has been related to the fatigue **resistance** of this detail. Both the response and resistance of the detail will be discussed in the following sections in the context of the AASHTO LRFD (2010) Fatigue limit state.

Response

Field measurements confirm that the response of the RD is driven by individual wheels of trucks. Stress ranges that are transverse to the rib wall in the deck plate and/or in-plane stresses (due to axial and bending stresses) in the rib wall contribute to the cyclic stress range producing fatigue damage.

Although a small proportion of the stress range cycle is due to global effects, such as global compression of the OSD cross-section or FB deflection resulting in bending of the entire deck, these effects are generally small in comparison to the local bending that occurs directly under the applied wheel loading (behavior Systems 1 and 2 discussed in Section 4.2). Furthermore, these other loads primarily produce shear, tension, or compression stresses that are parallel to the weld axis (i.e., parallel to the weld toe and weld root). Hence, they do not generally present a concern for the Fatigue limit state. The local, and more critical, behavior is shown in Figure 4-13, which was taken from measurements made on the Williamsburg Bridge (Connor and Fisher, 2001). The gages, identified as CH90R and CH90D were installed on the rib and deck plate respectively, immediately adjacent to and perpendicular to the longitudinal RD weld. The individual axles of this five-axle truck are clearly seen in the data.

It is also important to recognize that very often the total stress range cycle is the result of multiple vehicles in series. Note the data in Figure 4-14, which were obtained from a strain gage installed on the rib wall perpendicular and immediately adjacent to the longitudinal RD as two random trucks and one other vehicle passed (Connor and Fisher, 2001). It can be clearly seen that each truck produces a unique, but different, stress range cycle; the first truck produces a tension stress range, while the second truck produces a compression stress range. (The small stress cycle after the first truck is believed to be a passenger car.) The same observation was true for gages mounted on the deck plate or rib wall and measured stresses were either tension or compression, depending on the transverse position of the wheel load. In fact, for the second truck in Figure 4-14, the individual axles of the tandem can be made out. However, the front axle is not readily apparent, again illustrating the sensitivity of the detail to transverse position.

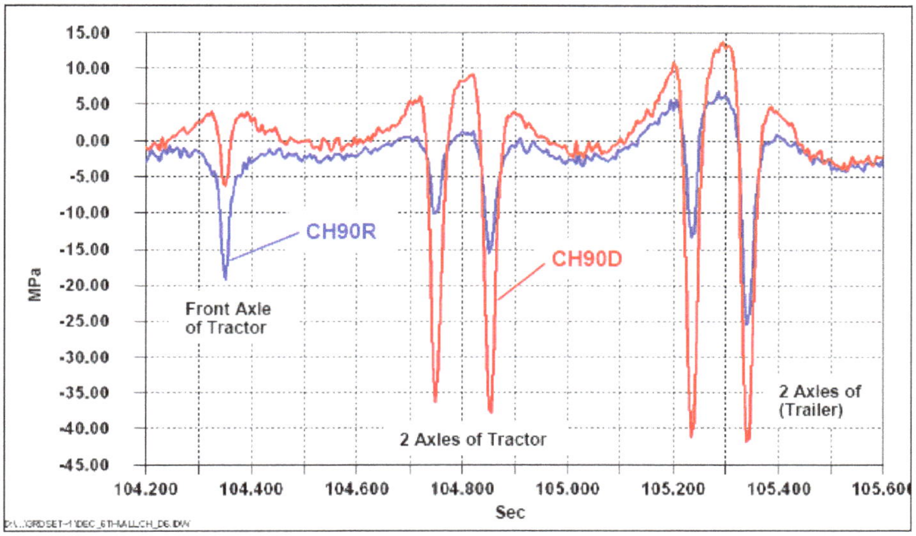

Figure 4-13 Measured Strain Response of Two Gages on the Rib and Deck Plate as Random HS Type Truck Passed (Note that Each Axle Produces an Individual Stress Cycle).

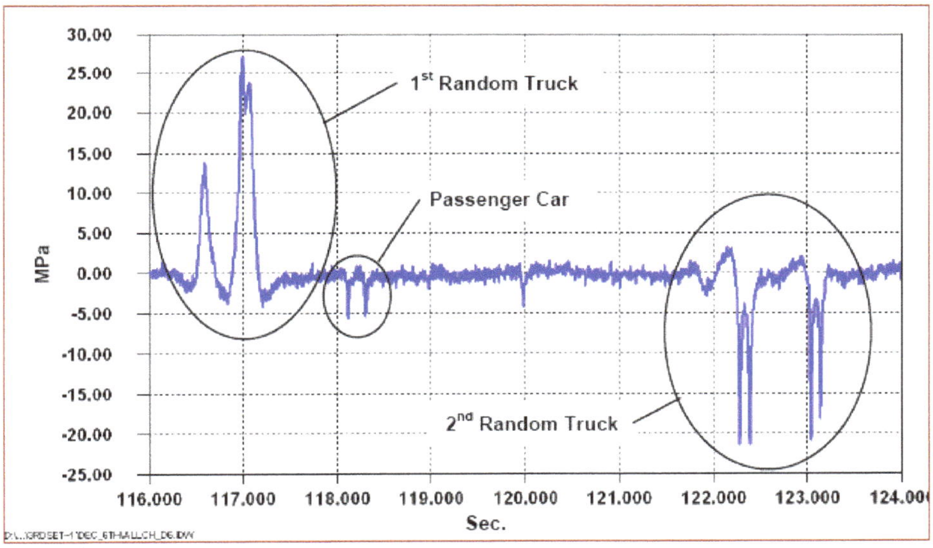

Figure 4-14 Measured Strain Response at of One Strain Gage Adjacent to a Rib Due to the Passage of Two Random Trucks (Note that the First Truck Produces a Tension Stress Range while the Second Truck Produces a Compressive Stress Range)

Observed Cracking

Fatigue cracks have been observed to initiate at the weld toe located on the rib wall and deck plate, depending on the relative stiffness of the rib/deck plate system. In addition, cracking has been observed to grow out of the lack of fusion zone found at the RD (see Figure 4-15). Of the two types of cracking (toe or root), root cracking has the potential to result in the poorest fatigue resistance. To prevent root cracking, the degree of penetration and the fit-up gap are the most important parameters controlling the performance of the connection. Large lack-of-fusion zones where the deck and rib wall come in contact effectively act as initial cracks within joints. Since these cracks grow from the inside out, once observed, they have already grown through thickness. In some cases, the cracks can be quite long (parallel with the longitudinal axis of the weld) once visible. In addition, due to the complex stress state at the connection, it has often been observed that root cracks will propagate into the deck plate, even though they initiated in the weld root, as the stress range in the deck plate is typically greater than in the rib wall. New research indicates that if the fit-up gap prior to welding is controlled, root cracking can be prevented (Wright, 2011).

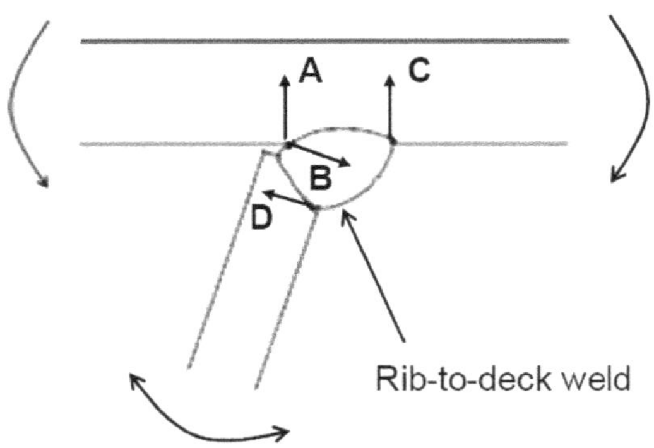

Figure 4-15 Rib-to-Deck (RD) Weld showing potential Locations for Cracking as noted by locations A though D in Response to Stresses in the Plates

4.5.2. Rib and Deck Splices

Rib and deck splices can be bolted or welded. Welded splices have a lower fatigue resistance, but are used to provide the most economical solution. The most common practice in the United States has been to use a bolted rib splice and a welded deck plate splice.

Response

In terms of live load, the ribs themselves primarily carry the moments and shears from passing wheel loads. The splices are generally located some nominal distance, roughly one to three rib depths, away from a FB/diaphragm. Measurements made near splices indicated that the rib and deck plate essentially behave as bending members, though some level of torsional moment also exists. The response of an individual rib is heavily dependent on the transverse position of the truck, as expected. For a gage installed on the bottom flange of a closed rib, the effects of individual axles and axle groups can be discerned. However, the individual axles of a tandem axle do not produce two unique and independent stress range cycles at this location. The same is not true of the response of the deck plate splice, which is subjected to direct wheel loads, similar to response of the RD connection discussed earlier. (That is, a three-axle dump truck would likely produce one primary stress range cycle due to the rear axle tandem and one smaller cycle due to the front axle at the bottom of the rib, while three individual cycles would be observed in the deck plate. This response is often referred to as a “camelback” response due to the two humps in the time history data”).

In many structures where an OSD is supported by a separate, primary FB (such is often the case in situations where the deck is installed on an existing structure) the flexibility of the FB must be included in the analysis. This is necessary since modeling a FB as a rigid support will not accurately reflect the response in the rib. Field and laboratory measurements have shown that due to the flexibility of the FB, tensile stresses (i.e., positive moments) could be expected in the bottom of some ribs due to the deflection of the FB. Thus, a rib splice may not only be subject to negative moments, as would be predicted assuming a rigid internal FB.

The reason for this can be illustrated by considering the example of a continuous beam (rib) on elastic supports (behavior System 3 as discussed previously). If a point load were applied directly over the centerline of a flexible support (FB), a positive moment would be produced in the beam (rib) over the support (FB). However, on the deck, the load is moving and is directly over the FB for only an instant. Hence, as the load travels longitudinally, both a negative moment and a positive moment (due to the deflection of the FB) are generated in the beam (rib). The addition or superposition of these two moments produces the final stress range in the rib.

It is also noted that the relative magnitude of the positive and negative moments are influenced by the FB stiffness, gross vehicle weight (GVW) and proportion of wheel live loads. Hence, heavy and light trucks will likely produce different proportions of negative and positive moments. The effects of FB flexibility are especially important to consider in situations where the OSD is installed on a cantilever overhang.

Although the response of the lower portion of the rib is primarily driven by axles and axle groups as described above, the deck plate splice is subjected to stress ranges produced by individual wheels, as well as the more "global" bending of the rib/deck beam. Since the deck plate acts a top flange to the rib, the latter component is generally small due to the rather large area of the deck plate (i.e., the neutral axis is rather high). As a result, the fatigue design of the deck plate splice is generally controlled by the local load effects from individual wheel loads (behavior System 1).

Observed Cracking

Bolted splices have higher fatigue resistance than welded splices. For welded splices, both the deck plate splice and rib splice most often involve a one-sided full penetration weld, with backing bar often left in place. Focus is given here to deck plate splices since these are subjected to more severe loading from direct wheel loads. Rib splices have similar fatigue resistance, but stresses are primarily in-plane only.

For welded deck plate splices, there are effectively two locations where fatigue cracking is a concern and each should be considered separately. Both are illustrated in Figure 4-16. First is throat cracking of the deck plate weld, whether or not the backing bar is left in place at the full-penetration transverse groove weld. Second is toe cracking of the deck plate at the end of the longitudinal weld between the deck plate and the rib wall. Both of these details are subjected to a combination of in-plane and out-of-plane stresses. It is noted that if the backing bar is to be left in place, it is critical that any splices in the backing bar be made using CJP welds. There have been cases where cracks have initiated out of the lack of fusion zone present between backing bars that simply butted together (Kolstein, 2007) Ceramic backing bars have also been used with mixed performance, though generally, the observed problems were related to poor adhesion of the backing bar or improper execution. It is noted that if a ceramic backing bar is used, small vertical misalignments of the deck plate cannot be corrected as easily when a steel backing bar is used and strongbacks may be required.

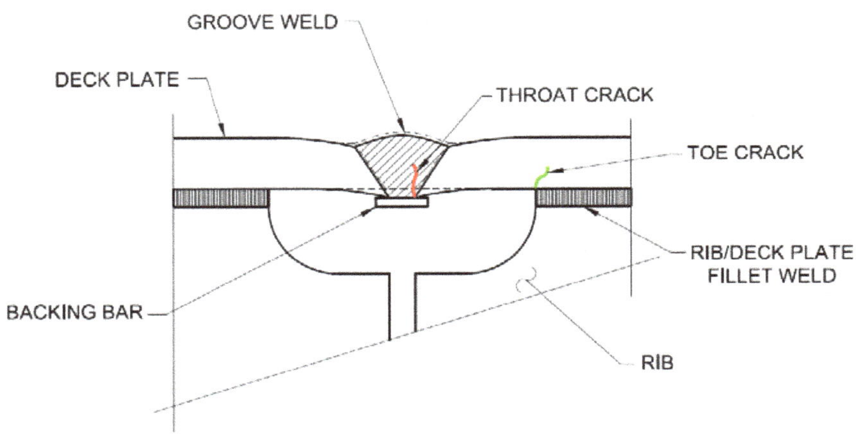

Figure 4-16 Detail at Transverse Groove Weld of the Deck Plate Illustrating the Two Types of Potential Cracking that Can Occur (Deflected Shape of the Welded Region is Exaggerated)

The full-penetration groove weld is continuous across the width of the deck and is subjected to the vertical forces applied from individual wheel loads as each axle produces a single cycle. Hence, each passing wheel produces local out-of-plane bending stresses in the deck plate. The deck plate is also subjected to longitudinal in-plane stresses generated by the longitudinal bending of the OSD panel between FBs and the global response of the entire girder span. Of these three stress components, the local out-of-plane bending stress has been observed to dominate the stress cycle (Connor and Fisher, 2001).

Figure 4-16 indicates the orientation of potential crack growth at the weld toe termination near the cope hole. Also indicated in Figure 4-16 is the exaggerated deflection of the deck plate due to wheel loads. The deck plate in this region acts as a beam with a span equal to the clear distance between the ends of the rib cope holes. At the edges of the cope hole, restraining moments are developed. The stresses produced by these moments dominate the stress cycle and may ultimately produce cracking at the weld toe if the Fatigue limit is exceeded. The presence of the weld toe at the termination of the RD aggravates the condition.

4.5.3. Rib-to-Floorbeam (RF)

The behavior, fatigue resistance, and evaluation of stresses adjacent to the rib-to-floorbeam connection (RF) are also well researched for OSDs. There have been many details that have been used to make this connection. Details with and without cut-outs, with and without internal bulkheads, various weld types, and plate sizes have been used. In addition, where cut-outs have been used, the geometry of the cut-out itself has been studied extensively.

It is important to recognize there is not a single solution that will work in every situation. One type of connection, though successful in one bridge, many not work in another due to differences in detailing, overall geometry, stiffness, and of course, loading. Nevertheless, there have been some concepts used that have such poor characteristics that they would likely result in premature failure in any bridge.

Response

Field studies have confirmed that the FB web plate is subjected to a unique stress cycle due to in-plane and out-of-plane forces, as shown in Figure 4-17. Stresses are the result of behavior Systems 4 and 5 described previously. The same is observed in laboratory and analytical studies where boundary conditions and loads have been properly simulated. In many cases, the in-plane component dominates the stress range cycle, but this may not always be the case. For certain, the proportion of the out-of-plane stress-range can vary substantially from rib to rib, depending on the transverse position of the truck and the stiffness of the OSD. Likewise, in decks where the rib and/or the FB spacing is large, the out-of-plane component will be larger and may exceed the in-plane component at some ribs.

The proportion of the out-of-plane stress range in the FB plate is related to the rotation of the longitudinal ribs. Ribs directly under a wheel load rotate nearly the same amount at the FB regardless of the transverse position of the rib. This is because the relative stiffness and boundary conditions for each rib are essentially the same (excluding external ribs). The same is not always necessarily true regarding the in-plane component of the stress range cycle. Thus, although the magnitude of the out-of-plane stress range may often remain about the same from rib to rib, its contribution in the total stress-range cycle will vary.

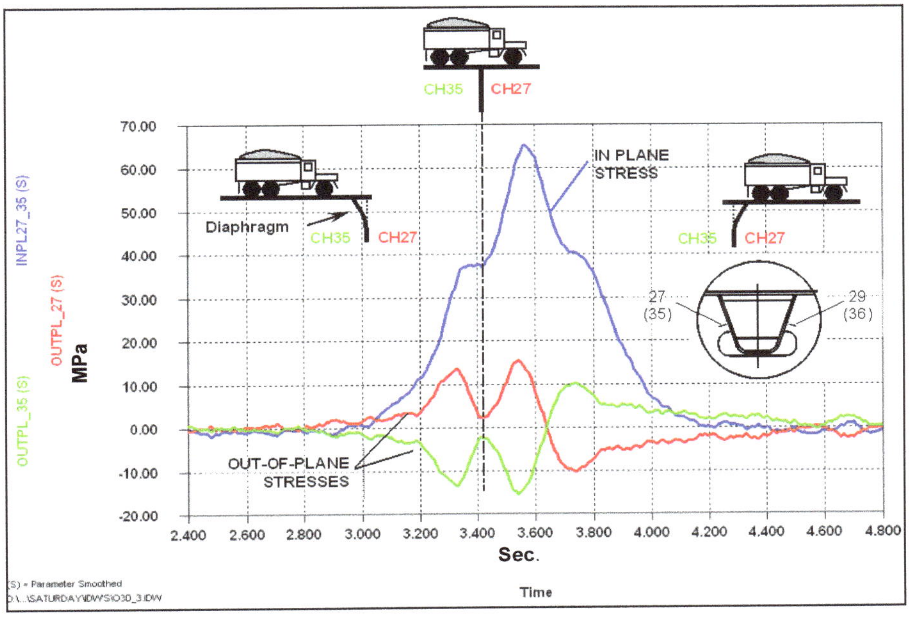

Figure 4-17 In-plane and Out-of-plane Stresses Measured at the Cut-out Near a Rib with Wheel Load Passing above, Demonstrating the Effect of the Truck Position on the Response (Connor and Fisher, 2004)

Another interesting observation can be made by examining Figure 4-18, which is data obtained as a random five-axle HS series truck passed. Shown in the figure are stresses measured in the deck plate near the FB, which clearly shows the individual stress-range cycle produced by each axle (CH_88D in Figure 4-18). However, contrast this response from the gages installed adjacent to the cut-out at two separate ribs. (i.e., CH_27 and CH_29 on Rib 5, CH_45 and CH_46 on Rib 7; Rib 5 was directly beneath a wheel load). Although just a few inches below the deck plate, the effects of the individual axles are already distributed such that their individual effects are not apparent. Thus, although this truck produced five individual cycles in the deck plate, only one primary stress cycle is produced by the truck (one smaller secondary cycle is present, but of much smaller magnitude). The behavior below is often referred to as a camelback response due its resemblance to the humps of a camel. The same type of observation has been observed through field measurements on other bridges (Connor and Fisher, 2001) (Connor and Fisher, 2004).

This is an important observation since it indicates that regardless of the configuration (i.e., H or HS type truck) the passage of a truck will produce in a single stress range cycle at the FB. The smaller secondary cycle produced during the passage of HS series trucks will produce essentially minimal damage. For example, the HS truck in Figure 4-18 produced a single primary stress cycle of about 60 MPa (8.7 ksi) at channel CH_27. The smaller secondary cycle is only about 20 MPa (2.9 ksi), or 33 percent of the primary cycle. In terms of cumulative damage using Miner's rule, the smaller secondary cycle contributes much less damage than produced by the primary cycle. Hence, it is clear that the secondary cycle contributes very little to fatigue damage.

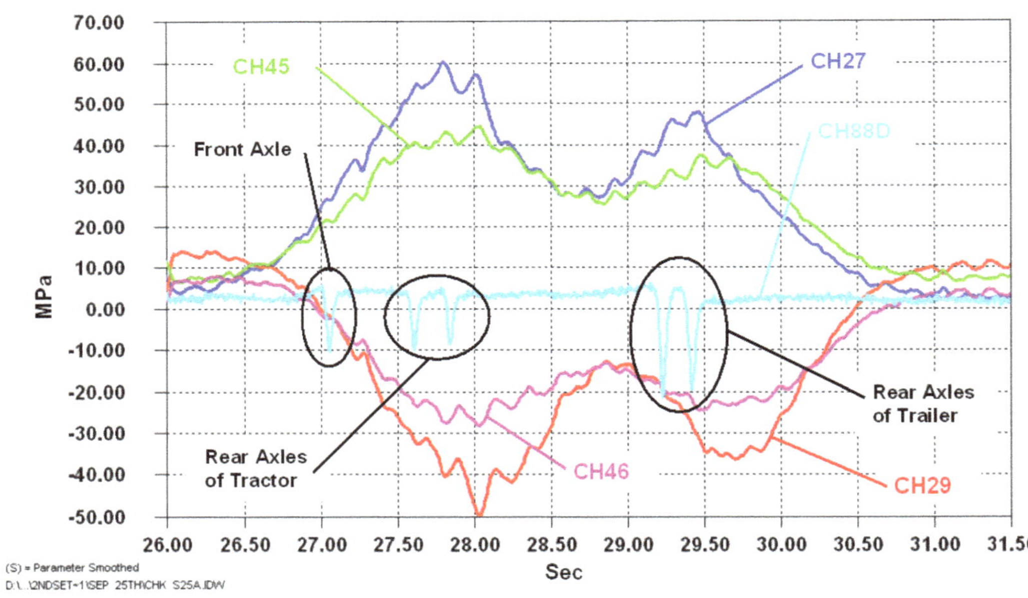

Figure 4-18 Detail of Response on Floorbeam Plate Due to Passage of a Long Random 5-Axle HS Series Truck (Channel CH88D is Located on the Deck Plate) (Connor and Fisher, 2004)

It is also clear that the primary stress range cycle is essentially driven by a single axle or group of axles spaced closely together (e.g. a tandem or tri-axle configuration). Thus, for long vehicles, such as tractor trailers, it is not the gross vehicle weight of the entire truck that controls, but rather the groups of axles. Of course, this depends on the spacing of the FB and the distance between axle groups. However, for typical trucks and common FB spacing, the above is a reasonably representative statement. Therefore, the assumption that each truck will produce a single primary cycle is appropriate for the FB cut-out in many cases. The same observation would be true for RF connections where there is no cut-out present. For short "H" series trucks, the group of axles comprising the truck will control and hence the GVW of the entire vehicle would need to be considered.

4.5.4. Rib-to-Deck at Floorbeam Joint (RDF)

Where the rib-to-deck (RD) weld crosses the FB, a concentration of stress is created in the RD weld due to a local rigid support from the FB web, and from the distortions that occur by in-plane flexure of the FB (System 5). Simple models such as a fixed-fixed beam solution have been proposed to assess the local stresses in the deck plate due to a wheel load placed directly over the RDF, however, this only treats deck plate flexure component of stress and the effective width of deck plate is not defined. Further, it is the distortional stress that may cause tension in the weld root which is thought to be more detrimental to the fatigue performance. Arguably, this is the least understood of all the OSD details and there may be need for future research to develop better analytical techniques.

As with any detail on the deck plate, each passing axle produces an individual stress-range cycle. Hence, the weight of individual axles and not GVW will control the design. Clearly, the detailing of the joint drastically affects the response. For example, details with copes (rat holes) in the FB verses those with no cope will obviously behave differently, with potential cracking potential at different locations. Cope details have been applied in some countries to avoid intersecting welds, but there is no known evidence of problems arising directly from this.

In the United States, the non-coped detail has been favored and the performance of the joint seems to have been adequate. It would also appear that from the literature, the non-coped detail has generally outperformed those where copes have been used. However, it is very important to point out that regardless of the detail used, most of the RDF cracking reported in the literature, particularly in Europe, has been observed on bridges where very thin deck plates were specified. For example, deck plates as thin as 9 mm to 11 mm (3/8 inch to 7/16 inch) were commonly used. In reality, it is not surprising that fatigue cracks appeared in these decks in a relatively short period of time (7 to 10 years).

4.6. REFINED ANALYSIS FOR FATIGUE ASSESSMENT

In all fatigue evaluations, a calculated (or measured) stress range is compared to some permissible stress range. The processes by which each of these parameters is established ranges from the very simplistic to the extremely complex. As with most analytical procedures, those that are the most complex require input that is often difficult to obtain. In addition, slight changes to the input can sometimes have drastic effects on the output, such as calculated life. However, the more simplistic methods, though "easier" to implement, may result in overly conservative and uneconomical designs. Even more importantly, simplified analysis can lead to unconservative designs.

Obviously, any analysis method must be capable of achieving the desired goal of the fatigue evaluation; that is, preventing premature fatigue cracking of the various components of the OSD. In order to achieve this goal, there are a few qualities the methodology must possess. These are listed, in no particular order:

- **Accurately accounting for the load effects from in-plane and out-of-plane forces.** The model used to obtain these load effects must accurately simulate the actual 3-D behavior of the complete OSD and supports and how these forces enter the connection.
- **Accurately accounting for the effects of subtle changes in weld type and profile.** It is well known that small changes in the type of weld and weld profile used at this connection can substantially alter the fatigue performance of the joint. This is especially true if a cut-out detail is used. Hence, the procedure must be capable of accounting for these parameters in the fatigue life prediction model. Note that this is different than the changes in stress range due to changes in the overall geometry of the detail or the system behavior discussed in the item above. For example, assume a cut-out detail was being evaluated for the Fatigue limit state. If a CJP weld were used, it would have a greater fatigue resistance than had the same geometry been used, but a fillet weld was used instead. Furthermore, if the CJP joint were used, but the welds were not ground smooth, the fatigue resistance would also be different for each condition. These subtle differences must be accounted for in some fashion in the model.
- **Producing consistent fatigue stress predictions.** Ideally, the method will be applicable to all geometries, weld types and configurations for the joint and produce consistent estimates that are user independent.
- **The accuracy of the method should be verified.** The ability of the model to accurately predict fatigue life must be verified. Predicted fatigue lives, crack locations, and other important parameters predicted by the method should be verified by comparing to experimental data. The data could be new or existing test data found in the literature could be used. The data would not need to be from the exact geometry to be used in the design. It should be from specimens that are very similar so that all important variables are included in the model and database.

Such information can only be obtained using 3-D finite element modeling. Although the global behavior of selected components of the OSD system (i.e., the ribs or FBs) can be reasonably predicted with approximate methods such as Pelikan–Esslinger, none of the approximate methods provide ability to quantify the distortional stresses and displacements at critical details, like the RF connection. However, every documented performance failure of an OSD has been the result of localized stresses and not global response (Connor, 2002) (Kolstein, 2007).

Generally, there are two established methods for numerical fatigue evaluation that are viable for use on OSDs: 1) a nominal stress S-N curve approach (e.g. AASHTO [2010]) and 2) a local structural stress approach (e.g. IIW [2007]). Each of the methods has advantages and disadvantages, which are discussed in more detail in Chapter 5. While one method may be appropriate for a certain type of detail, the same approach may not be the best or most efficient choice for another detail.

4.6.1. Influence Surface Analysis

Accurate and efficient refined analysis of OSDs demands the use of influence surface based techniques. The stress response at critical details is often sensitive to the precise location of the wheel loading. Because of integral nature of the OSD and the complicated in-plane and out-of-plane effects that exist at many details, the governing positions for loading are not always apparent by judgment. The total stresses at critical details are typically the result of multiple behavior systems as described previously, all of which may not have maximum response when the load is in the same position. Furthermore, fatigue evaluation requires quantification of the full range of response from the passage of trucks. Development of the influence surface reveals to the analyst where the wheel loads should be placed to maximize or minimize total response.

An example of an influence surface at a critical orthotropic detail from Jong, et al (2004) is shown in Figure 4-19. This is the influence surface for strain in the topside of the deck plate near the location of the RDF from a wheel load of 50 kN (11.24 kip), with patch size of 50 cm. The model is a seven-span OSD panel with six ribs in the cross-section. The longitudinal position is measured from crossbeam 1 and the transverse position is measured from the left wall of Rib 2. As can be observed from the surface, it is only when the loading is located within 300 mm (11 13/16 inch) transversely and 500 mm (1.64 ft) longitudinally to the point of interest that any significant response is observed. Also, the slope of the surface is very steep in all directions and contains a severe peak. Thus, for analysis, the wheel patch loading must be placed as close as possible to the precise location indicated by the peak to obtain the maximum response.

Influence lines can also be used when the controlling position of loading is known in one direction. Influence lines require much less computational effort than influence surfaces, and should be used when possible. For example, local stresses at deck plate splices and RD can be assessed using transverse influence lines. Figure 4-20 shows the influence lines for stress at deck plate splice and RD weld from a dummy load of 44.5 kN (10 kips). The panel section consists of 305 mm (12 inches) deep trapezoidal ribs spaced at 610 mm (2 ft) and span of 4.57 m (15 ft), with 19 mm (¾ inch) thick deck plate. This is the same panel geometry that is the subject of design Example 1 (see Chapter 12 for more details). From these lines, it can be seen that the maximum stress at the RD weld is in the deck plate at the toe of the weld. The shape of this influence line indicates that when a typical size wheel load is near the rib wall, this will maximize response. Although there is a small region of reversal for the portion of load at the weld, integration of the 0.51 m (20 inch) wide wheel patch superimposed on the influence line will show maximum net compressive stress when the load is nearly centered over the rib wall. As with the deck splice, reversal can occur from multiple trucks in series in different transverse positions.

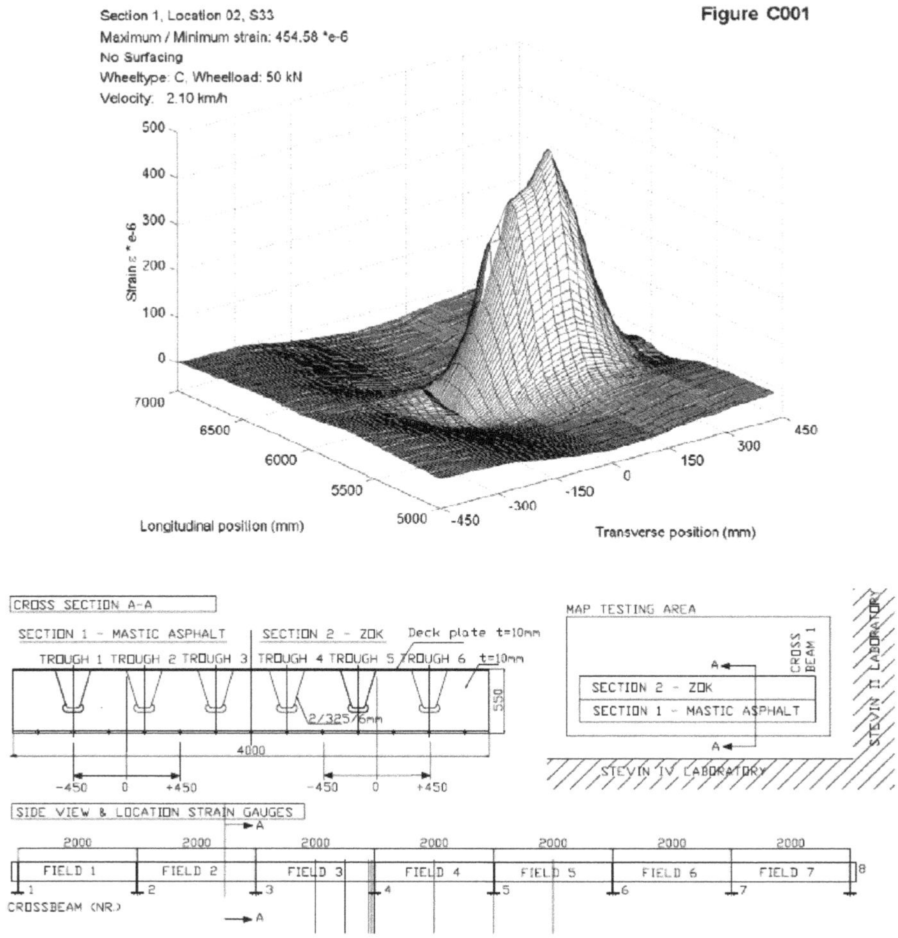

Figure 4-19 Three Dimensional Plot of an Influence Surface for Deck Plate Strain at Rib-to-Deck at the Floorbeam (RDF) Detail in Test Deck Panel showing the Sensitivity of Response to Wheel Load Position (Jong, et al, 2004)

The influence line for bottom surface stress at the deck plate splice indicates that maximum response occurs when the wheel load is located midway between the two ribs as expected. Thus, a smaller wheel patch width that fits between two adjacent ribs will maximize the response, which can occur in a single tire steering axle. Also note that the total range of stress that is possible at this detail is the result of two trucks in series passing over this location in different transverse positions--one truck with wheel directly over the splice and a second truck 0.610 m (24 inches) to the right or left. Calculation of maximum possible stress range must account for this scenario.

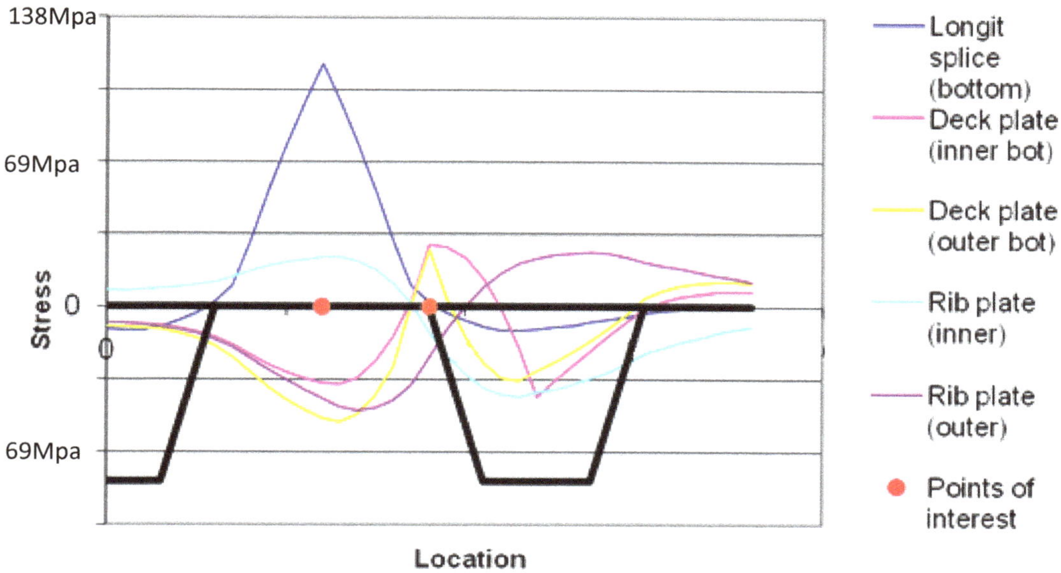

Figure 4-20 Influence Lines Running Perpendicular to the Rib for Stress at Deck Plate Splice and Rib-to-deck Weld (RD) as shown for Various Locations of Concern in the Rib and Deck Plate

Analysis may also use influence line development from longitudinal truck paths in fixed transverse positions. Figure 4-21 shows two influence lines for a stress concentration of a deck with an intermediate FB. The influence lines that are found by FEA contain St. Venant and distortion effects and all combined stresses. The paths "A" and "B" represented by the two lines in the figure are for two different transverse positions of the wheel (from the HS notional truck). It can be seen in Figure 4-21 that each position produces a different stress signature. The analyst may explore a number of influence lines (i.e. transverse positions) generated in this fashion and use the one that generates the maximum range.

The present method of obtaining a maximum design stress range in the FB is to use the stress range of a single wheel position relative to the center of the rib (the position that gives maximum effect) and by multiplying this by the design load factor. The basis for this factor is discussed in more detail in Chapter 5. For more accurate calculation of the effective stress range at sensitive details, the Monte Carlo Technique can also be used. When a truck weight spectrum representative of the expected traffic on the bridge is adopted along with a distribution model of the wheel transverse positions, a simulation of the stress environment can be obtained from which a complete stress range spectrum is derivable. For further information on derivation of the stress range spectrum, see NCHRP Report 299 (Moses et.al., 1987).

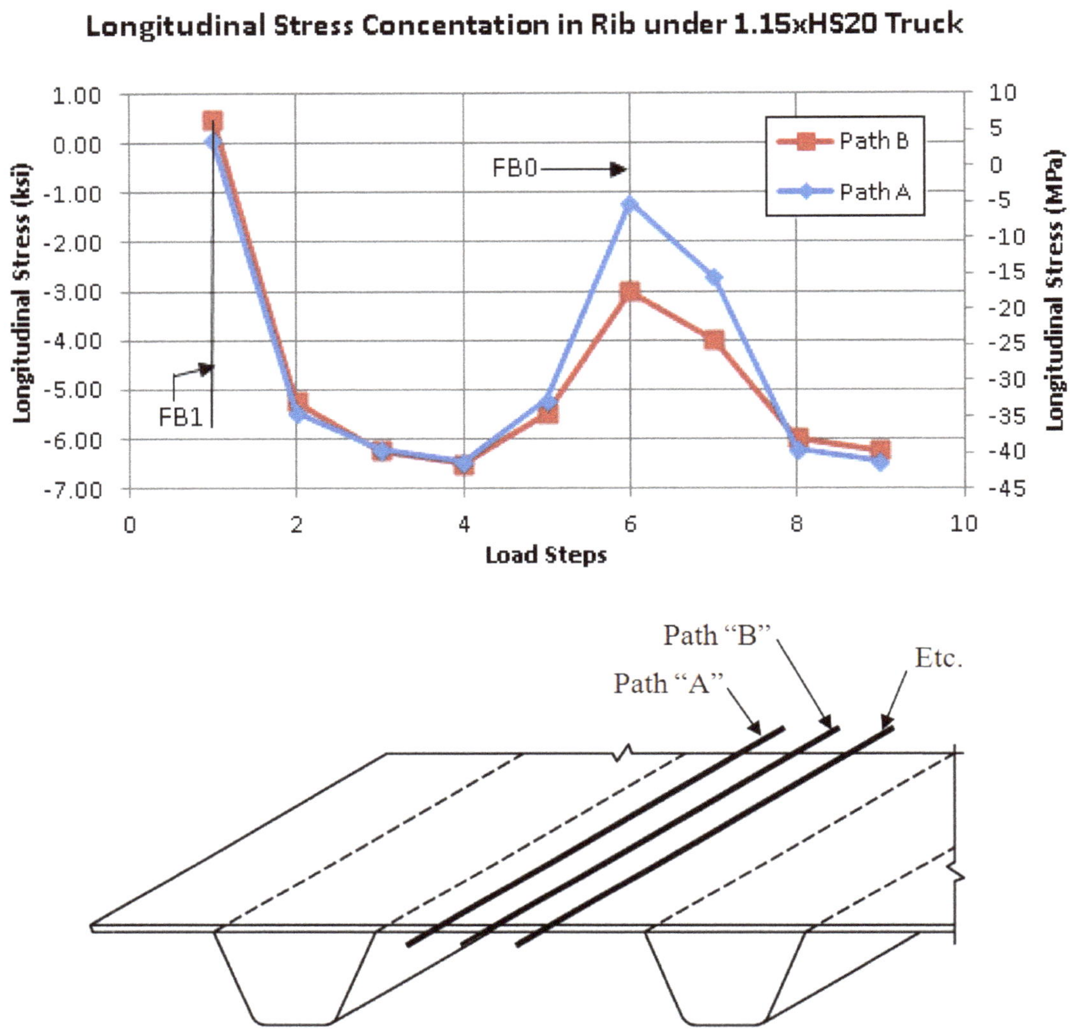

Figure 4-21 Comparison of Influence Lines (Parallel to the Rib) for Stress in the Rib-to-Floorbeam (RF) connection at Different Transverse Wheel Positions

4.6.2.Evaluation of Stresses

At the end of the 1970s, fatigue problems of welded structures were better understood and the "nominal stress" approach to evaluating fatigue resistance became standard assessment procedure. However, simplified techniques were used to evaluate stresses at the FB cut-out (i.e. those longitudinal flexural effects derived from moments found by the Esslinger-Pelikan method). These techniques are incapable of providing distortion and other local stresses that must be added to the primary, MC/I effects. They are a source of grave underestimation of the longitudinal (horizontal) rib plate effects at the FB, while the local vertical effects in the rib stem that are funneled into the FB termination are completely ignored. There are no simplified techniques that allow the practitioners to assess distortion stresses in this case.

Hence, there are two ways of correctly assessing stresses in continua in such a condition as a rib going through a FB, the FEA method and strain-gaging (laboratory testing or in-situ). The first is widely accepted, broadly used and highly efficient. The latter is equally effective, but must be used on a prototype faithfully fabricated and loaded. Strain-gaging is typically used on existing structures, or for field testing of design prototypes. At the planning as well as final design stages the FEA technique is indispensable. This perspective is supported by many codes, including: the International Institute of Welding (IIW) Recommendations; the Det Norske Veritas (DNV); the British standards; the proposed AWS provisions for fatigue evaluation.

In welded structures, fatigue propagates from two characteristic locations, assuming the weld is sound and not internally damaged:

- Toe cracking
- Root cracking

The aim of the engineer is to check in all locations of high stress concentration due to postulated geometry and calculated behavior (both primary and secondary effects), and how the stresses at the weld compare to the presumed resistance of the same. The critical locations for evaluation are presented in detail in Chapter 5, Design.

When fatigue analysis is used for evaluation of expected life, it may require a distinction as to whether the stresses are uniform across the plate thickness. Stresses can be uniform (in-plane) through the thickness, variable due to flexure, or are a combination of both. The techniques here described will give all the possible combinations that the engineer may want to evaluate.

Stress evaluation at weld toes presents technical problems to the analyst that uses the FEA technique, in that the angle the weld makes with the base material represents a sharp discontinuity. At that precise point, the discontinuity will cause the model to show a sharp increase in stress that is not in any way similar to the calculated stresses used to produce the fatigue database that created the widely adopted S-N curves. Thus, the concentration part of the stress that is treated as a black box in the baseline data provided in the AASHTO nominal stress provisions must be eliminated from the assessment of the stress for such data to be of any use. It follows that to any modeling technique, there must be associated a calibration method to bring the estimated stress in line with the procedure used to evaluate stresses for the production of the database.

Finite element meshing can be accomplished using shell plate elements or brick elements. It is the consensus of the industry at this time that shell plate elements are adequate to characterize structures and local effects that are typically composed of relatively thin material with small effects in the through-thickness direction. This does not prohibit analysts to use brick element techniques, should they so desire, as calibration techniques are available for both.

There are two accepted techniques that provide this calibration. Both use databases that are dissimilar to that adopted by AASHTO. They are:

- Extrapolation.
- The Battelle Structural Stress (BSS) (Dong, 2006).

The extrapolation method is more widely accepted and is discussed further below.

Extrapolation

Figure 4-22 shows diagrams of conceptual connections for the purpose of illustrating concentrations at weld toes in abrupt transitions and in continuous fillets. In most cases, shell element sizes t•t are standard for determining toe of weld stresses, where "t" is the plate thickness of the smaller element. Brick elements are often invoked at abrupt terminations with t•t•t as illustrated in Figure 4-22(b). These graphics are from the Recommendations for Fatigue Design of Welded Joints and Components (IIW, 2007) published by the IIW, and chaired by A. Hobbacher.

Figure 4-23 shows that points on the finite element are selected at an appropriate distance from the weld toe for type "a" and type "b" details illustrated in Figure 4-22 and how the stress is linearly (or to the second power) extrapolated to the weld toe to obtain the structural stress.

The table in Figure 4-23 also gives appropriate mesh sizes. Figure 4-23 is also extracted from the Recommendations for Fatigue Design of Welded Joints and Components (IIW, 2007).

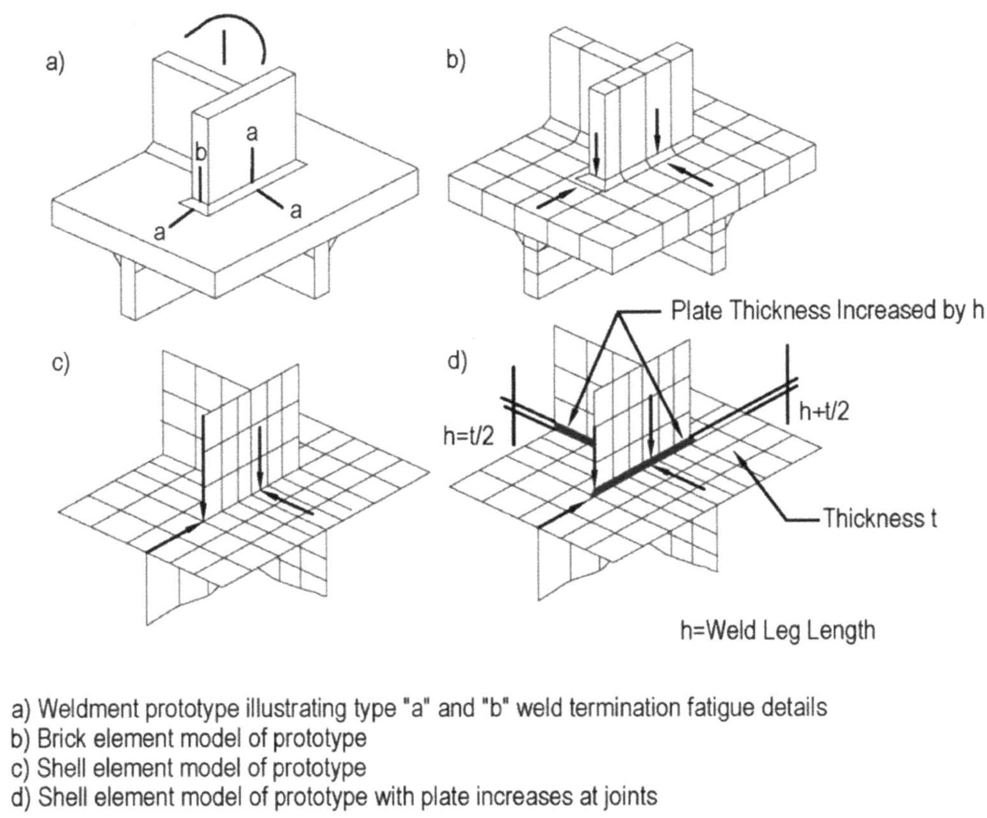

Figure 4-22 Modeling Guidance for Evaluation of Stress Concentrations by Extrapolation: (a) Weld Prototype for a Plate Welded Perpendicularly to another Plate; (b) as Discretized into FEA Brick Elements; (c) as Discretized into FEA Shell Elements; (d) with Increases in the Shell Elements to Account for Thickness at the Joints (adapted from IIW, 2007)

The validity of the locations of the extrapolation points is based on the authors' affirmations that those locations give best fit to the baseline data. Other codes, such as the DNV, or the newly proposed AWS fatigue criteria, have alternate extrapolation criteria, indicating that for shell elements the weld at the toe be extrapolated to the mid plane of the shell. This is a more conservative approach, but it is accepted practice.

Type of model and weld toe		Relatively coase models		Relatively fine models	
		Type a	Type b	Type a	Type b
Element size	Shells	t x t max t x w/2*)	10 x 10 mm	≤0.4 t x t or ≤0.4 t x w/2	≤ 4 x 4 mm
	Solids	t x t max t x w	10 x 10 mm	≤0.4 t x t or ≤0.4 t x w/2	≤ 4 x 4 mm
Extra-polation points	Shells	0.5 t and 1.5 t mid-side points**)	5 and 15 mm mid-side points	0.4 t and 1.0 t nodal points	4, 8 and 12 mm nodal points
	Solids	0.5 and 1.5 t surface center	5 and 15 mm surface center	0.4 t and 1.0 t nodal points	4, 8 and 12 mm nodal points

*) w = longitudinal attachment thickness + 2 weld leg lengths
**) surface center at transverse welds, if the weld below the plate is not modelled (see left part of fig. 2.2-11)

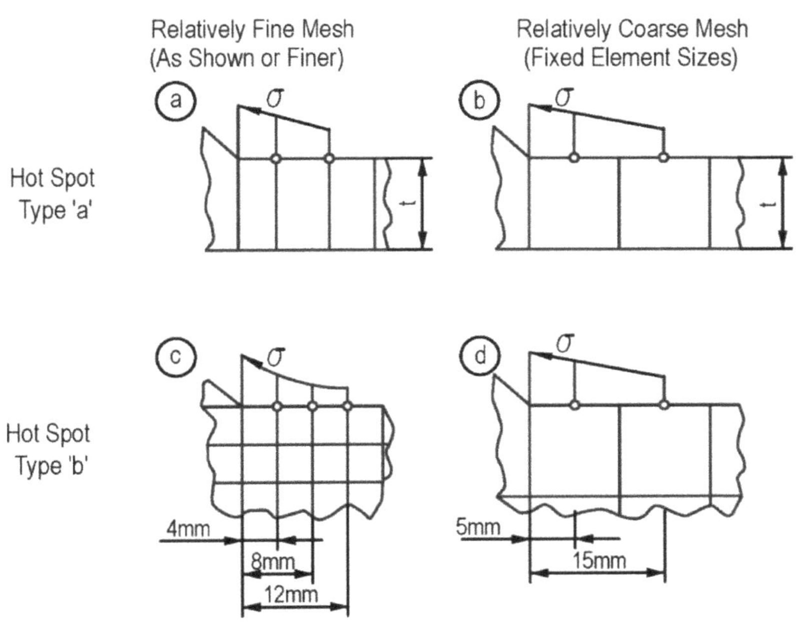

Figure 4-23 Recommended Mesh Sizing and Extrapolation for Fine and Course Meshes to be used when Performing Analysis of Stress Concentration (adapted from IIW, 2007)

4.7. COMPOSITE BEHAVIOR WITH DECK SURFACING

Research involving both laboratory tests and site measurements has clearly shown that wearing surfaces can reduce the displacements and stresses in the composite deck system of OSD bridges. This can be an important aspect of behavior, particularly for the accurate assessment of fatigue stresses at critical steel details. Experiments carried out by De Jong and Kolstein (2004), performed on OSDs with different surfacing materials at different temperatures, show that the stress range compared to an OSD without surfacing is reduced by a factor of 1 to 6. However, traditional bituminous or polymer surfacing materials are generally viscoelastic or plastic and provide rigidity at lower temperatures only. Furthermore, these materials have a tendency to crack or debond at highly stressed locations in service, which can reduce or eliminate the stiffening effect. (See Chapter 9 for more detailed information related to wearing surface material properties.) Such materials are not typically used as the basis for safe structural design in highway bridges. However, the potential for cost savings in design of the steel components can be significant since Fatigue limit states often control.

Conventional methods of composite analysis can be applied to assess the stresses, with the use of an effective deck plate thickness calculated based on the modular ratio between steel and the surfacing material, $n = E_{steel} / E_{WS}$. This ratio should be calculated with consideration of variability in the modulus of the surface material as a result of service temperature range and loading velocity. Consideration must also be given to the influence of the shear "slip" that occurs at the soft bonding layer. If no relevant data are available, refined analysis methods, as shown in Seim and Ingham (2004), or experimental testing may be required to accurately assess the composite stiffness developed by the wearing surface. See Chapter 9 for additional information on wearing surfaces.

4.8. STABILITY

Since the orthotropic panel is often integral with the primary bridge superstructure, the stability of the panel must be evaluated to ensure that buckling does not degrade the overall strength of the bridge when subject to axial and/or flexural demands. This is especially true when the OSD is part of a bridge superstructure primarily intended to resist global compressive loads, such as in cable-stayed and continuous box girder bridges. In flanges of box girders, global flexural demands on the bridge superstructure can result in nearly pure axial compressive stress on the flange plate components. For the case of global plus local demands, the rib may be subject to varying stress gradients with possible shear interaction. The potential stability-related limit states that must be evaluated in the orthotropic plate include:

1. Local buckling of the deck plate between ribs
2. Local buckling of the rib wall and
3. Buckling of the orthotropic panel between FBs.

Testing has indicated that failure of the rib in a stiffened panel is critical because it can produce a sudden collapse of the entire panel (Grondin et al., 1998).

For simplified evaluation of local plate stability, the accepted practice in the United States is to limit the width-to-thickness ratio (b/t) to a value that prevents local buckling. When these limits are exceeded, the local buckling causes a loss of stiffness and redistribution of stress and portions of the width become ineffective (SSRC 1998). The nominal critical buckling stress is reduced by an empirical reduction factor to account for this post-buckling behavior. This effective width concept has been used in design specification for many years (AISI, 2001) (AISC, 2005).

Stability can also be evaluated by more rigorous methods, including FEA. There are two common strategies for conducting a buckling analysis using FEA:

1. Eigen value buckling analysis
2. General non-linear incremental collapse analysis that traces the entire equilibrium path to the critical load and beyond.

The Eigen value buckling analysis is more attractive in that it requires substantially less computational effort, since it only needs to employ an Eigen value extraction routine on the global stiffness matrix, rather than conduct matrix inversion many times. This type of analysis is relatively simple to execute with most commercially-available software codes. However, this solution strategy is limited to problems where the precollapse displacements are relatively small and any changes in material properties do not significantly violate the assumption of linearity. This is often referred to as "elastic," "bifurcation," or "column type" buckling. This analysis also neglects any residual stresses and imperfections that exist, resulting in an overestimation of the true buckling load. This error can be small for plates with relatively high slenderness, but becomes more pronounced as the slenderness is reduced and buckling is coupled with inelastic behavior. The general non-linear incremental collapse analysis is more robust in that full consideration of residual stresses and imperfections can be considered. However, this demands a higher level of understanding in the FEA method by the practitioner in terms of definition of nonlinear elements, mesh imperfections, initial stresses, and incremental solution controls. No matter what solution technique is employed, the finite element model must always contain sufficiently refined mesh to describe the buckled configuration of the structure.

4.8.1. Local Buckling

Local buckling can occur in orthotropic plates in the deck plate between the connecting points to the ribs and in the rib walls, depending on the slenderness of each component, either a/t_d, e/t_d, or h/t_r (see Figure 4-24). This limit state has been observed both numerically and experimentally (Chou et al 2006) (Grondin, et al 2002). The problem of local buckling in OSDs was initially addressed in the Design Manual for Orthotropic Steel Plate Deck Bridges (AISC, 1963). The method proposed was based on elastic stability analysis of simple plate elements with varying loads and boundary conditions. The consideration of inelasticity was approximately accounted for by use of the stiffness reduction factor (τ) applied to the elastic solution. Post buckling behavior was not considered.

More recently, Yarnold et al (2007) conducted analytical parametric studies on the local buckling behavior of trapezoidal ribs in OSDs. Their studies recognized that typically the rib wall is the most slender element, and chose to investigate this element for both pure axial compression and combined axial plus bending. Their numerical analyses neglected imperfections and residual stresses, but the adequacy of this simplifying assumption for the problem at hand was confirmed by comparison of the results against physical testing. The work of Yarnold et al. (2007)

demonstrated clearly that local buckling will reduce the strength of the panel as a function of plate slenderness, and there exists a limiting b/t ratio at which the full yielding capacity of the panel can be achieved.

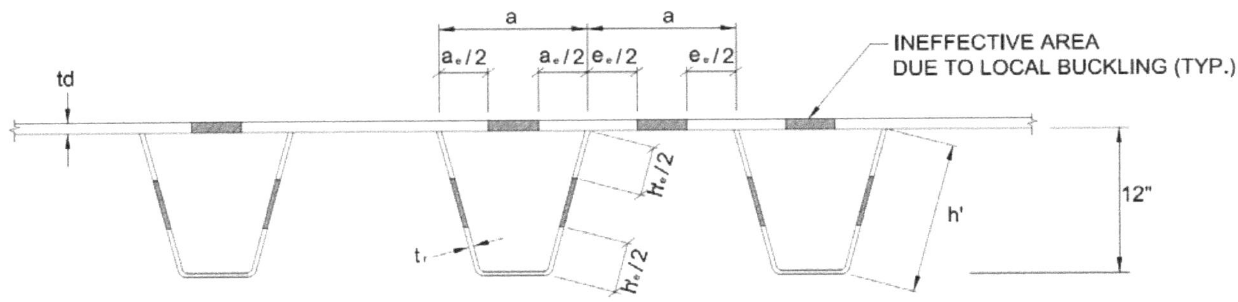

Figure 4-24 Deck Panel with Trapezoidal Ribs Showing the Effective Panel Section for Consideration of Local Buckling

The recommended method for quantification of strength reduction resulting from local buckling is the method as given by Specification for Structural Steel Buildings (AISC, 2005)for slender stiffened elements, which is based on the effective width approach and consideration of post buckling behavior (AISC, 2005). The method provides for simple calculation of the reduced critical stress by use of the slender element reduction factor (Q). This method is based on tests results from (Winter, 1947) and is also the basis of the North American Specification for the Design of Cold-Formed Steel Structural Members (AISI, 2001). The deck plate and the rib walls are considered "stiffened" elements since both longitudinal edges have support. When the width to thickness ratio of the deck plate between ribs or the rib wall exceeds the limit of:

$$\mathbf{b/t \geq 1.4\sqrt{\frac{E}{F_y}}} \tag{4-7}$$

Then the reduced effective width of the plate element is taken as:

$$\mathbf{b_e = 1.92t\sqrt{\frac{E}{f}}\left[1 - \frac{0.38}{b/t}\sqrt{\frac{E}{f}}\right] \leq b} \tag{4-8}$$

where:

b_e	=	reduced effective width of the plate
t	=	thickness of the plate element
b	=	width of the plate element
E	=	modulus of elasticity
f	=	applied stress (may be conservatively taken as F_y)

If all elements in the cross-section have width to thickness ratios less than Equation (4-7), then local buckling will not reduce the compressive strength of the panel. If multiple elements exceed Equation (4-7), then the effective width must be calculated for each individually. It is noted that this equation is conservatively based on stiffened plates, where the sides provide little rotational restraint to one another as is the case in hollow box sections (AISC, 2005). The applicability of equation (4-8) to the OSD is based on the assumption that the residual stresses and imperfections are of similar magnitude to those found in rolled or welded structural shapes.

Once the effective width for each slender plate element in the cross-section is determined, then the reduction factor for slender elements (Q) is determined by:

$$\mathbf{Q} = \frac{\mathbf{A_{eff}}}{\mathbf{A}} \tag{4-9}$$

Where:

A_{eff}	=	summation of the effective areas of the cross-section based on the reduced effective widths
A	=	total cross-sectional area

This slender element reduction factor is then incorporated in the calculation of the panel buckling strength, as described in the next section. This method allows for direct calculation of the panel buckling strength accounting for local buckling reduction.

It is noted that the Q Method as proposed will produce reasonable results for determining buckling strength of panels with rib wall slenderness, as found in most typical orthotropic cross-sections. However, for wall slenderness exceeding a value of approximately 60, then this method produces unconservative results and more rigorous analysis is required. This is due to the fact that local buckling of rib walls degrades the overall buckling strength of the panel to a larger degree than a typical rolled steel shape.

The methods for evaluating local buckling were presented here for closed ribs. Open ribs are not presented since the evaluation is similar, uses the same theory, and is less complex overall.

4.8.2.Panel Buckling

Buckling behavior of stiffened plate panels is a complicated problem due to the two-way orthogonal stiffening behavior and partially restrained boundary supports on four sides of the panel. A summary of relevant historical research on this subject is provided in the text by Troitsky (1977) and in (SSRC, 1998). Generally, the intermediate FBs and bulkheads in orthotropic plate structures are stiff enough to be considered as a pinned boundary support for the containment of buckling within the panel. Similar to stiffened plate elements, reserve post-buckling strength in the panel exists beyond the point of initial buckling and can be quantified by use of the local effective width approach.

As proposed by Horne and Narayanan (1977), a simplified approach to estimate the buckling strength of the stiffened panel is to analyze the panel as a series of isolated column struts comprised of a stiffener and the associated effective width of plating (see Figure 4-25). Basic column theory can then be employed. This approach conservatively neglects the bending and membrane stiffness of the panel in the transverse direction and the torsional stiffness of the closed rib sections.

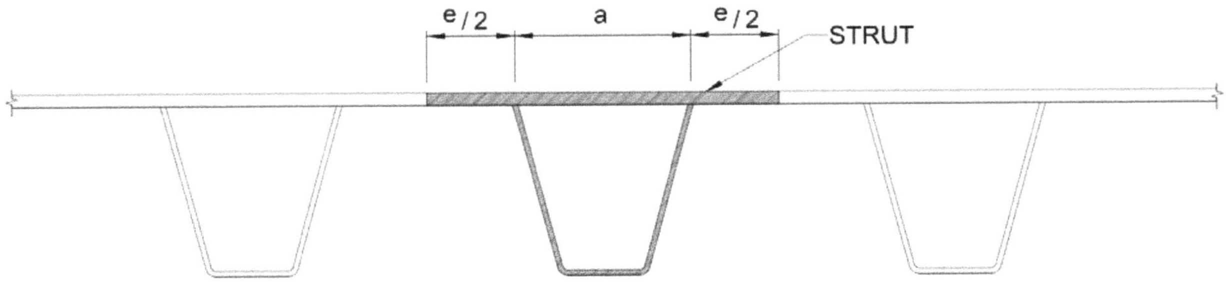

Figure 4-25 Deck Panel with Trapezoidal Ribs showing Axial Strut for Simplified Analysis of Buckling Strength in Orthotropic Panel.

Applying the strength equations of Specification for Structural Steel Buildings (AISC, 2005) with consideration of local buckling reduction from the slender elements, the critical buckling stress is determined as follows:

(a) when $\frac{\mathbf{KL}}{\mathbf{r}} \leq 4.71\sqrt{\frac{\mathbf{E}}{\mathbf{QF_y}}}$

$$\mathbf{F_{cr}} = \mathbf{Q}\left[0.658^{\mathbf{QF_y}/\mathbf{F_e}}\right]\mathbf{F_y} \qquad (4\text{-}10)$$

(b) when $\frac{\mathbf{KL}}{\mathbf{r}} \geq 4.71\sqrt{\frac{\mathbf{E}}{\mathbf{QF_y}}}$

$$\mathbf{F_{cr}} = 0.877\mathbf{F_e} \qquad (4\text{-}11)$$

where:

F_e	=	is the elastic critical buckling stress = $\pi^2\mathbf{E}/(\mathbf{KL}/\mathbf{r})^2$
Q	=	is the slender element local buckling reduction factor
K	=	is the effective length factor
L	=	is the span length of the panel in the direction of compressive stress (between FBs)
r	=	is the radius of gyration of the strut

For more accurate analysis of the panel buckling strength with full consideration of the orthogonal stiffening behavior, the FEA method is recommended. An isolated panel with idealized boundary conditions can be considered for simplicity, or multiple spans that account for the continuity can be analyzed for improved accuracy. It is noted that panel buckling in OSDs has been determined analytically to be sensitive to the residual stresses (Chou et al 2006). Therefore, the FEA modeling strategy should employ nonlinear incremental collapse analysis with modeling of the initial stresses and consideration of inelastic material behavior, in situations when yielding is expected prior to buckling (i.e. when the panel slenderness is relatively low). Alternatively, inelastic behavior can be approximated by modification of elastic buckling analysis results with a stiffness reduction factor (i.e. use of a "tangent" modulus) in the model or by decreasing the resistance factor.

4.8.3. Residual Stresses

Residual stresses that exist in orthotropic steel panels are created primarily by the cooling of the longitudinal RD weld. The maximum tensile residual stress at a weld or in a narrow zone adjacent to a flame cut edge is equal to or greater than the yield strength of the plate (Bjorhovde et al., 1972). For consideration of the panel stability, it is the longitudinal residual stresses in the deck plate and rib that are of primary concern. Limited data exists for the magnitude and distribution of these stresses; however, a distribution similar to that shown in Figure 4-26 has been found by Grondin et al. (2002). It is noted that the magnitude of residual stresses in orthotropic panels (F_y max, $0.25F_y$ min) is near that observed in typical hot rolled or welded columns (Bjorhovde et al., 1972). However, the distribution is quite different. In the case of the orthotropic panel, stiffness of the cross-section can be lost quickly once yielding is initiated, depending on the plastic material hardening that is provided by the steel. Thus, for stocky panels, equation (4-10) may be unconservative, as was observed in the research by (Chou et al., 2006). Until future research more accurately quantifies the inelastic buckling strength of an orthotropic panel in the full domain, rigorous incremental nonlinear FEA of the panel buckling would provide a more accurate result in this case.

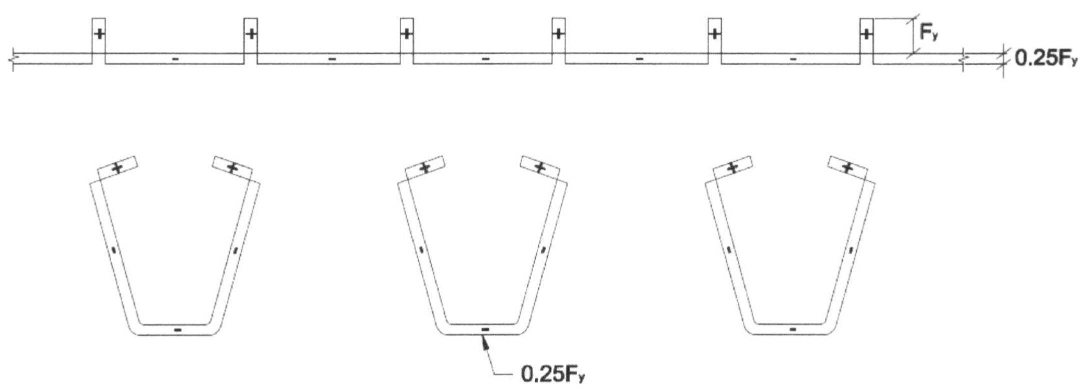

Figure 4-26 Approximate Residual Stress Pattern in Orthotropic Panel in the Deck and Rib showing Locations of Tension and Compression Fields.

4.8.4. Imperfections

The behavior of an out-of-straight or warped panel changes the stability problem from one of bifurcation to one of plate bending from eccentricity of the axial load. Imperfect geometry will reduce the buckling strength of the panel, although it has been found to be much less influential than the presence of residual stresses (Chou et al., 2006). The imperfection that will exist in the finished panel is related directly to fabrication technique and the quality control measures that are employed during construction (see Chapter 10). The analysis assumptions must be consistent with the fabrication tolerance allowed in construction. For development of the AISC Specification equations, a value of L/1000 was utilized, based on the upper limit of what is acceptable for actual delivery of structural members (SSRC 1998). This magnitude of imperfection is also applicable to bridge construction in general, which supports the applicability of the AISC equations to orthotropic panels. For modeling of imperfections in nonlinear FEA, a half-sinusoidal shape with maximum amplitude of L/1000 at the center is considered conservative and reasonable, where L is the span length in the direction of the compressive stress.

4.8.5.Second Order Effects

Orthotropic panels, when used as decks to support direct traffic loading, are typically subjected to combined bending plus axial compression. Local wheel loads cause out-of-plane deflection of the panel, which results in second-order moments from eccentricity of the axial load. The magnitude of the live load deflection is typically limited to increase the longevity of the wearing surface. However, as the compressive load approaches the elastic critical buckling stress, the second order P-δ effect can become significant and should not be neglected. This can be assessed directly by FEA that employs **geometric** nonlinearity; where loads are applied on the deformed structure. Note that proper discretization of each structural element is necessary to capture P-δ effects. Alternatively, the Specification for Structural Steel Buildings (AISC, 2005) method of "Second Order Analysis by Amplified First Order Analysis" may be employed. This method allows for simple estimation of the total moment in the panel as follows:

$$\mathbf{M_u} = \mathbf{B_1 M_o} \qquad (4\text{-}12)$$

$$\mathbf{B_1} = \frac{\mathbf{C_m}}{1 - \mathbf{f_u} / \mathbf{F_e}} \qquad (4\text{-}13)$$

where,

M_u = is the maximum factored moment including second order effects,

M_o = is the maximum factored moment based on first order analysis with the transverse loading,

B_1 = is an amplification factor to account for additional moment caused by lateral displacements in the panel (P-δ),

f_u = factored pure axial stress from compressive loading only,

and C_m is the equivalent moment factor calculated as follows:

$$\mathbf{C_m} = 1 + \psi\left(\mathbf{f_u} / \mathbf{F_e}\right) \qquad (4\text{-}14)$$

$$\psi = \frac{\pi^2 \delta_o \mathbf{EI}}{\mathbf{M_o L^2}} - 1 \qquad (4\text{-}15)$$

where δ_o is the maximum deflection due to transverse loading

Alternatively, C_m = 1.0 can be used as a simple conservative approximation for all cases involving transversely loaded panels.

5. DESIGN

This chapter outlines the approach to design of orthotropic steel bridge superstructures by evaluation of applicable limit states using the Load and Resistance Factor Design (LRFD) methodology. The discussions related to design specifications, including loads and factors, are put into context with the AASHTO LRFD Bridge Design Specifications (AASHTO LRFD). For determination of nominal resistance, references to other sources such as American Welding Society (AWS), the European Committee for Standardization (ECS), and other internationally published literature are shown as applicable.

5.1. GENERAL DESIGN APPROACH

Orthotropic steel panels are employed in a variety of different ways. One form is as an independent (floating) deck system where the design need only consider local effects as the deck spans between points of intermediate support from the global superstructure (such as a truss or cable support). A second form is a panel that also serves as an integral flange or web to a built-up steel plate or box girder where the design must consider both local effects as well as global demands imposed from the superstructure. A third form is as an integral deck rigidly attached to a supporting global bridge superstructure system where the deck is used for redecking an existing bridge with rigid attachment. In this last case, the demands on the panel from global response of the existing structure need to be carefully considered to assess all the complex interactions.

The criteria provided in this chapter are primarily applicable to the case of orthotropic panel functioning as a bridge deck subjected to direct wheel loads. By comparison, when the orthotropic panel is used for a bottom flange or web of a box girder, the design is greatly simplified because no traffic loading is applied directly to the panel. In this case, the panel is primarily subjected to simple in–plane axial loading.

Regardless of the design condition, the individual components of the panel such as the deck plate, ribs, FBs and their connections need to be evaluated for all applicable limit states.

5.1.1. Design Level

For OSDs to become more accepted as a common bridge deck solution, design verification requires a new approach. Since many of the controlling aspects of OSD panel design are local rather than global demands, a well-designed and detailed panel has the potential to become a standardized modular component that can be used in multiple future applications that are sufficiently similar. If such a deck panel can be developed and verified, the required effort for design would be much less for these types of bridges. In contrast, those bridges that have unique characteristic will continue to require a more rigorous analysis. Thus, the Design Level is determined depending on the application and the test data available to the designer. The Levels of design are summarized as follows (see Section 5.6 for more detail on each level).

- **Level 1 Design** - Level 1 Design is based on little or no structural analysis, but is accomplished by selection of details that are verified to have adequate resistance by experimental testing (new or previous). When appropriate laboratory tests have been conducted for previous projects or on specimens similar in design and details to those proposed for a new project, the previous tests may be used as the basis for the design on the new project. All details must provide a measure of reliability consistent with the AASHTO LRFD Specifications. Previously verified Level 1 designs may be used as the basis for design on new projects without additional testing, subject to approval by the owner.

- **Level 2 Design** - Level 2 Design is based on analysis of certain panel details that can be evaluated with sufficient accuracy by simplified 1-D or 2-D analysis techniques. Calculations consider only nominal stresses, and not local concentrations. Acceptable techniques include transverse strip models, the Pelikan/Esslinger Method, the Vierendeel Model (as described in Chapter 4), or other methods that are properly calibrated to experimental test data. This design Level can also include incremental improvement of previously tested details by comparative analysis. Not all orthotropic panel details can be designed by Level 2.

- **Level 3 Design**- Level 3 Design is based on refined 3-D analysis of the panel to quantify the stresses to the most accurate extent reasonably expected (from a qualified design engineer experience in refined analysis) for all components and connections. Calculations consider local stress concentrations at fatigue-prone details. This may require a detailed sub-model of the panel within a global model of the bridge superstructure system.

If no previous test data is available for a panel, new testing (Level 1) or refined analysis (Level 3) is required unless it can be demonstrated that the local distortional mechanisms (FB distortion and rib distortion) will not lead to fatigue cracking. Strength, Service, and Constructability generally only require a Level 2 design. For design of panels for bridge redecking applications, design Level 3 should always be used.

Standardized panel details for use in Level 1 design have been developed and are promulgated in specifications worldwide, such as the Eurocode (ECS 1992) and the Japanese Bridge Specifications (JRA 2002). However, additional standard panels have not yet been developed and tested in the United States at the time of this Manual's publication. Recent domestic full-scale prototype tests (Connor 2004, Tsakopoulos 1999) were special designs conducted for redecking of existing bridges, and they are not considered optimum solutions for standardization. Future research and testing should provide additional data that can be used in Level 1 design. For implementation, owners should consider adoption of a verified design and incorporation into their standards.

Appendix C demonstrates the application of Level 1 Design of an OSD based on available experimental test data from a previous project. The basis of design is the prototype OSD for redecking of the Bronx-Whitestone Bridge, which was tested at Lehigh's ATLSS Research Center in 2002 (Tsakopoulos, 2002). This full-scale laboratory test simulated 4.1M cycles of 2.25 times the AASHTO HS20 Fatigue Truck, plus additional 2M cycles of three times the HS20, producing effectively 239M cycles of the fatigue loading. Fatigue cracks were not found in any of the primary connections, which demonstrates very good fatigue resistance and verifies the design performance for the given conditions.

5.1.2.Design Life

OSDs are often handicapped by their own expected success. While other competing deck systems are considered to be disposable (i.e. expected to only last 30-50 years), the OSD is often required to provide extended service life. In the past they have often been rejected because of uncertainties about potential fatigue cracking and maintenance. As the technology has improved and more experience gained, fatigue problems have been greatly reduced or eliminated, and have decreasingly become a concern. As such, OSDs are economically viable and highly competitive on a life cycle cost basis, despite their higher initial costs. Life cycle analysis does require that certain assumptions be made about service life. **Designs made according to these provisions can be expected to perform very well and meet the design life as per AASHTO LRFD. There is no reason to expect that an OSD should not last as long as the other more common steel bridge members subjected to the same heavy traffic and environmental conditions.**

5.2. GENERAL DESIGN APPROACH

The applicable limit states for the design of orthotropic panels include Strength, Service, Fatigue, and Constructability. All limit states need to be considered for complete design, but, as previously stated, it is generally the Fatigue limit state that will control the majority of design details. Note that extreme-event limit states are beyond the scope of this manual and are not covered.

5.2.1. Strength Limit State

Strength limit states maintain the load-carrying capacity governed by geometry and material properties. Thus, yield strength and/or geometric properties, such as loss of stability, must be considered in the design of orthotropic panels. Global and/or local geometry may govern stability considerations.

Testing has shown that OSD panels can have tremendous reserve strength for lateral loading beyond the yield strength, due to membrane stiffening. This reserve, however, is dependent upon the boundary support conditions. For simplicity, the approach to Strength design should conservatively limit stresses to the specified minimum yield strength or critical buckling stress.

Strength design must consider the following demands: rib flexure and shear, FB flexure and shear, and axial compression. The rib, including the effective portion of deck plate, must be evaluated for flexural and shear strength for its span between the FBs. The FB, including the effective portion of the deck plate, must be evaluated for flexural and shear strength for its span between primary girders or webs. The reduction in FB cross-section due to rib cut-outs must be considered by checking flexure and shear where the portion of web is removed. When the panel is part of a primary girder flange, the panel must be evaluated for in-plane compressive strength based on stability considerations.

The Strength limit states that include live load and dead load as the primary loads in the combinations govern the design of orthotropic panels in most cases. In AASHTO LRFD, these are the Strength I and Strength II limit states. These Strength limit states must be satisfied for both buckling and yielding. The Strength I load combination is applied in conjunction with the HL-93 notional live-load model representing random traffic, while the Strength II load combination is applied with owner-specified permit loads (for example, the Caltrans P-15 permit load model).

Often, a bridge design specification will include Strength limit states for special situations. The AASHTO LRFD includes the Strength IV limit state to allow the specifications to be applied to long-span bridges where the dead load is predominant over the live load. This limit state governs where the dead-load effect is seven or more times the live-load effect. As such, it will likely only govern the design of OSDs when they are made integral with a long-span bridge superstructure.

5.2.2. Service Limit State

Service limit states exist to provide checks for maintaining the service life of the bridge. These limit states should also be considered as means to minimize maintenance costs and traffic disruptions for repairs. Additionally, these limit states could include elastic and plastic deformations and other forms of service–induced deterioration, such as debonding or cracking of the wearing surface of an OSD.

The basic Service limit state applies load factors equal to 1.0 to each significant component of load. In the AASHTO LRFD, this is the Service I limit state. For OSDs, the Service I limit state must be satisfied for overall deflection limits for the deck plate (span/300) and the ribs (span/1000) and relative deflection of adjacent ribs (2 mm [0.10 inches]). These deflection limits are intended to prevent premature deterioration of the wearing surface.

Another applicable Service limit state is the Service II limit state for the design of bolted connections against slip in the overload scenario. This should be considered for the design of rib and FB splices. The remaining Service limit states III and IV are for tensile stresses in prestressed concrete sections under vehicular live loads, and tensile stresses in prestressed concrete substructures under wind loads, respectively. Thus, neither of these additional Service limit states is applicable to OSDs.

5.2.3. Fatigue Limit State

Two types of designs are possible within the context of the AASHTO LRFD Specifications for fatigue: infinite-life and finite-life design. As such he AASHTO LRFD introduces two Fatigue limit states: Fatigue I for infinite-life design and Fatigue II for finite-life design. Because OSDs are governed by wheel loads, they experience millions of repetitive cycles of wheel loads and thus will most often be required to be designed for Fatigue I. By comparison, other code-writing bodies acknowledge other fatigue life prediction concepts instead of the infinite life concept used in the AASHTO LRFD. For example, Eurocode (ECS, 1992) specifies fatigue-resistance curves with merely a decreased slope below some threshold value of stress range, instead of AASHTO's horizontal threshold (the constant amplitude fatigue limit [CAFL]) for variable amplitude loading. Other specifications use the infinite life concept for constant amplitude fatigue, while relying on a bi-linear fatigue life curve for variable amplitude curves for certain life estimates. Fatigue II finite life design may produce more cost-effective proportions when the traffic volume is not excessively high.

5.2.4. Constructability

The strength and stability of the orthotropic panel and the integrity of the wearing surface must be maintained during all stages of construction, including handling, storage, shipping, and erection. Very often, orthotropic panels are shipped by ocean-going vessels, which may cause the controlling loading scenario for the panel. There have been reports of wearing surface failure due to stresses applied during erection of pre-topped panels.

5.3. LOAD FACTORS AND COMBINATIONS

State-of-the-art bridge design specifications are reliability-based with a partial-factor format that mimics deterministic design methodology. The design equation for such specifications can be generalized as:

$$\sum_i \gamma Q_i \leq \phi R \tag{5-1}$$

Where,

Q_i	=	force effect,
γ_i	=	appropriate load factor,
R	=	nominal resistance,
ϕ	=	appropriate resistance factor.

In the partial-factor method of reliability-based design, load and resistance factors are specified for application to the load and resistance sides of the design equations, respectively, to achieve desired levels of reliability or safety. These factors are specific to the nominal loads and resistances specified in a particular design specification and must be applied together. The load factors from one specification are not necessarily appropriate for application with the resistance factors, or the nominal loads and resistances of another design specification.

The Strength limit states of AASHTO LRFD are calibrated to achieve a target reliability index of 3.5, which results in a probability of failure just above to the average of the past specifications (Nowak, 1999).

In the case of OSD design, the nominal force effects are typically stresses or deformations, with the nominal resistances of limiting stresses (for example, yield strengths or buckling strengths) or limiting deformations, respectively.

Sets of load factors, called load combinations, are calibrated to achieve the target reliability under various combinations of loads (Nowak, 1999) (Kulicki et al, 2007). The magnitudes of the load factors in a combination reflect the uncertainty of the loads and the probability of the simultaneous occurrence of the loads represented in the combination.

The load combinations of typical design specifications are categorized as Strength limit states or Service (or serviceability) limit states. Strength limit states are those intended to maintain load-carrying capacity. Service limit states are those intended to maintain service life.

The applicability of the live-load load factors for the Strength I and Strength II limit states, 1.75 and 1.35 respectively, to the design of OSDs governed by wheel loads is an extension of the original calibration of AASHTO LRFD. These load factors were not derived specifically for OSDs, but the derivation did consider other bridge systems and components governed by wheel loads. These strength load factors are likely somewhat conservative due to lower uncertainty in axle loads as compared to gross vehicle weight (GVW). However, the nominal design axle loads may be somewhat low, as described below, and may offset some additional safety margin.

The appropriateness of the resistance factors of AASHTO LRFD for OSDs has not been rigorously established through research. These resistance factors were derived by considering the uncertainty of force effects in girders subject to truck loads. It is noted that the magnitude of residual stresses and the tolerances for fabrication are consistent with other more conventional fabricated structural steel members. Until future research more accurately quantifies the reliability index provided for deck systems, the current AASHTO LRFD resistance factors are considered acceptable to achieve safety in the design of OSDs.

The load factors for fatigue are dependent upon the nominal fatigue load. Using the AASHTO LRFD provisions for fatigue design of OSD components and connections, the Fatigue load factors are taken as $\gamma_I = 1.50$ and $\gamma_{II} = 0.75$. **There is an exception to this for connections to the deck plate and details around the FB cutout where the Fatigue I load factor should be increased to 2.25.** The increased Fatigue I load factor is based on stress range spectra monitoring on both the Williamsburg Bridge (Connor and Fisher, 2001) and the Bronx Whitestone Bridge (Hodgson and Bowman, 2008), which indicate that the standard Fatigue I load factor, which was developed for girders, FBs, truss members and other "global" components is unconservative for the design of certain OSD components. These studies indicate that the ratio of maximum stress range to effective stress range is increased as compared to standard bridge girders. This is attributed to a number of factors such as occasional heavy wheels and reduced local load distribution that occurs in deck elements, as opposed to a main girder for example. The influence of the enhanced load distribution, which is not accounted for in the AASHTO distribution factors is apparent as this ratio is in fact more consistent with the original findings of NCHRP Report 299 (Moses et.al., 1987). This is increase is accomplished simply by using an additional modifier of 1.5 for the appropriate orthotropic details. Thus, 1.5 x 1.5 = 2.25 for Fatigue I. These stress-range load factors limit the stress ranges exceeding the constant-amplitude threshold to a rate of 1 in 10,000. (This rate is comparable to that observed for typical fatigue-sensitive details on girders.)

5.4. PERMANENT LOADS

The permanent loads to be considered in the design and evaluation of OSD are the dead loads of the steel deck and its wearing surface. The AASHTO LRFD includes a separate load factor for wearing surfaces (termed DW) of 1.50, which is greater than the load factor for other dead loads (termed DC) of 1.25. This increased load factor acknowledges the uncertainty of future asphalt wearing surfaces and can be reduced at the discretion of the designer, considering the well-controlled thicknesses of the wearing surfaces on OSDs.

5.5. LIVE LOADS

Live load demands on the OSD can include both local and global effects. Global effects result from the OSD participating as an integral part of the bridge superstructure. Local effects result from the application of the wheel loads directly on the panel. Both effects need to be superimposed for certain design conditions to calculate the total force effect in the deck components and connections. Different live load components govern the global and local effects. It is most often the local demands (due to cyclic stress ranges from wheel loads), however, that control the OSD design details. Thus, the complete HL-93 notional live-load model of AASHTO LRFD should be used to determine total force effects.

5.5.1. Design Truck or Tandem Load

5.5.1.1.Application

As demonstrated in Section 4.5, local effects in OSDs are governed by wheel loads, and the maximum response can be sensitive to how the wheel loads are applied. The simplified approach of using concentrated wheel loads should only be considered for Level 2 design, although it may result in more conservative designs. For Level 3, the wheel load must be distributed over the tire contract area yielding a uniformly distributed pressure to be applied to the OSD contact surface. This pressure may be subsequently distributed downward through the wearing surface and deck plate to the mid-plane of the deck plate, assuming a 45° angle of distribution.

5.5.1.2. Orthotropic Steel Deck Refined Design Truck

The magnitude of wheel loads specified in AASHTO LRFD is 71 kN (16 kips) for the HL-93 design truck and 55.5 kN (12.5 kips) for the HL-93 design tandem. The 142 kN (32 kip) axles (consisting of two 71 kN [16 kip] wheels) of the three-axle HL-93 truck, date back to the H20 load from the 1930s and represent a design expedience developed for components other than OSDs. The 142 kN (32 kip) axles in the design truck represent two closely spaced tandem axles of greater weight than 71 kN (16 kips) (of a five-axle tractor trailer truck). A history of live-load model development is included in Kulicki and Mertz (2006). More recently, Nowak (2008) found by extrapolation procedure on weigh-in-motion (WIM) data from 13 bridges in Michigan that the 75-year mean maximum axle weight is 196 kN (44 kips), based on the same procedure used to develop AASHTO LRFD live loads. This value is somewhat larger than the current HL-93 truck in AASHTO LRFD; however, the load factor is deemed to be sufficiently conservative when applied to axle loads, as described previously.

For OSDs, it must be recognized that the AASHTO LRFD-specified 145 kN (32 kip) truck axle actually represents a tandem consisting of two 71 kN (16 kip) axles spaced at 1220mm (4ft.). Thus, each wheel of the 71 kN (16 kip) axle is properly modeled in more detail as two closely spaced 45 kN (8 kip) wheels, 1220mm (4ft.) apart to accurately reflect an actual Class 9 tractor-trailer with tandem rear axles. Further, these wheels are distributed over the specified contact area of 510 mm (20 inches) wide by 250 mm (10 inches) long for rear wheels and 250 mm (10 inches) square for front wheels, which approximates actual pressures applied from a dual tire unit (Kulicki and Mertz, 2006) (Nowak and Eamon, 2008).

The refined HL-93 Design Truck is shown in Figure 5-1. This loading should be positioned both longitudinally and transversely on the bridge deck ignoring the striped lanes to create the worst stress, stress range, or deflection, as applicable. Note that the smaller 255mm x 255mm (10 in. x 10 in.) front wheels can be the controlling load for fatigue design of some OSD details since the patch width will often fit within the dimension of the rib wall spacing.

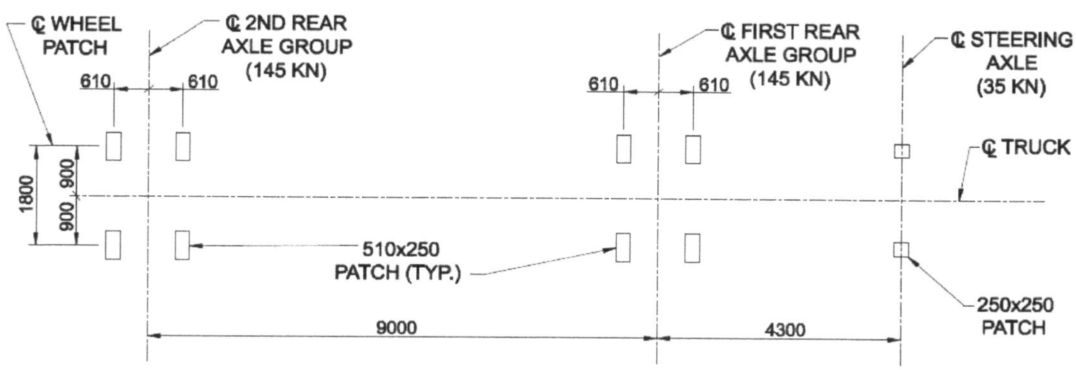

Figure 5-1 Refined Design Truck Footprint for Level 3 Design of Orthotropic Decks with Tandem Rear Axles and Individual Wheel Loading Patches

5.5.1.3. Fatigue Application

The fatigue-load specification includes configuration, magnitude, movement, and frequency. The configuration and magnitude of the fatigue wheel load is identical to the one shown in Figure 5-1 and should be modeled accordingly. For fatigue design, the live load analysis must consider moving loads to quantify the full range of stress at details from an individual truck passage since many details are subject to stress reversal when the loading is placed in adjacent spans (or even adjacent ribs). This can be done by considering a straight travel path for the design truck at a set transverse position on the bridge deck. This approach has the effect of neglecting larger stress ranges that may result from the passage of two trucks in series or side-by-side in different transverse positions, but these events are covered by design with the fatigue load factors given above.

The frequency of loading is critical for finite life design in OSDs. The Average Daily Truck Traffic (ADTT) and cycles per truck passage (n) both influence the total number of cycles for design. For components and connections of the OSD subjected to direct wheel loads, the number of cycles for design is governed by the number of axles expected to cross the bridge. Conversely, it is the number of truck crossings that equate to fatigue cycles for the main load-carrying members. For the refined tandem-axle truck, this results in 5 cycles per truck passage. However, the work by Connor (2002) found that other components such as the rib and FB typically experience only one primary stress cycle per truck passage. Thus, for design of all welded connections to the deck plate use n = 5.0 and for all others use n = 1.0. Additionally, since axle (and wheel) loads are variable for design trucks, it is a matter of variable stress range loading. The force effect (Δ_f) can be conservatively taken as the worst case from the five wheels, or by application of Miner's Rule to determine the effective stress range from the group of wheels.

5.5.2. Multiple Presence Factors (MPF)

Multiple-presence factors (MPFs) account for the reduced probability associated with pairs or multiple trucks being as heavy as individual trucks for evaluation of Strength limit states. The multiple-presence factors contained in AASHTO LRFD table 3.6.1.1.2-1 were developed from the observation of the effects of gross-vehicle weights (GVWs) of trucks. Although truck weights do not govern the design of OSDs, these values are appropriate for use with the wheel loads for OSD design. The single-lane MPF of 1.2 accounts for particularly heavy wheel loads.

5.5.3. Dynamic Load Allowance (IM)

Bridges are typically analyzed for static live load with the resultant static stresses amplified by dynamic load allowances to estimate dynamic stresses. The dynamic load allowances (IM) of AASHTO LRFD, 0.33 for strength and service and 0.15 for fatigue, were originally derived by examining the responses of bridge components to vehicle loads and assuming application to the HL-93 notional live-load model. (If applied to vehicles alone instead of the superposition of vehicles and lane load as for HL-93, the value for strength and service could be reduced to 0.25.) The specified values of IM consider surface roughness to be predominant and beyond anticipation by the designer. Further, they assume the potential surface roughness associated with potholes in typical reinforced concrete decks.

The dynamic load allowances specified in AASHTO LRFD may be reduced at the discretion of the engineer for OSDs considering the well-controlled nature of the wearing surfaces proposed in this Manual. Special consideration for the need for higher impact factors in the regions of expansion joints or other details that may result in amplified loads may be considered by the engineer.

5.5.4. Site Specific Live Load Models

At the discretion of the engineer, a site-specific live load model may be developed to achieve a more cost-effective design or more accurate structural assessment. Researchers have found that analysis of local or regional traffic conditions can justify changes (either increases or decreases) to the loads and load factors from the standard values found in AASHTO LRFD (Pelphrey and Higgins, 2006). This approach may be particularly applicable to the design of OSDs due to the relative lack of research on reliability and their sensitivity to fatigue design. Development of site-specific loading requires understanding of structural reliability theory and knowledge about some of the basic criteria used for calibration of AASHTO LRFD. Any site specific load modeling and calibration must provide a safety index that is consistent with current AASHTO philosophy. This may involve statistical analysis of data collected at the site or by use of existing data that is considered representative of the traffic loads anticipated on the bridge.

A good source for axle load spectra data is available in the Mechanistic Empirical Pavement Design Guide (NCHRP, 2003). Since pavement design is controlled by axle loads and not GVW similar to OSDs, this data is considered the most relevant for development of a site-specific load model. The NCHRP Guide provides a method for determining live loads for pavement design, which is based on WIM data collected from 134 sites across the country and from varying roadway functional classes. The method allows for varying levels of input for load development, depending on the data available to the designer. The more data available about the traffic conditions at the project site, the more accurate (and often less conservative) the load model can be developed. The axle load data summarized in the NCHRP Guide may be used to develop a site-specific "effective fatigue truck" for fatigue design and/or the maximum design truck for strength design.

5.6. ANALYSIS

Traditionally, the analysis approach for the design of highway bridges is one of component analysis and not system analysis. In a component design approach, the girders, FBs, stringers, and the deck are designed independently of one another. The approach is predicated on the principle that the assumed behavior of the individual component is fully compatible with the actual behavior once placed in the complete system. In most situations, these assumptions lead to conservative and acceptable designs with respect to global behavior. Unfortunately, most fatigue problems in steel bridges are often the result of a lack of consideration for stresses produced by local or secondary behavior at connections. These local stress components almost always arise from the "real" three-dimensional behavior of the system that is not accounted for in component design.

Experience has shown that component design of OSDs does not appear to be acceptable in all cases. Generally, the local stresses cannot be properly accounted for during a component design of the deck system. For example, the FB web plate of OSDs is subjected to a complex stress-range cycle comprised of both in-plane and out-of-plane stress cycles. In-plane stresses are analogous to membrane stresses within the FB plate. Out-of-plane stresses are produced by the rotation of the rib where it passes through the FB. These rotations are the result of bending of the rib under moving loads. The proportions of in-plane and out-of-plane stresses adjacent to the cut-out are dependent on the geometry of the cut-out and ribs, stiffness of the FB and ribs, and the type of internal bulkhead if used. The transverse position of the rib, i.e., toward the edge of the deck (external) or toward the middle of the deck (internal) also has a significant influence on the behavior adjacent to the cut-out.

Since the complex interactions of the various components of the orthotropic bridge cannot be accurately quantified in a component design, simplified analytical models may not provide sufficient or accurate information at all fatigue-sensitive details. As a result, the current AASHTO nominal stress approach routinely and successfully used for fatigue design in highway bridges cannot always be used for all details in OSDs. In some cases, other more refined methods must be used to ensure an adequate design. Although commonly used in other industries, these methods are not well known to many bridge engineers in the United States.

As such, the updated design approach for OSD bridges is based on the following:

1. The current AASHTO nominal stress approach for fatigue design cannot be directly used to evaluate the Fatigue limit state at all critical details due to the complex stress field present. Hence, more refined fatigue evaluation techniques are required for many of the details;

2. Although previous experience is useful and important to consider in any design, adequate performance cannot be guaranteed by implementing simple detailing requirements. This is due to two primary factors. First, the lack of tested and established standard deck panel details makes it difficult to recommend a single "all-purpose" solution. Second, research has shown that in cases where the deck is installed on an existing structure, there are often subtle, yet very influential factors that affect the actual stress response at a given detail which can't be addressed without analytical modeling or testing. In other words, it cannot be assumed that a detail that was successfully used on one bridge will perform similarly on another bridge. Also, publishing standard details in the governing national specifications would hinder future development of improved designs;
3. Although the above are true, it is also recognized that refined analysis for new designs will add engineering cost and potentially limit use for routine span arrangements. Hence, a standard OSD design, which includes all details, must be developed for use on "typical" applications in order for this system to be widely used; and
4. Verification testing of every design adds unnecessary cost and has the potential to delay construction.

Hence, design verification of OSDs requires a different approach than what is used for more common steel bridge members. Since many of the controlling aspects of OSD panel design are local rather than global demands, a well-designed and detailed panel has the potential to be reused in future applications and become a standardized modular component. Therefore, the required effort for design can vary depending on the application and available test data. As summarized previously in Section 5.1.1, these different levels of required effort for design or "Design Levels" are as follows:

Level 1

Level 1 is based on full-scale laboratory testing and may be completed without consideration of Levels 2 and 3. The test must appropriately represent or be a prototype for the design to be used for the structure. That is, all structural components and details must be verified as providing sufficient resistance to test loads. Test loading should be equivalent to the maximum truck load, and stress ranges at details should accurately simulate expected in-service demands and should have accurate boundary conditions. For finite fatigue life design, the resistance shall provide 97.5 percent confidence of survival and the constant amplitude fatigue limit (CAFL) should be exceeded no more than one in 10,000 cycles (0.01 percent). A full-scale test should include a minimum of two rib-spans with three FBs. The number of ribs required will depend on multiple factors. A minimum of five ribs is recommended but more may be required in order to resemble the crossbeam in terms of correct bending and shear. All details must provide a measure of safety consistent with the AASHTO LRFD Specifications. Guidance on testing procedures is provided in Chapter 10.

Additionally, it is allowed that Level 1 designs that have been previously verified by laboratory testing may be used as the basis for design on new projects without additional testing, subject to approval by the owner. It is anticipated and acceptable that if two structures have similar geometry and loading conditions and the same orthotropic detailing is to be used, additional testing is not required to reprove the adequacy of the repeated design. However, analysis must be performed to verify that the boundary conditions and loading conditions are equal and that distortional stresses will not be impacted by the new application.

Level2

Level 2 is based on simplified 1-D or 2-D analysis of certain panel details for which there is little experimental data and acceptance of certain details that are similar to previous tested details as described in Level 1. Calculations need only consider nominal stresses and not address local stress concentrations. Level 2 Design is primarily intended to allow incremental improvement of previously tested details, as demonstrated by Level 1. It is a reasonable assumption that small variations in certain aspects of a bridge may affect the global forces affecting the OSD, yet will not greatly impact the locally-loaded, fatigue-prone details. Careful consideration must be used by the designer to ensure that any changes do not impact the previously verified details, that OSD fabrication techniques will be the same, and that the calculation of nominal stresses is sufficient to ensure the durability of the bridge.

Furthermore, it has been argued by some that certain designs may be verified by long-term observation of existing decks where the details have sufficiently performed. Details that have been proven effective by Level 3 designs and long-term observation while subjected to the appropriate loads may also be verified by Level 2 (i.e. considering only nominal stresses with simplified analysis). Again, great care must be taken to ensure that the detailing and fabrication techniques are identical, and that the loading conditions are nearly exact. For example, OSDs that have been used successfully in Europe, China, and other locations are subjected to different design, and in-situ, axle, and wheel loads and configurations. Thus, the same detail resisting those loads may react differently to loads in the United States in a negative fashion.

Approximate analysis of both open rib and closed rib decks may be based on the Pelikan-Esslinger method presented by Design Manual for Orthotropic Steel Plate Deck Bridges (AISC, 1963) and Troitsky (1987). This method gives conservative values of global force effects in the OSD supported on longitudinal edge girders. Load distribution of adjacent transversely located wheel loads on decks with closed ribs is discussed in the Design Manual for Orthotropic Steel Plate Deck Bridges (AISC, 1963).

Level3

Level 3 is based on refined 3-D finite element modeling of the panel and the supporting bridge superstructure (where applicable). Localized stress concentrations at sites of potential fatigue initiation are quantified for use in the fatigue design. The mesh refinement and stress calculations must follow the guidelines provided in Chapter 4. Meshing must be sufficiently detailed to perform extrapolation of stresses at weld toes and for resolving the wheel patch pressure loading with reasonable accuracy.

Except for special analyses of stability or composite interactions with the wearing surface, structural modeling techniques that utilize the following simplifying assumptions can be applied:

- Linear elastic material behavior
- Small deflection theory

- Plane sections remain plane
- Neglect residual stresses
- Neglect imperfections and weld geometry

Level 3 analysis for structural details is an extension of current AASHTO LRFD methodology for fatigue evaluation by nominal stresses. A similar methodology is applied by the American Petroleum Institute (API) and American Welding Society (AWS 2004) and is well documented by the International Institute of Welding (IIW 2007). It is used extensively for the fatigue evaluation of tubular structures and plate-type structures with complex geometries by various industries, where there is no clearly defined nominal stress due to complicated geometric effects, conditions very similar to orthotropic deck details. This approach recognizes that fatigue damage is caused by stress raisers that exist at details and attempts to quantify them by refined analysis rather than classification into general categories.

Design Level Additional Comments

In most cases, the Design Level will be dictated by the requirements to provide reliability for Fatigue limit states. If no test data is available, Design Level 3 is required unless the designer can verify that local distortional mechanisms (Systems 5 and 6 from Chapter 4) are not expected to cause fatigue cracking. Limit states with respect to Strength, Service, and Constructability generally only require a Level 2 design. For design of panels for bridge redecking applications, Design Level 3 should always be used due to complex interactions that can occur with an existing bridge structure unless an exception is approved by the owner.

A design flowchart is provided in Figure 5-2 to help guide the designer to select the appropriate design level for OSD details.

Composite Stiffness

Design of the steel components of the OSD should be conservatively based on the noncomposite (steel only) stiffness in most cases, for both analysis and calculation of stresses. Alternatively, at the discretion of the engineer, fatigue design may be based on consideration of composite stiffness when a bituminous or cementitious wearing surface is utilized and the required properties are proven. Since the stiffness behavior of bituminous surfacing is strongly influenced by temperature, loading velocity, and the composition of the system, the stress-reducing effect in the steel components cannot easily be described in a design specification. The engineer must develop and apply a rational approach to design accounting for the operating temperature ranges and daily/seasonal thermal cycling. This often requires a damage accumulation approach to fatigue evaluation.

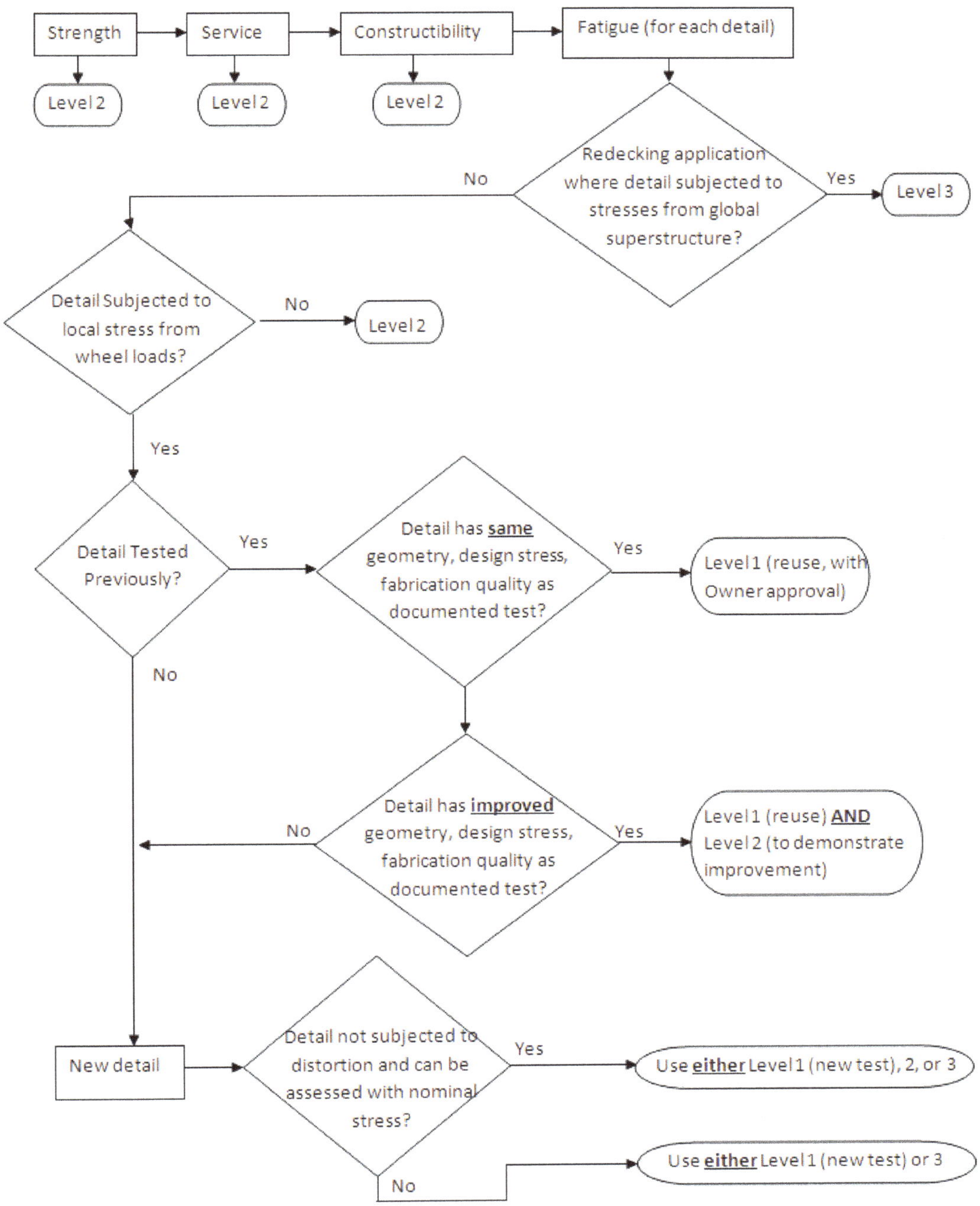

Figure 5-2 Flowchart for Selecting the Appropriate Design Level for an Orthotropic Steel Deck Bridge Detail

5.7. FATIGUE RESISTANCE

Depending on the detail and Design Level applied, there are different approaches for the determination of fatigue resistance at details for design. Level 1 Design is based on proof of resistance by experimental testing, as described previously. Level 2 Design is based on evaluation of nominal stresses near critical details. (This is the philosophy of the current

AASHTO LRFD method for fatigue design, with which most designers in the United States are familiar.) However, this method has not been fully extended to the design of all orthotropic panel details and is only applicable to Level 2 design for certain details. For Level 3 design, a resistance model must be employed that is compatible with local stresses obtained by refined analysis. In this case, evaluation of the local stresses is required. A discussion on each method is provided below.

5.7.1. Analytical Resistance Models

5.7.1.1. Nominal Stress S-N Curve Approach for Level 2 Design

In the nominal stress approach, various details (usually connections) are separated into different categories with similar fatigue resistances. This "detail category" accounts for several parameters that are highly variable and difficult to quantify in practice (for example, local stress concentrations and initial discontinuity size). Full-scale laboratory testing of representative details is used to completely define the fatigue resistance of the detail. Typically the data generated are subject to a considerable amount of scatter, and therefore a statistically significant number of identical tests must be carried out. The data generated consists of the magnitude of stress range and number of cycles to failure when subjected to a particular constant amplitude loading. It has been observed that the logarithm of the number of cycles (N) to failure is approximately normally distributed at a particular stress range (S_r) (Fisher et al., 1974). In the lognormal approach, the mean S-N curve is found using a linear regression analysis, minimizing the error in log N using the method of least squares with the log S_r as the independent variable. The data are then plotted on a log-log scale with the result referred to as an S-N curve, as shown in Figure 5-3. The exponential equation of the line is:

$$N = C / S_r^m \qquad (5\text{-}2)$$

Corresponding to

$$\log N = \log C - m \log S_r \qquad (5\text{-}3)$$

where:

N	=	Number of cycles to failure,
C	=	Constant dependent on the detail category,
S_r	=	Applied constant amplitude stress range,
m	=	the inverse of the slope.

It is important to recognize that the detail category accounts for the local stress concentration effects present at the detail, as well as the variability in discontinuities. This is because it was the complete detail that was tested, with same geometry and fabrication procedures used in the real construction. Hence, only the nominal stress range at the detail need be calculated. This nominal stress is determined in the same manner as that used in strength design and is readily calculated using member properties and simple mechanics principles for most common bridge components. Because this method uses simple, straightforward procedures to determine the applied stress range, the approach lends itself to typical design office practice. However, it is noted that it is absolutely critical that the stress calculated in the fatigue assessment be consistent with that used to develop the details classification(s) provided in the given specification. For example, it would be inappropriate to use the stress results from FEA, which include the effects of local stress risers at a detail, when comparing to an allowable stress for a detail category that developed based on the nominal stress approach.

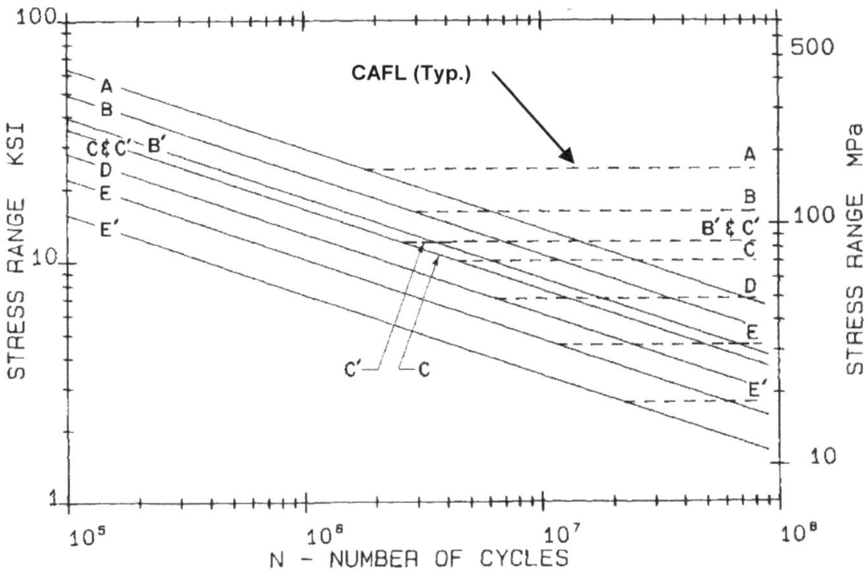

Figure 5-3 AASHTO S-N Curves for Fatigue Design by Nominal Stress

Unfortunately, unique S-N curves obtained from testing are theoretically required for every type of detail and all combinations of geometry, orientation, and fabrication method. However, because of the large amount of scatter in the data, the subtle effects of many of these factors are not readily apparent and many similar details appear to have the same fatigue resistance. Often though, some parameters cannot be ignored and are accounted for by increasing or decreasing the category. For example, the AASHTO LRFD Specification classification system assigns transversely loaded, fillet-welded attachments with welds parallel to the direction of stress and end welds ground smooth (see Figure 5-4) as a Category B connection, if a transition radius of 610 mm (2 ft) is provided. If the transition radius is less than 610 mm (2 ft), the fatigue resistance is reduced. If no radius is provided, the fatigue resistance is reduced to category E or E', depending on the thickness of the plates. The nominal stress calculation would remain the same as the geometric effect is accounted for by the detail, as determined through full-scale testing.

The fatigue curves presented in AASHTO LRFD (Figure 5-3) were developed through extensive research and testing of typical full-scale details known to be fatigue sensitive. These curves provide a designer with an estimate of the number of cycles to failure for a given stress range. In order to provide a certain amount of confidence, a statistical analysis was made of the data during the development of the AASHTO LRFD. The curves shown in Figure 5-3 represent the mean minus two standard deviations of the data, resulting in a 97.5 percent confidence of survival. In other words, there is a 2.5 percent chance a given detail will develop fatigue cracks if designed according to the curves in Figure 5-3, ignoring load variability.

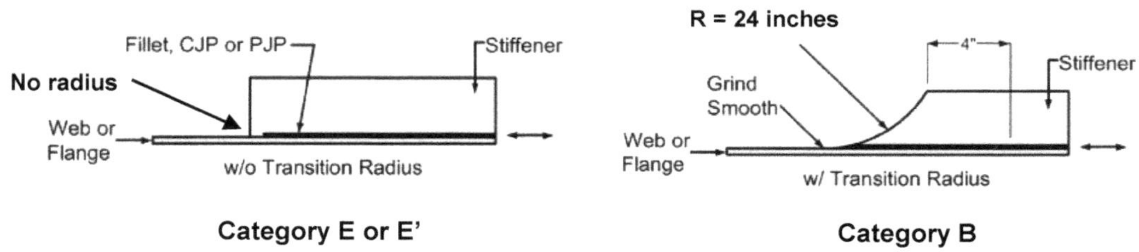

Figure 5-4 Illustrative Example of How Detail Geometry Affects Fatigue Category Assignment at the Termination of a Longitudinal Attachment Using the Nominal Stress Approach

Test data also suggest that there is a limiting stress range below which fatigue crack growth will not occur under constant amplitude loading. This limit is known as the Constant Amplitude Fatigue Limit (CAFL), or Fatigue Threshold, and differs for different details. Though not explicit in the nominal stress range approach, in reality the CAFL also reflects the different stress concentration factors and/or inherent initial discontinuity sizes associated with different details. For an infinite life design approach, the designer simply must detail the components such that all stress ranges are below the constant amplitude fatigue limit of the detail or that it is exceed only a very small number of times. An accepted exceedance interval is one cycle in 10,000. Thus, as long as the CAFL is not exceeded more than 0.01 percent of the time, infinite life can be assumed.

For finite life design, the equation presented above can be used to estimate the number of cycles to failure, if the stress range is known. It should be noted that the current AASHTO LRFD does not require an infinite life design for fatigue. Designers must, however, verify that the detail has sufficient fatigue resistance to meet the design life specified by the owner.

The use of the nominal stress approach lends itself for use in evaluating some details found in OSDs where the nominal stress range is easily defined and local distortions do not occur. For example, the evaluation of longitudinal bending stress in a rib at a splice is appropriate. However, the approach cannot be applied at others such as at the RF detail, where distortions occur and a nominal stress can't be defined. In these cases, one must rely on existing laboratory test data (Level 1 Design) or perform refined analysis (Level 3 Design).

5.7.1.2.Local Structural Stress Approach for Level 3 Design

Unlike the nominal stress range approach, the local stress approach involves use of a more refined analysis to evaluate local stresses at welds that are prone to fatigue. In this case, the resistance model is also different, based on calibration between refined stress analysis and test data. The local stress approach as defined herein is similar to methods applied by the American Petroleum Institute (API) and American Welding Society (AWS) and is well documented in worldwide publications and readily available from the International Institute of Welding (IIW). It is used extensively for the fatigue evaluation of tubular structures and plate-type structures with complex geometries by various industries, where there is no clearly defined nominal stress due to complicated geometric effects. The local stress approach recognizes that fatigue damage is caused by stress raisers that exist at details and attempts to quantify them by more refined analysis (Figure 5-5) rather than classification.

This approach is based on assessment of the surface stress precisely at the weld toe of the joint. The local structural stress (σ_{lss}) at the concentration includes all stress-raising effects of a structural detail, excluding all stress concentration due to the local weld profile itself (Figure 5-6). Since the stress gradient is very high in the vicinity of the weld toe due to the notch effect, the extrapolation procedure must be used to evaluate the structural stress as described in Chapter 4.

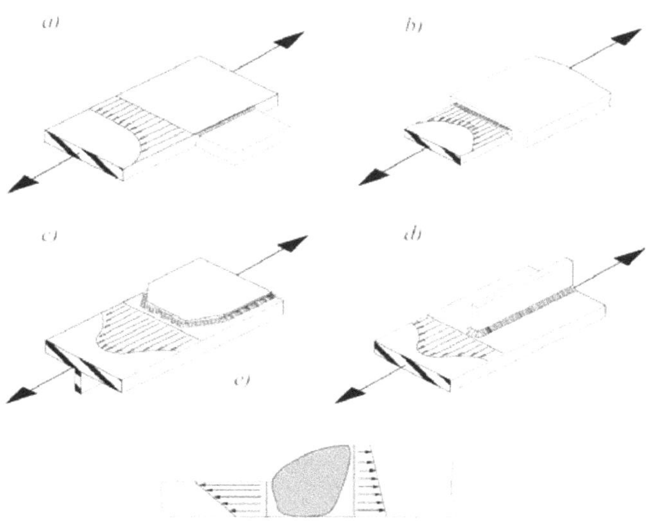

Figure 5-5 Local Stress Profiles for Welded Details for (a) a welded attachment (b) a Plate Size Transition (c) a Cover Plate Termination (d) a Longitudinal Stiffener Termination and (e) a Plate thickness Transition in Section (adapted from IIW, 2007)

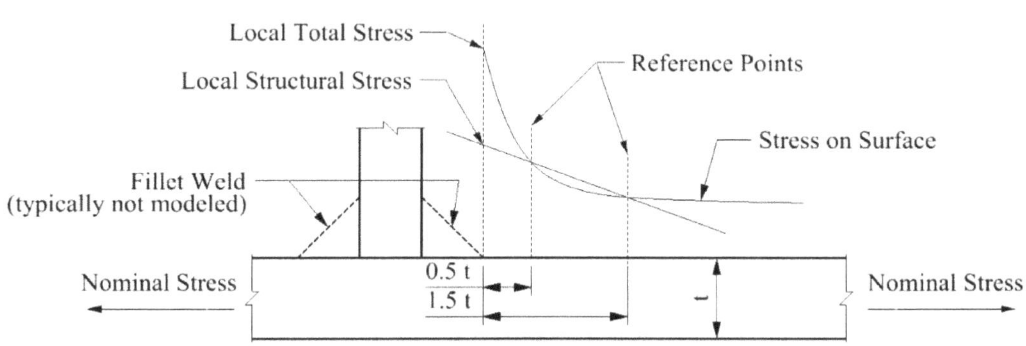

Figure 5-6 Derivation of Local Structural Stress (LSS) using Extrapolation from Reference Points Based on the Finite Element Modeling of the Connection

In the local stress approach, a nominal stress range (σ_{nom}) can be considered to be modified with a stress concentration factor (SCF) denoted as K_s yielding a local or "hot-spot" stress range (σ_{hs}) at the detail:

$$\sigma_{\mathbf{hs}} = \mathbf{K_s}\sigma_{\mathbf{nom}} \tag{5-4}$$

Similarly, through extrapolation, the local structural stress can be derived by:

$$\sigma_{\mathbf{lss}} = 1.5\sigma_1 - 0.5\sigma_2 \tag{5-5}$$

where σ_1 and σ_2 are the surface stresses at the locations 0.5 t and 1.5 t from the weld toe, respectively.

Although there are standard SCFs published in the literature, it important to recognize that these can only be used if they are consistent with that defined with the local structural stress approach. A modifier (i.e., SCF) that is consistent with the approach is often referred to as K_s to denote its compliance with the method, according to IIW (2007). Unfortunately, limited data is available on SCFs in orthotropic panel details for simplified evaluation.

In cases where both bending and axial stress components are present, $K_{s,a}$ and $K_{s,b}$ are used to modify the nominal axial and bending stresses separately as the gradients may be different for each stress component. In this case, the equation takes the form:

$$\sigma_{\mathbf{hs}} = \mathbf{K}_{\mathbf{s\,a}}\sigma_{\mathbf{nom,a}} + \mathbf{K}_{\mathbf{s\,b}}\sigma_{\mathbf{nom,b}} \tag{5-6}$$

An advantage to this method is that a reduced number or even a single S-N curve is all that is required. This approach, commonly used in fatigue evaluation of tubular structures, generally utilizes a baseline S-N curve. The baseline S-N curve is associated with butt weld or fillet weld details in a nominal stress field. In this method, the stress concentration factor accounts for effects associated with global geometry, and any local discontinuities and flaws are incorporated into the S-N curve. As discussed above for the nominal stress approach, the same statistical method of analyzing the test data and providing a lower bound estimate is employed.

If a single S-N curve is used, a major assumption is that local fatigue failure is independent of detail type, and differences in fatigue resistance are entirely incorporated into each SCF. The SCF is assumed to account for stress increases near the weld toe, but all details are assumed to possess the same initial flaws and local stress concentrations as the base-line specimen. A major limitation with this assumption is that using one baseline curve also leads to one CAFL for all details. However, it is known that different details may provide different CAFLs at a wide range of stress range levels, as well as fatigue lives.

Another complication is that near the weld toe, stress gradients are rather steep, thus the maximum stress used to determine the SCF will be influenced by the mesh size of the FE model used in analysis and the strain gage location and length used in the experiments. As a result, SCF determined at different distances from the weld toe will each require a compatible baseline S-N curve. Unfortunately, where to determine the hot-spot stress at the detail varies in design codes and research recommendations (Fisher et al., 1993). Nevertheless, IIW does provide guidance on how to ensure consistency in the application of the technique, both analytically and experimentally (IIW, 2007).

Because the local stress approach requires calculation of a stress concentration at the weld toe, the method cannot typically be applied to details having internal flaws. For example, it is recognized that full and partial penetration butt joints ground smooth exhibit different fatigue characteristics. However, the effects of the internal flaw present in the lack of fusion zone at the root of the partial penetration joint cannot be defined by the SCF. Both strain gage data and FEA would indicate there are no significant stress gradients that affect the internal discontinuities. The SCF for either joint is undefined because there are no surface stress concentrations present. One could assume the full penetration joint as the base S-N curve. But, since the SCF for the partial penetration joint would be calculated as 1.0, the method would indicate no adjustment in the nominal stress is needed. As a result, both joints would be assigned the same SCF and subsequently identical fatigue lives. (It is emphasized that the previous statement is only true if specific experimental data for each joint do not exist.)

Recent research has demonstrated that evaluating the local structural stress perpendicular to the weld toe and evaluating the stress range with the AASHTO Category C provides a reliably conservative assessment of the weld toe cracks at OSD panel welded joints subjected to distortional stresses. As previously mentioned, the AASHTO Category C curve is similar to the curves provided in the Eurocode (ECS, 1992) and the IIW (2007) for local stress evaluation of welded details. Furthermore, research by Dexter et al. (1994) found that the AASHTO Category C curve provides the 97.5 percent survival lower bound for welded details on flexible plates subjected to combined in-plane and out-of-plane stresses in all cases where local stress measured 5 mm (3/16 inch) from weld toe was used for the fatigue life stress range. The work by Connor and Fisher (2006) also found similar results. **Therefore, for Level 3 Fatigue Analysis, the AASHTO Category C curve (Figure 5-3) is conservatively used in conjunction with the LSS approach.** This is predicated on the modeling and stress analysis being conducted by the prescribed methods given in Chapter 4.

5.7.2. Rib-to-Deck (RD) Weld

This connection can be designed and detailed using either a one-sided fillet weld or a partial penetration weld, depending on the application. When the panel is being used as a deck subjected to direct traffic loading, the partial penetration weld should always be used. In other cases, the fillet weld can provide cost-effective design. Provided that sufficient penetration exists and the gap is sufficiently small to ensure that root cracking does not occur, the fatigue strength of the weld will be controlled by traditional weld toe cracking in the deck plate or rib wall, and should be treated as an AASHTO Category C detail with CAFL = 69 MPa (10 ksi). The stresses in this joint are typically dominated by transverse bending from behavior systems 1 and 2 as described in Chapter 4, and thus can be assessed by either Level 2 or 3 Analysis.

In order to use the nominal stress approach for Level 2 Design, several simplifying assumptions need to be made regarding how the wheel load is distributed to this joint and hence, calculate the nominal stress range at the weld toe. Experimental tests indicate that the effective width for transverse strip analysis can be conservatively taken to be the length of the wheel patch footprint (10 inches) plus any load dispersion through the wearing surface thickness.

In-service monitoring has shown that the front wheels of trucks are the ones that typically cause the maximum stress in this joint, even though the magnitude of the front axle loading is less. The single tire front wheel will typically fit within the distance between the rib walls, which maximizes transverse bending stress at the weld.

5.7.3. Rib Splice

Rib splices can be either bolted or welded. Welded rib splices can eliminate the need for internal sealing plates, but this approach obviously adds field welding to the construction and requires a one-sided full penetration weld with internal backing bar left in place. Welded splices also have the advantage of providing continuous ribs from end to end of the bridge, which offers benefits to inspection, and the possible double use as a dehumidification duct to circulate air. The fatigue performance of the various details used to splice ribs is mixed. Interestingly, most of the details that have performed the poorest would either not be permitted in the United States according to the current AASHTO LRFD fatigue provisions, or their failures could have been predicted had a basic fatigue consideration been made. Kolstein provides an excellent summary of the performance of many failed rib splice details in Europe, Australia, and Japan (Kolstein, 2007). However, the review is focused on welded splice details, as they have been more common abroad. These have also been plagued most with fatigue cracking. However, in the United States, the bolted rib splice has been the preferred method in the last 10 to 15 years and will likely remain as such. The performance of this connection has been excellent due to its inherent high fatigue resistance.

Based on the measurements made in the laboratory and the field, the stress range in the rib can be calculated by Level 2 or 3 Analysis. The fatigue resistance of the high-strength bolted connection can be classified as AASHTO Category B, with CAFL = 110 MPa (16 ksi) when the bolts are fully pretensioned. The fatigue resistance of the welded connection can be classified as AASHTO Category D, with CAFL = 48 MPa (7.0 ksi) when the weld root gap is at least the thickness of the rib wall.

5.7.4. Deck Plate Splice

Deck plate splices are typically made in the field using a complete joint penetration groove weld from one side, with or without backing bar removed. Bolted splices have been used in some applications, but welded splices are generally preferred. **A thicker wearing surface is required in cases where a bolted connection is utilized to sufficiently overlay the bolt heads.** A number of deck plates splice details that have been used with apparent success are shown in Chapter 6. It would seem that these joints provided adequate stiffness and resistance to local bending.

The fatigue stresses at this detail can be evaluated using Level 2 or Level 3 Analysis. Most deck splices are subjected to wheel loads causing local response; for other cases, the design need only consider in-plane effects. Longitudinal splices should be located away from primary wheel paths to minimize the number of stress cycles from passing wheels. If local bending due to the wheel loads is substantial, it may be appropriate to conduct testing or apply other weld improvements to ensure long-term performance.

5.7.4.1. Transverse Splice

These splices are subjected to both in-plane stresses from the deck panel spanning between FBs and local bending stresses from the deck plate spanning across the opening in the ribs. The weld toe at the termination of the rib can be protected by minimizing the width of the opening as described in Chapter 6. The rib and weld is terminated to permit access for installation of the backing bar.

In other regions when the weld is subjected to longitudinal in-plane stresses in the deck plate alone (e.g. between ribs) the joint could be classified as Category D, C, or B, depending on the level of non-destructive testing (NDT), and if the weld toes and backing bars are removed. The joint would be similar to a girder flange splice subjecting to minimal stresses, as classified by AASHTO. The fatigue resistance of the deck plate splice subjected to in-plane stresses, without the backing bar removed, is classified as a Category D detail in the previous versions of the AASHTO LRFD. This is consistent with full-scale fatigue testing of very similar details on beams containing cope holes at splices with backing bars subjected to longitudinal stress ranges (Dexter and Kelly, 1997).

The local out-of-plane bending stress field is dominant, as demonstrated by field measurements. Although the actual fatigue resistance of this detail has not been fully established through specific testing, it is proposed that it can reasonably be considered an AASHTO Category C detail with respect to out-of-plane bending stresses at the weld toe, based on tests conducted on stiffeners subjected to out-of-plane deformation and other similar details susceptible to weld toe cracking (Fisher et al., 1990). However, as stated, the fatigue strength of this detail subjected to out-of-plane bending stresses, produced by the vertical wheel loads, has not been well established.

5.7.4.2.Longitudinal Splice

For most of their length, these splices are dominated by local bending in the deck plate as it spans transversely between ribs. At the FBs, the splice is also subjected to in-plane stress. Similar to transverse splices, this detail should be considered Category D for in-plane stresses and Category C for out-of-plane stresses. Locating these splices away from primary wheel paths, and applying finite life design, is an effective way to achieve the most durable design.

5.7.4.3.Bolted Splice

Assuming adequate stiffness (i.e., resistance to local bending from wheel loads) is provided, the fully tensioned bolted connection could be evaluated using the nominal stress approach and classifying the detail as AASHTO category B with CAFL = 110 MPa (16 ksi).

5.7.5. Rib-to-Floorbeam (RF)

There are two accepted approaches to detailing the intersection of the rib to the floorbeam (RF):

1. Providing a stress relieving cut-out in the FB web below the rib
2. Providing a tight fit between the continuous rib and FB with a continuous weld all-around.

At the time of this writing, standardized procedures to evaluate this connection with respect to the Fatigue limit state have not been fully developed and validated. Even a quick literature review will reveal that there are several proposed methods available. While some have been calibrated with full-scale experimental work, other methods, though used successfully in other industries, have not been compared to data obtained from testing details specific to OSDs. Regardless of the method used, it is clear that the RF has the greatest potential for cracking.

Considering the above requirements, it is clear that the Level 2 nominal stress approach as contained in AASHTO LRFD Specifications can only be used for evaluation of rib primary flexural stresses. For stresses in the FB web, cut-out, and rib wall distortion, the "nominal" stress range cannot be calculated using a simplified approach. Therefore designers must rely on alternative analysis methods, such as refined structural analysis (Level 3 Design) where the local structural stress can be evaluated with Category C as shown in Section 5.7.1.2, and/or by direct strain measurements on an appropriately loaded prototype specimen (Level 1 Design) to track the actual S-N behavior of the local structural stress.

Detailed guidance for evaluation of the likely fatigue prone areas is summarized below.

5.7.5.1. Cut-out Detail

The fatigue resistance of the RF connection with cut-out first depends on the type of weld that is selected. This weld can be either a two-sided fillet weld or a CJP weld with termination ground smooth. The fillet weld is the option with least fabrication cost, but the CJP offers superior fatigue strength for rib wall distortional cracking. Some designers have also utilized a transition from fillet to CJP to minimize weld cost but gain the smooth termination. To utilize the fillet weld, the rib may require stiffening or the FB spacing may require reduction. The designer must consider these options carefully before making a selection.

Rib Cracking

At the RF weld, fatigue cracks may initiate in the rib from primary flexural stress in the longitudinal direction. This is considered a traditional weld toe crack at a transverse stiffener, and can be considered an AASHTO Category C detail with CAFL = 69 MPa (10 ksi). This can be evaluated by either Level 2 or Level 3 analysis. Although much of the stress cycle at the bottom of the rib is compressive, this site will typically see stress reversal due to flexibility in the FB.

Rib Wall Distortional Cracking

When the cut-out detail is used, fatigue cracks may develop in the rib wall at the cut-out termination due to distortion of the rib cross-section. In almost all cases, the cracks initiate at the weld toe, even though it may have been removed and ground smooth. Even in the presence of an internal bulkhead plate, rib wall distortions can be expected, thereby producing local transverse bending stresses in the rib wall at the weld toe. It is noted that these local stresses are additive to the longitudinal bending stresses. Thus, any evaluation procedure must take into account both stress components to ensure an adequate design is achieved.

The CJP with ground termination serves to reduce the stress concentration in the rib wall as compared to the fillet weld option, but at an increased weld cost. Both options have performed satisfactorily and should be considered. Since no simplified methods exist to evaluate this detail and nominal stress is undefined, Level 3 design by the local stress approach should be used with resistance of AASHTO Category C with CAFL = 69 MPa [10 ksi], or by testing and Level 1 design.

Floorbeam WebCracking

Fatigue cracks may also initiate in the FB web at the weld toe. This location is subjected to both in-plane and out-of-plane stresses, and can only be assessed by Level 3 analysis. Based on his review of the existing data, Kolstein suggests using European Category 71, which effectively corresponds to AASHTO Category D in the finite life regime up to a limit of five million cycles (Kolstein, 2007). It is believed that this lower recommendation is based on the fact that his database included some details possessing a lower fatigue resistance (e.g., sharper radius transitions, etc.) than included in the database developed at Lehigh University based on years of testing. Kolstein also suggests that the stress range be calculated using the local structural stress approach, but notes that more data are needed to better define the appropriate design category to confidently use this approach.

Based on the testing conducted in the United States, the fatigue resistance of the full-penetration RF weld can be characterized as an AASHTO Category C detail (Tsakopoulos and Fisher, 1999, 2002). However, the use of Category C is predicated on the assumption that the stress range at the detail is calculated in a manner that is consistent with how the stresses were determined during the fatigue tests (Connor and Fisher 2006). As an alternative, a preliminary review of the data obtained from the Lehigh University tests suggests that using Category C with the local stress approach should result in conservative designs.

Root Cracking

Fatigue is also possible in the root of the RF weld, especially for details utilizing the internal bulkhead, depending on the proportions of the fillet or PJP welds and the thickness of the bulkhead plate. Also, both terminations are difficult to fabricate for achieving physical transitions that do not have inherent asperities.

For cases where continuity of the FB and bulkhead are intended by design (creating a cruciform detail), the AASHTO LRFD uses the Frank-Fisher formula (which can be found as equation 6.6.1.2.5-4 in the AASHTO LRFD Bridge Design Specifications [AASHTO, 2010]) that includes three parameters: a) plate thickness "t_p"; b) internal notch length "2a"; and c) weld leg "H". Figure 5-7 shows the cruciform detail fully aligned and with the modification of category C (toe cracking) to a degraded category (root cracking). The modification formula shown in Figure

5-7 presumes no termination effects that may develop at the top of the bulkhead. Because the stress concentration is highly dependent on the stress pattern internal to the FB, increments in stress concentrations are envisioned. Eccentricities between the bulkhead and FB web are considered a fabrication quality issue and are not typically accounted for in design (see Section 6.5). **For these reasons, as well as for associated costs, designs should start without the use of a bulkhead, if other alternatives are available.**

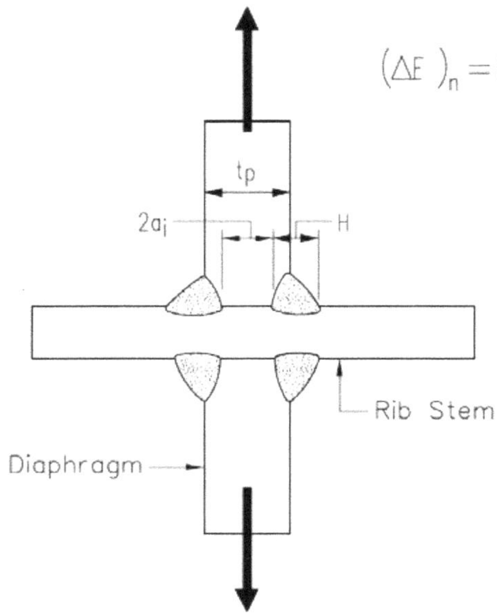

$$(\Delta F)_n = (\Delta F)_n^c \left\{ \frac{0.71 - 0.65\left(\frac{2a_i}{t_p}\right) + 0.79\left(\frac{H}{t_p}\right)}{1.1\, t_p^{1/6}} \right\}$$

Figure 5-7 Fatigue Resistance for Cruciform Detail to account for Root Cracking

Base Metal Adjacent to the Cut-out

The material at the inside edge of the cut-out is subjected to high stress ranges due to the interruption in the stress field in the FB plate. Out-of-plane stresses also contribute to the stress range at this detail. There are no details included in AASHTO LRFD that directly apply to this detail. Since this location is subjected to both in-plane and out-of-plane stresses and nominal stress is undefined, it can only be assessed by Level 3 analysis.

Since there are no welds involved, the surface or edge of the cut-out can be considered a base metal condition if properly finished (i.e., ground). It could be argued then, that the edge of the cut-out can be characterized using the category for base metal (i.e., Category A). However, Category A applies to base metal that is ground smooth and was developed from rolled sections with straight edges with the principal direction of stress parallel to the edge under consideration (Fisher et al. 1974). If the surface of the cut-out is inspected and ground smooth, only the stress concentration effect due to the geometry of the cut-out need be addressed in order to apply Category A. (As per the AASHTO LRFD Bridge Design Specification [2010], a 'smooth' flame-cut edge must comply with ANSI/AASHTO/AWS D1.5 and have smoothness of 1,000 u-in or less.) Using the finite element method or experimental measurements, the stress range along the cut-out can be determined and a fatigue analysis performed.

The procedure to design and analyze the base metal at the cut-out is proposed as follows:

1. The stress range along the edge of the cut-out must be determined by either refined analysis (Level 3) or laboratory tests (Level 1). If refined analysis is used, a sufficiently refined mesh must be used. The stress range used for evaluation must be tangent to the edge of the cut-out.
2. It may be necessary to experimentally measure or calculate the stress a defined distance from the free edge of the cut-out for calibration. For example, if experimental measurements are used, it is very difficult to make strain gage measurements exactly at the edge of a plate subjected to in-plane and out-of-plane stresses. Finite element results would also be more consistent a small distance from the edge of the mesh. Based on the laboratory data (Tsakopoulos and Fisher, 1999, 2002) and the above reasons, a 6 mm ($^1/_4$ inch) offset is proposed. This corresponds to a location that is consistent with the strain gage placements in laboratory and field (Tsakopoulos and Fisher 1999, 2002; Connor and Fisher 2001, 2004, 2006).
3. The stress range would be compared to the CAFL for Category A (CAFL = 165.6 MPa [24 ksi]) for infinite life designs. For finite-life calculations, the S-N curve for Category A should be used. The edge preparation of the cut-out must be consistent with the requirements prescribed for Category A, as contained in AASHTO LRFD.

5.7.5.2. Weld All Around Detail

The second viable alternative for connecting the rib and FB does not utilize a cut-out at the bottom of the rib. In this case, the rib is continuous through the FB, which is fitted tightly into the FB "window." Once fit up is complete, a weld is placed around the outer perimeter of the rib on both sides of the FB web plate, typically a fillet weld. If only a fillet weld is used, a lack-of-fusion plane will exist where the FB web is not fused to the rib wall. Cracks may develop from the root of such welds, primarily driven by the in-plane stress component in the FB. However, toe cracks in the rib and/or FB plate will most often control. Obviously, if this detail is selected, a very clear understanding of the demands placed upon the joint is required. It should also be noted that the all-around weld detail places a Category C detail at the bottom fiber of the rib for primary flexure. The associated weld toe is located where the longitudinal bending stresses would be greatest and where additional out-of-plane stresses may be present due to distortion of the rib wall. Nevertheless, there are examples where bridges with this detail have demonstrated excellent in-service performance.

Generally speaking, for the fully welded detail to be successful, a flexible FB plate is essential in order to accommodate the distortion from the rib rotation. Hence, this detail may not be feasible in cases where a very shallow FB is required, as is often the case where a shallow deck is placed on existing bridge FBs.

Floorbeam WebCracking

It is the opinion of the authors that this connection should always be detailed such that sufficient weld penetration is provided to ensure that root cracking does not control, especially at the bottom of the rib. Assuming this is guaranteed, weld toe cracking will then control. Hence, the detail could be evaluated using the same methods proposed for the cut-out detail described earlier. However, rather than focusing at only one location, as can generally be done with the cut-out detail, the entire perimeter of the connection must be examined for the Fatigue limit state. Note that the components of in-plane and out-of-plane stress must both be considered and stresses perpendicular to weld toe should be used.

Consideration of Compression and Shear Stress Cycles at Concentrations

In the design of this detail, careful consideration must be given to the treatment of compressive stress cycling because the majority of the stresses generated by wheels over this location will be compressive. Generally, compressive cycling may cause initial damage at welded details where tensile residual stresses exist, but will eventually be unable to cause further crack propagation as the stress field is adjusted, and only tensile stresses from the spectrum that overcome compression from dead load would drive continued crack growth. AASHTO LRFD states that fatigue design provisions shall only be applied to details where the compressive (permanent) stress is less than twice the maximum tensile live load stress (AASHTO, 2010). Conversely, certain codes dealing with welded structures consider compressive cycling to be equally damaging as tensile cycling in all cases. In OSDs, the point at the bottom of the rib is subject to several stress components (Figure 5-8), which make applicability of AASHTO LRFD questionable. As such, it may be prudent to consider compressive cycling to be fully damaging because the rib material is thin and any propagation by one mode could continue into a different mode.

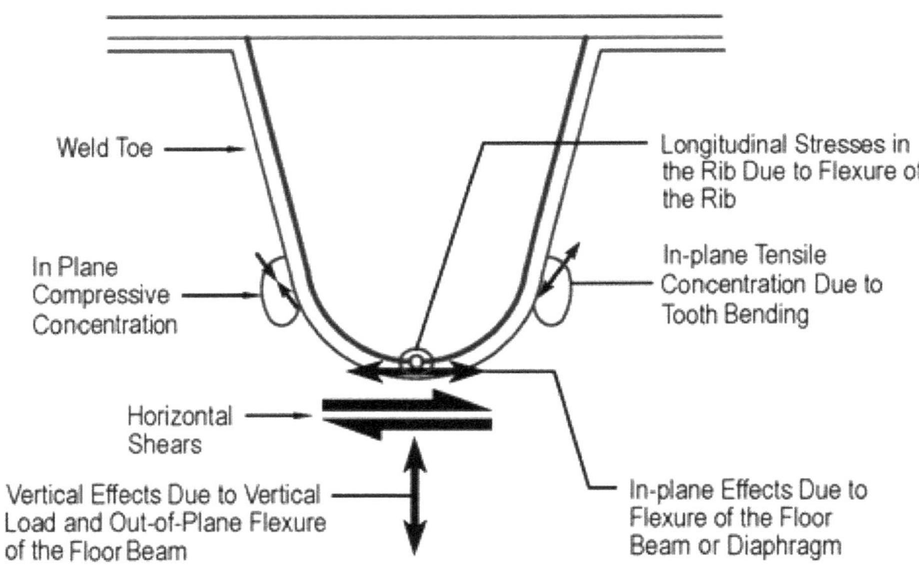

Figure 5-8 Stress Conditions at Base of Rounded Rib subjected to Shear Stresses between the Deck and the FB Web and Vertical Effects from Out-of-plane Flexure of the FB

At the bottom convergence of the rib and FB there are orthogonal stresses and shears. The vertical effects in the FB and the longitudinal stresses in the rib are not in phase with the shears in the FB and with the in-plane effects due to floor beam bending. This is called non-proportional loading and detrmining the range of stress will most likely require an influence surface based approach (see Section 4.6.1).

There are also concentrations in the FB web on both sides of the belly at the end of the tangent part of the rib where it meets the curved part. The principal stresses there are not perpendicular to the weld toe. Stresses perpendicular to the weld toe are to be used for the fatigue design verification, but many software tools will not have the ability to develop the stress influence surface at a specified orientation. The evaluation criteria necessary for this detail have not been fully developed and validated and it may be prudent to use more conservative criteria.

5.7.6. Rib-to-Deck at the Floorbeam (RDF)

Fatigue cracking at the rib-to-deck at the floorbeam (RDF) welded joint has received relatively little attention, compared to some of the other fatigue prone details used on OSDs. As with other details, there is a variety of methods that have been used to make this connection. The fact that the state of stress at the detail has received relatively little attention is most likely due to the fact that the joint is complicated to address analytically, especially with respect to the Fatigue limit state. Much of the limited experimental data for the connection was reviewed by Kolstein (2007) and recommendations were made as to how to best evaluate the joint. However, he clearly states that the data used to make the recommendation are limited.

This connection has been detailed using several different geometric configurations. One of the most obvious differences between the details is in the use of a cope hole in the FB web. Early designers believed that intersection of the rib and FB welds was not preferred, and chose to detail a cope hole in the top corner of the FB to allow uninterrupted passage of the RD weld. However, the termination of the FB to deck weld was found to initiate fatigue cracks in the deck plate, while the non-coped detail performed well in service. The intersecting weld is believed to be acceptable since there is minimal triaxial restraint and plate components are relatively thin. The present recommended detail is the non-coped approach. The overall field performance of the joint suggests it offers superior fatigue resistance. Since there is limited experimental data available, any recommendations with respect to the design of the joint are based on engineering judgment and rely upon existing methods of evaluation. Kolstein seems to suggest a similar approach, i.e., relying somewhat on the performance of joints that appear to be performing well in service when selecting the configuration of this joint (Kolstein, 2007).

A local stress concentration is created in the RDF weld root at this location due to the local "rigid" support from the FB web and the distortions that occur in the deck plate from in-plane flexure of the FB. Root cracking is controlled in other typical locations of the RD weld by limiting the gap tolerance to achieve closure of the root after welding. It is not certain whether such gap control alone is sufficient to prevent root cracking at the RDF in all cases. Tensile stresses at the RDF root may be large enough to open the gap and cause cracking in some cases.

Clearly, Level 2 Analysis cannot be applied to this detail since there is no nominal stress that can be determined and stresses are primarily distortional. This detail can only be verified by Level 1 or 3 Design approach. Unfortunately, application of Level 3 Analysis to this detail is beyond the limits of the method as described earlier (i.e. does not apply to root cracking). It is not known how effective the local stress approach with AASHTO Category C limit will be at prediction and control of RDF root cracking since the notch condition is different from that at the weld toe. Also, little known research has been done to quantify the performance of the detail and parameters of influence. This should be the subject of future research efforts. Until then, Level 3 can be applied, but with additional verification to best extent reasonably possible.

5.7.7. Summary

A summary of the fatigue design requirements for fatigue-prone orthotropic panel details, including illustrative examples, is shown in Table 5-1.

Table 5-1 Summary of Fatigue Design Requirements for OSD Details

Detail	Illustrative Example	Description of Condition	Fatigue Category Detail	Allowable Design Level
Rib-to-deck (RD) Weld	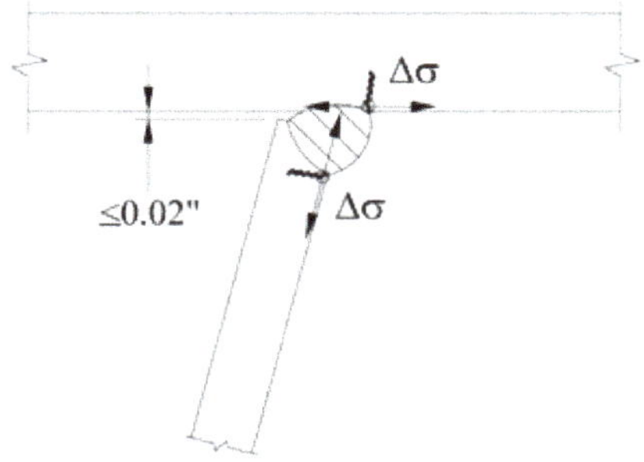	One-sided 80% (70%min) penetration weld with root gap ≤ 0.5mm (0.02") prior to welding	C	1,2, or 3
Rib Splice (Welded)	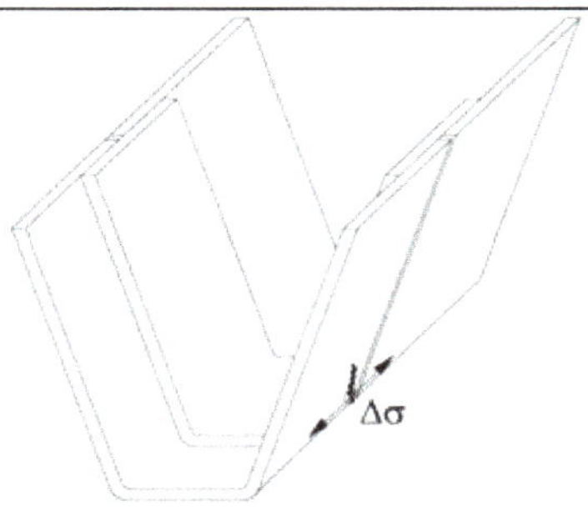	Single groove butt weld with permanent backing bar left in place. Weld gap > rib wall thickness.	D	1,2, or 3
Rib Splice (Bolted)	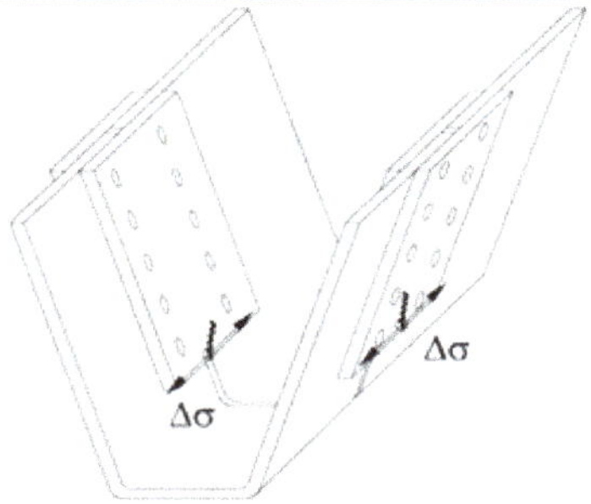	Base metal at gross section of high strength slip critical connection.	B	1,2, or 3
Deck Plate Splice	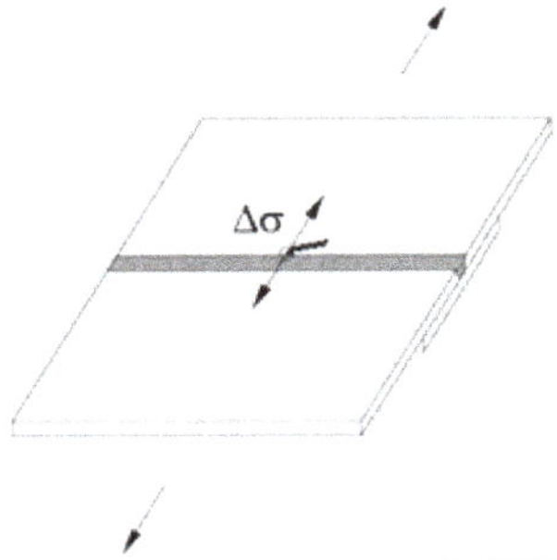	Transverse or Longitudinal single groove butt splice with permanent backing bar left in place.	D (in plane) C (out of plane)	1,2, or 3

Rib to FB (RF) Weld (Rib)	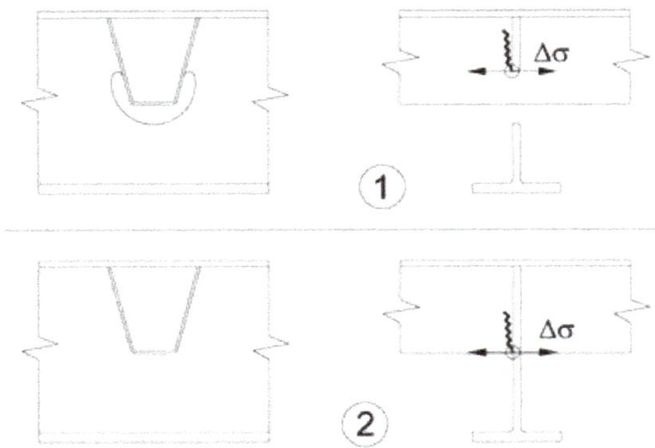	Rib wall at rib to FB weld (fillet, PJP, or CJP).	C	1,2, or 3
Rib to FB (RF) Weld (FB Web)	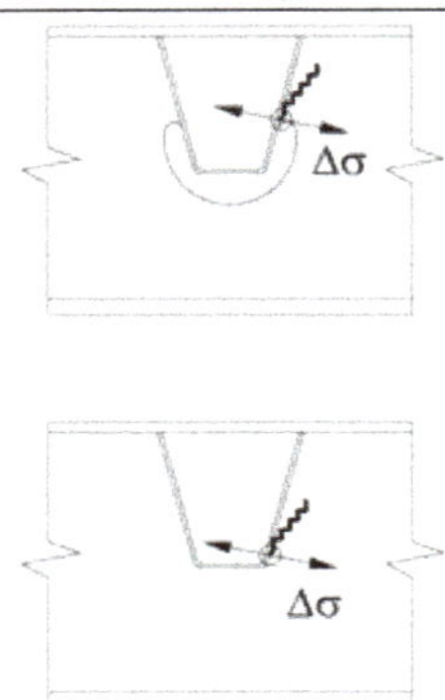	FB web at rib to FB weld (fillet, PJP, or CJP)	C*	1 or 3
FB Cut-out	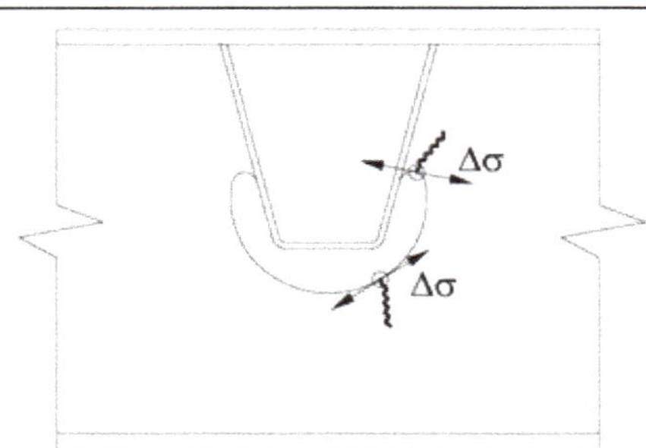	Base metal at edge with "smooth" flame cut finish as per AWS D1.5.	A	1 or 3
Rib Wall at Cut-out	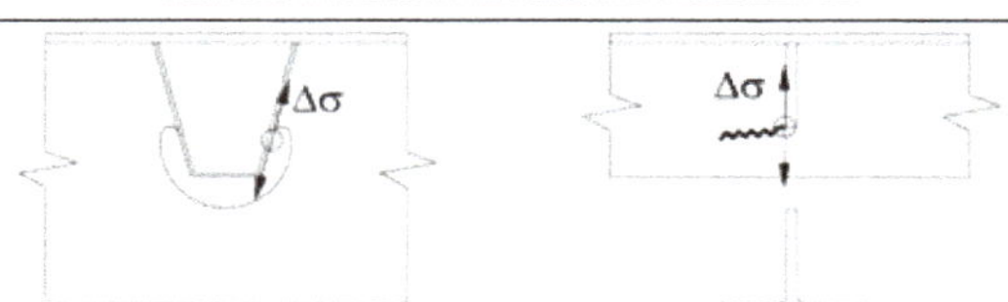	Rib wall at rib to FB weld (fillet or CJP).	C	1 or 3
Rib-to-deck Plate at FB (RDF)	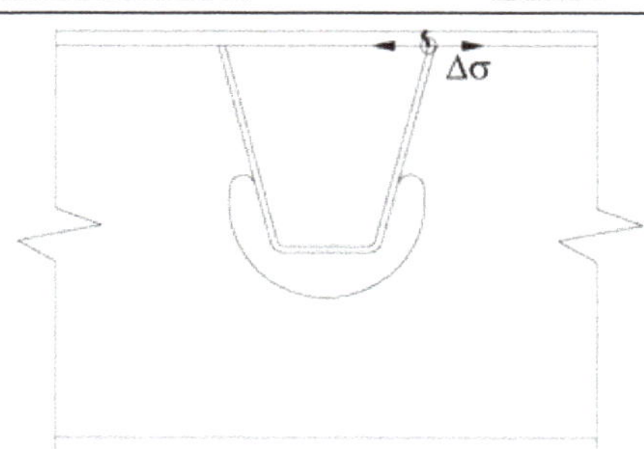	Base metal of deck at root of rib-to-deck fillet weld	C	1 or 3

* Where stresses are dominated by in-plane component at fillet or PJP welds, Figure 5-6 shall be considered. In this case, stress should be calculated at the mid-thickness and extrapolation procedure need not be applied.

5.8. FRACTURE CONSIDERATIONS

Fatigue and fracture are two related, albeit different, limit states that need to be checked in the design and evaluation of all dynamically loaded steel structures. In its most simple form, fatigue is the crack growth of a flaw due to cyclic or repetitive loading. Generally, the Fatigue limit state is defined or met when a crack grows to through-thickness of the component. However, in many structures, this is really just a serviceability limit state and does not necessarily mean that the structure is in danger of fracture or collapse. This is especially true in the case of OSDs, as they are highly redundant systems. There are numerous examples of cracked OSDs, some with very large cracks, which show no signs of distress in terms of load carrying capacity.

Although there are no known failures attributed to brittle fracture in ODSs, fracture is a unique limit state that is distinguished from fatigue. Since many engineers incorrectly use the terms interchangeably or always in the same sentence, there has been some confusion as to the difference of the terms. In its simplest form, brittle fracture, as referred to herein, is the rupture in tension in conjunction with rapid, unstable extension of a crack, leading to gross deformation, complete separation of the component, and usually loss of function or serviceability. Fracture is addressed in the AASHTO LRFD Specifications through material selection to achieve specified levels of fracture toughness.

Although the possibility of brittle fracture of a given component of an OSD exists (e.g. fracture of a rib at a welded detail), fatigue has traditionally been the limit state that controls and is the focus of design. OSDs are not fracture critical on their own. A bridge with an ODS would only be classified as a Fracture Critical Member (FCM) if the girder portion was a non-redundant member subjected to tension. Hence, in the preceding discussions, the emphasis has been placed on the Fatigue limit state. However, there are several good texts that can serve as a starting point for more in-depth study of brittle fracture (Barsom and Rolfe, 1987) (Anderson, 2005).

5.9. REDECKING CONSIDERATIONS

Orthotropic panels for the redecking of existing bridges can be designed and detailed as a "floating" system with weak, flexible connections to the existing framing or as an integral system with rigid connections. Generally, the latter is the preferred approach, since this allows for the reduction or elimination of deck joints and adds rigidity to the existing framing. However, this requires careful Level 3 Design of the composite structure, including evaluation of any new concentrations that may be created in the existing framing. In all cases, connections must be designed and detailed to address the interaction stresses.

For the construction of an analytical model, rigidities must be assigned to all connections, which must have strong similitude to the actual ones. There must be accurate modeling of at least one connection, from which all the rigidities of other connections are represented by idealized ligaments, and all are then correctly represented.

The available depth over the existing floor beam or floor truss is the greatest determinant in selection of crossbeam spacing (over stringers), and of rib-diaphragm design options in redecking projects. Various options are covered in more detail in Chapter 6.

6. DETAILING

Detailing is a critical step in the development of a cost-effective and safe design of an orthotropic bridge. This is especially important for orthotropic steel deck (OSD) bridges since much of the design is controlled by Fatigue limit states, which are sensitive to localized effects. Detailing involves selection of components and connection geometries to make the system successful, balancing the requirements for structural performance, economy, and constructability. This demands a full understanding of the controlling parameters from an engineering perspective, as well as close communication with fabricators and contractors with experience in the construction of orthotropic bridges. Some detailing decisions, such as material selection, can be made somewhat independently, but others, such as the rib-to-floorbeam (RF) connection, must be made with careful consideration of impacts to the rest of the system.

Selection of details must rely heavily on the past record of performance and test data (when available). As outlined in Chapter 5, the Design Level can vary depending on the availability of test data on similar details (see Section 5.1.1). This demands that the designer be familiar with the worldwide literature related to orthotropic construction. Two of the most important detailing decisions that must be made are as follows:

- Welded or bolted splices.
- Cut-out or weld-all-around RF connection.

Each of these options has advantages and disadvantages, which have been discussed throughout this Manual. All have proven successful. Ultimately, the designer must select a detailing approach that addresses criteria for design, construction, and maintenance. This chapter provides general rules for appropriate structural details related to steel components and connections for inclusion in the contract plans.

6.1. MATERIALS

Structural Steel

Current practice is to use ASTM A709 Grade 345 (Grade 50) material in most cases, consistent with other common steel bridge structural members. Grades of lower strength are possible, as design is fatigue driven, but they tend to be of decreased availability. Uncoated weathering steel is not recommended in conditions where the underside of the deck is exposed to salt spray or other environmental conditions that would negatively affect performance. HPS material could work well but the added expense is likely not justified. Because OSDs are highly redundant, standard toughness is typically specified. OSDs would only be classified as a Fracture Critical Member (FCM) if the deck was part of a non-redundant girder subjected to tension.

Generally, in OSD construction, mill plates have the direction of roll oriented the direction of the ribs and bridge axis. It has been argued that the elongated grains, in the steel, running transverse to the ribs, and in the direction of the floorbeam (FB), would retard crack propagation in critical areas. The extent to which this might occur is not known, and is in most cases impractical.

Consumables

American consumables fall under AWS A5 criteria, but must be validated by Welding Procedure Records (WPR). Also, the usual practice is to fabricate customized mock panel tests in which macroetching and other tests are used to validate the weld process (see Chapter 7). In the past, consumables using the automated submerged arc welding SAW process for the rib-to-deck (RD) weld have faired best. For FB welding, Shielded Metal Arc Weld (SMAW) and Flux Core Arc Weld (FCAW) are the processes more often used. Gas Metal Arc Weld (GMAW) can be used, however, to produce smooth welds of high quality, but a greater degree of welder experience is necessary.

As many fabrications of OSDs occur abroad, specifications typically require that consumables conform to American standards or be produced by American manufacturers.

6.2. CORROSION PROTECTION

The corrosion control strategy should be selected in a manner similar to other common steel bridge structures, with options including use of: 1) material resistance 2) coating or 3) dehumidification. Uncoated weathering steels and other common bridge coatings can be expected to perform very well on the OSD since there are very few deck joints and few places for debris to collect. Currently there are no known examples of any OSDs with a zinc coating of any kind. This may not be viable due to limitations in hot-dip baths and concerns about further reduction in fatigue resistance at welded details.

Addressing corrosion may require different strategies depending on area being protected. Four areas of corrosion protection need to be addressed:

- Top of the deck plate.
- On the exposed bottom of the deck.
- Inside closed ribs.
- Inside of box girder (when applicable).

Deck Plate Protection

The bonding agent that serves as a substrate to the wearing surface also serves as waterproofing for the OSD. When an asphalt wearing surface is used, a waterproofing membrane that bonds to the steel is used. Two types are predominant in the United States: a) a methacrylate compound, on the order of 3 mm (1/8 inch) thick, with grit cast on it, or b) an epoxy-asphalt mixture with silica sand, about 25 mm (1 inch) thick. The first is sprayed on and dries to a rubber-like consistency, the latter dries to a hard concrete. Both serve as waterproofing and as substrates to a wearing layer of asphalt. Consideration must be given to the surface preparation for adequate bonding of the wearing surface system. This is usually dictated by manufacturer's requirements, and often involves shot blasting prior to application of the initial surfacing layer.

Instead of asphalt, some bridge owners use polymer concrete as a wearing surface, especially in long suspension bridges. This is done to reduce suspended deck structure weight. Polymer concrete layer is in the order of 9 mm (3/8 inch) thick and serves as waterproofing as well. There are two types of installations, slurry or multiple-layer. The first method involves using slurry with mixed-in fine aggregate, which is applied with a squeegee onto a clean deck and coarse aggregate is applied prior to drying. The second method is constructed in layers of epoxy and cast aggregate. Both surface types contain epoxies with "flexibilizers," or urethanes that render them less brittle.

For further discussion of properties, advantages, and disadvantages of alternate substrates and of epoxy concrete, the reader is referred to Chapter 9.

Bottom of Deck

Protection of the bottom of OSDs is typically achieved with the use of a paint coating system. Painting technology for bridge structures is an ever-evolving science. The FHWA has ongoing studies at the Turner-Fairbank Highway Research Center that will provide key findings, not only on preparation and coating performance, but on corrosivity of environments, environmental compliance, toxicity and containment, etc. Preliminary findings are that three-coat systems in which zinc based primer, epoxy coat, and urethane finish coat work the best. Various state DOTs and bridge owners will derive from these studies their own standards and practices.

Paint systems are most often applied in the shop and touched up in the field where damaged by transportation and handling, or over field-welded areas. OSDs are optimal structures for resisting corrosion, as they have smooth surfaces and no perches, such as bottom flanges for moisture to pond or for birds to nest.

Rib Protection

The rib needs to be protected internally from corrosion due to condensation. The common practice in the U.S. is to place internal steel bulkheads (sealing plates) on each side of the bolted splice to completely seal the ribs from moisture penetration. These are welded all around or welded along the rib and caulked at the top. This prevents moist air from entering the inside of the closed rib. In the area of the splice, the ribs are painted internally.

Interior ofOSD Box Girder

Where orthotropic ribs stiffen the walls of a box girder, the interior of the box girder needs to be protected internally from corrosion due to condensation. This can be accomplished with a coating, or by a dehumidification system, which has been done successfully overseas in many cases. Since the entire interior space is protected by this system, the internal sealing diaphragms in the ribs at the splices are not needed and maintenance of the interior surfaces is not required. Inspection of welds is easier because the welds are not painted and the access is provided without disturbance of the traffic below or on top of the bridge.

6.3. GENERAL PROPORTIONS

Orthotropic bridge fabrication is labor intensive, and the design should not strive solely to minimize material. As discussed in previous chapters, this has been one of the leading causes of fatigue and wearing surface failure in early designs. Rather, the design should strive for repetition in details, lending itself to automation and advanced technology, such as automatic welding processes and robotics to maximize economy. OSDs require significantly more shop welding than any other steel bridge system. Strictly speaking, there is nothing difficult about fabrication of OSD panels, compared to other steel bridge construction, except for a singular welding issue for the RD connection and time consuming cut-out details, if they are required. Erection of OSD panels in deck replacements is now commonplace, even if strict field welding rules must be applied.

The recommended limits for panel detailing proportions are summarized in Table 6-1.

In box girders designed for suspension bridges, where low resistance to cross winds is necessary and sharp corners are part of the outer edges of the boxes, close attention must be paid to access for welding of stiffeners and shell plate corners.

Table 6-1 Recommended Limits for Orthotropic Panel Proportions

Detailing Dimension	Limit
Deck Plate Thickness	$t_d > 14$ mm (5/8 inch)
Rib Thickness	6 mm $< t_r <$ 12 mm (1/4 $< t_r <$ 1/2 inch)
Rib Spacing – direct wheel load	600 mm $< s <$ 762 mm (24 $< s <$ 30 inches)
Rib Spacing – no direct wheel load	600 mm $< s <$ 1000 mm (24 $< s <$ 40 inches)
Floorbeam Spacing	$L <$ 6000 mm (20 ft)
Ratio of Rib-to-Floorbeam Depth	$h_{rib}/h_{FB} < 0.4$
Floorbeam Web thickness	10 mm $< t_{FB} <$ 20 mm (3/8 $< t_{FB} <$ 3/4 inch)
Ratio of Cut-out to Rib Depth	$h_{cutout}/h_{rib} > 0.33$

6.4. RIB-TO-DECK PLATE (RD) WELD

Historically, the RD weld has been specified as a one-sided, partial penetration weld with 80 percent minimum penetration and limited melt-through. Because of inherent variations in components when ribs are thin, weld size has tolerances that are of a high magnitude relative to rib thickness. Achieving a minimum of 80 percent penetration without melt-through is difficult. Fabricators, fearful of melt-through, have chosen electrode angles of 60 degrees with the deck plate, without preparing a bevel in the rib, and have often failed to achieve 80 percent penetration. The primary factors that are known to influence the fatigue resistance of the joint are: penetration, melt-through, and root gap.

Penetration

It is not known exactly where the "80 percent rule" came from and hence, it appears somewhat arbitrary. A review of the literature suggests that it was more of a practical limit, i.e., the maximum penetration that could be achieved without regularly resulting in weld melt-through. Kolstein's review of existing test data suggested that the weld throat should be, at a minimum, of the same size as the rib wall and that the amount of penetration should be between 50 to 80 percent (Kolstein, 2007). This statement must be used with caution, however, as the influence of the lack of fusion zone size is not constant for different plate thicknesses with constant weld reinforcement size. In other words, 50 percent penetration in a 9 mm (3/8 inch) thick rib wall would result in a different fatigue resistance than a 50 percent penetration weld in a much thicker rib wall with the same weld reinforcement. However, for typical values of rib wall thickness used in highway bridges, the range would seem reasonable.

Although Kolstein suggests a lower limit of 50 percent, this can result in a rather large lack of fusion plane and greater penetration is recommended. Accordingly, penetration levels between 70 to 95 percent, with a target of 80 percent, seem more reasonable. The lower bound of 75 percent is from FEA analysis by Xiao, which indicated that increasing the penetration above 75 percent did not result in any significant improvement in fatigue resistance of the joint (Xiao et al 2008), though no experimental work was performed as part of the study. Nevertheless, all of the data suggest that a range for the degree of penetration is appropriate and that requiring 80 percent is not justified.

Melt-through

There are legitimate concerns that weld melt-through could create undesirable discontinuities at the root that could become the source of fatigue crack initiation, and therefore it should be avoided (Ya, 2011). Fatigue tests conducted by Sim and Uang (2007) at the University of California at San Diego, on behalf of Caltrans, however, indicate that melt-through is not a real problem, even when intentionally introduced after 80 percent penetration, to create abrupt transitions. This suggests that, using SAW, weld size limits between an 80 percent minimum penetration and a maximum amount of melt-through should be a workable and acceptable fabrication procedure, even for 8 mm (5/16 inch) rib thickness.

Root gap

Recent research has shown that fatigue resistance of the weld is clearly improved when the root gap is closed in the final condition (Wright, 2010). Shop experience indicates that using a tight fit prior to welding will also help prevent weld melt-through with a limit of 0.5 mm (0.020 inch).

As a result, the consensus is that the proposed detailing criteria for the weld between the webs of a closed rib and the deck plate be a one-sided, nominal 80 percent penetration, with 70 percent minimum and no blow-through, and with a tight fit less than 0.5 mm (0.020 inch) prior to welding. Additional fabrication techniques and tolerances are discussed in Chapter 7.

6.5. RIB-TO-FLOORBEAM (RF) CONNECTION

At the rib-to-floorbeam (RF) connection, fatigue cracks have been observed in the rib wall and FB at weld toes and weld terminations at cut-outs. The detailing must be made to prevent such crack growth. This detail is the most labor-intensive of all the details, and is where the designer is free to choose the geometry. Cracks at the RF are generally due to the presence of high out-of-plane stress ranges due to distortion of the rib wall. The distortion comes from the FB pumping the rib wall in and out-of-plane. Where cut-outs are used, the problem can be likened to web gap cracking at transverse stiffeners in a plate girder. In order to prevent such cracking, internal FBs or bulkheads have been used to ensure the geometry of the rib is maintained, and hence the distortion minimized or eliminated. However, bulkheads may create other problems (see Section 4.1 for more discussion).

There are generally three viable options to consider at the RF. In order of increasing cost these are: 1) no cut-out and welded all around 2) cut-out with fillet welding 3) cut-out with CJP and ground termination (Figure 6-1).

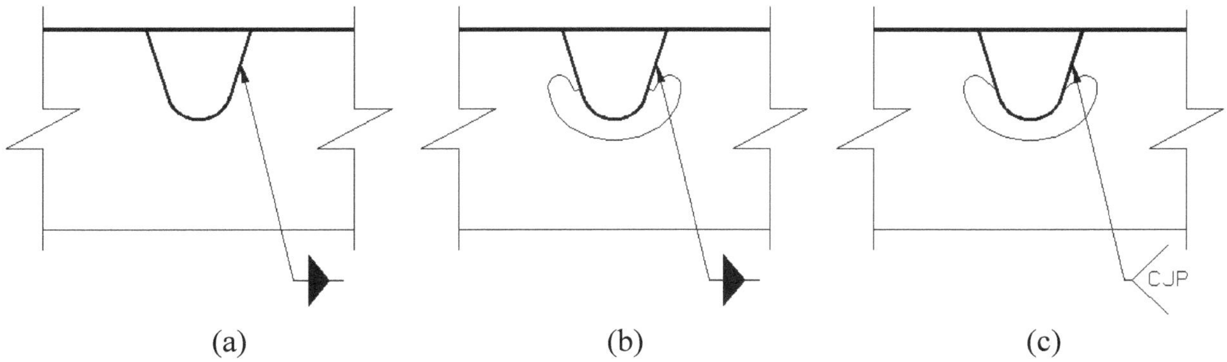

Figure 6-1 Three Viable Detailing Strategies for RF Connection showing (a) no cut-out and welded all around, (b) cut-out with fillet welding, and (c) cut-out with CJP and ground termination.

Cut-out Shape

The shape of the cut-out should be detailed with careful consideration of the fatigue stress concentrations created at the welds and the free edge of the cut-out (if used). The designer is free to choose the ideal geometry and there are few limitations since the cut will be made by CNC equipment in fabrication. Worldwide consensus has not yet been established for this. The optimum shape should satisfy, to the best extent possible, the following conditions:

- The radius of curvature in each point of the free edge of the cut-out must be large, so that the stress concentration at the free edge is low.
- The dimension of the FB web tooth between ribs must be large, so that the in-plane stresses due to shear and bending of the tooth can be minimized.
- The cut-out must terminate high enough on the rib wall to avoid the highly stressed area of the rib and provide out-of-plane flexibility for rib rotations.
- The cut-out must terminate low enough on the rib wall to prevent excessive distortion in the rib walls.

For example, Figure 6-2 shows two possible cut-out details. Figure 6-2(a) shows a non-preferred bulkhead detail that has stress concentrations due to abrupt weld terminations. Concerning the geometry effects, improvement of that detail is represented by Figure 6-2(b), where there is no bulkhead and a "large" radius is introduced to alleviate the concentration effects from the geometry of Figure 6-2(a) (as indicated at point "a" in Figure 6-2). To keep higher tooth rigidity and minimize flexure in the upper portions of the cut-out, the radius should quickly be reduced (at point b). Note that in Figure 6-2(b), the weld is formed and then ground to match the radius transition. As an alternate method, a fillet weld all-around the termination of the radius could be used, but the local stresses will need to be carefully analyzed to ensure root cracking of the fillet is avoided. The fillet welded detail may be considered to be of lesser quality with a higher potential for defect-induced cracking at the start/stop or wrap-around.

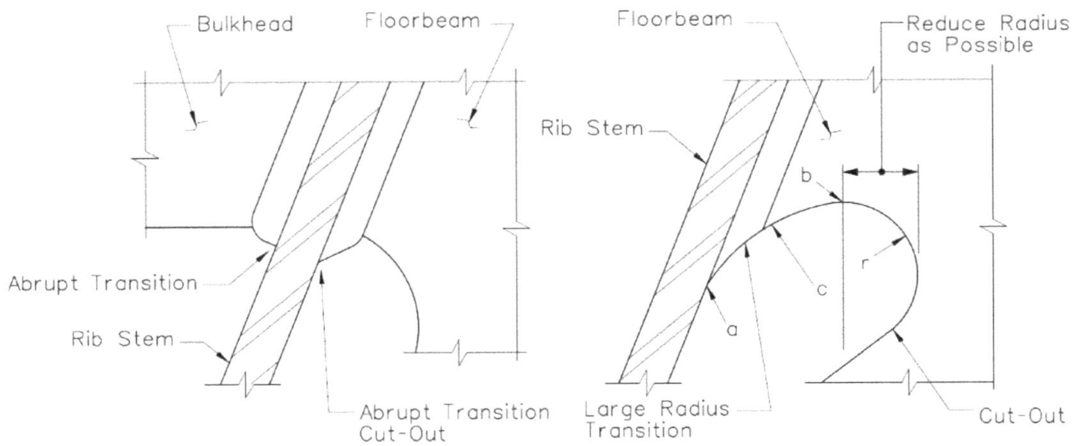

(a) (b)

Figure 6-2 Rib-to-Floorbeam (RF) Cut-out Detail showing Termination of the Floorbeam Web Utilizing a Stress Relieving Radius with (a) a Bulkhead, and (b) No Bulkhead

For in plane stresses, the detail in Figure 6-2(a) is treated using both the well-established weld termination fatigue category and toe of weld category. For the complex stress fields introduced by rib rotation, the connection is evaluated according the guidance in Chapter 5. The transition in Figure 6-2(b), as indicated in the figure by point "a", is equally as poor as the abrupt transition illustrated in Figure 6-2(a), **but** the stress transition is greatly changed. As such, the best approach to this fatigue evaluation is to use Level 3 finite element evaluation to find the stress at the approach of the transition accurately and optimize the cut-out geometry accordingly. At point "b" the accepted technique is to take the principal stress at the toe and cut-out surface and take the component of that stress perpendicular to the toe at point "c", where the resistance is presumed equal to that of the base metal (Category A) and the model should give the concentrations due to the curved geometry.

Cut-out Termination

The cut-out often requires that distortional stresses from the rib wall going into the RF weld do not encounter a notch inherent in a double-sided fillet weld detail. The area immediately at the transition and some length above can be a CJP weld, with abundant reinforcement on the sides. This transition is shown in Figure 6-3. The length of CJP extension depends on internal stresses to be quantified by FEA and compared to the experimental database.

The figure also shows grinding of the cut-out to the extent deemed necessary. Grinding of the side reinforcements is also beneficial, but not necessarily needed. Figure 6-3 shows that a transition of PJP between the CJP and fillet weld can be used to minimize welding. Grinding is required all along the surfaces where it is necessary to maintain a Category A in the base metal.

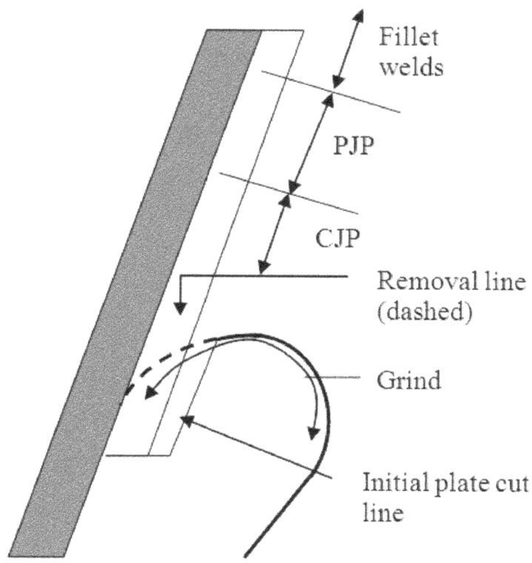

Figure 6-3 Weld Detail Transition from Fillet, to Partial Penetration, to Complete Penetration for Smooth Termination of Cut-out at the Rib-to-Floorbeam (RF) Connection

No Cut-out, Welded All-Around

When the OSD is fabricated without a cut-out at the FB, increased penetration can be considered around the rib's belly with suitable PJP transitions to fillet welds on the sides.

Details with Bulkheads

As mentioned, the use of an internal bulkhead presents additional considerations with respect to fatigue that must be addressed, not to mention the added fabrication costs and inability to inspect. Since the bulkhead provides a load path for continuity of FB forces through the rib, the welds used to attach it must be properly designed for fatigue. Measurements and analytical studies show that internal bulkheads primarily carry in-plane force, consistent with expected behavior. Since the geometry of the rib does not distort as it rotates or "pushes" the FB out-of-plane, distortion of the bulkhead would not be expected.

Laboratory testing has demonstrated that the use of fillet welds for attaching the bulkhead to the rib wall should generally be discouraged, unless confirmed by FEA and fatigue evaluation. This is due to cracks initiating in the weld root from the lack of fusion zone. Hence, CJP joints, or PJP joints with increased penetration, are generally required to ensure fatigue resistance that is compatible with the fatigue resistance of the exterior RF weld.

As previously mentioned, designs should avoid the use of a bulkhead when possible. When deemed appropriate or necessary for use, Figure 4-3 and Figure 6-2(a) illustrate that the bulkheads will have concentrations at terminations. This may require welds of increased penetration such as CJP or PJP groove welds. The type and extent are determinable from stress levels and the use of the Frank-Fisher formula (see Section 5.7.5.1). If a transition is needed in the bottom it will likely need a CJP there. As this is more difficult to fabricate, alternate solutions should be sought, if the stress concentrations are high at the abrupt bulkhead transition. Additionally, consideration must be made for proper bulkhead alignment during fabrication (see Chapter 7 for additional information on fabrication).

6.6. RIB-TO-DECK AT THE FLOORBEAM (RDF)

To minimize the stress concentrations at the leading edge of the tooth of the FB, the corner of the FB should be fully welded to the adjoining elements.

Practitioners in Asia and in Europe have been using a rat hole at the corner of the tooth at the RDF, presumably to avoid crossing welds, as they are believed to degrade the resistance of the weld crossed. Experiences in Japan have shown that the rathole introduces distortions when the wheel passes, causing crack propagation at the top of the FB from the toe of the weld on the rib side.

Therefore, it is recommended that the weld between the FB web and rib/deck plate in the corner be continuous on the outside. The internal bulkhead weld should be 100 percent penetration to ensure that failure can start only at the root of the longitudinal RD. The length required in each leg is determined by the stresses along it.

6.7. SPLICES AND DECK JOINTS

Mill plates are not typically sufficiently large to make up the full length of a box section or OSD panel and must be spliced by CJP groove welding. Groove welds for the deck panel splices are typically ground flush on the top surface, to provide continuity for the wearing surface. The bottom is either ground flush or not, depending on the stress environment to which it is subjected. Field splice locations that facilitate placement and removal of backing bars in transverse and longitudinal field joints must be designed into the panel layout.

For deck plate field splices, the longitudinal weld is typically made in an area away from a primary wheel path and between two ribs. At this location the maximum stress range perpendicular to it would likely be higher than its natural Category C, depending on deck plate thickness. Additionally, it may have to be ground smooth for a better resistance category (AASHTO Category B).

Transverse deck plate welds of field splices are typically placed within the range of the rib inflection point. The splice plates must be designed for shear and bending. Their fabrication has been achieved either with the backing bar remaining or with backing bar removed, depending on the stresses at the point where it is located. If left in place, the fatigue resistance of the weld is Category D. For this reason and for lack of consistency of performance in the trades, the current trend is to remove the backing bar.

Backing bar removal for the transverse butt weld, which is ground flush at the bottom, requires space for access in the rib work. Figure 6-4 and Figure 6-5 show bolted rib splices as practiced in the United States, for backing bar removed and backing bar remaining, respectively.

Alternately, some new field techniques use overhead welding first, to weld the bottom part of the plate and to produce inherent backing; then the welding is completed from the top.

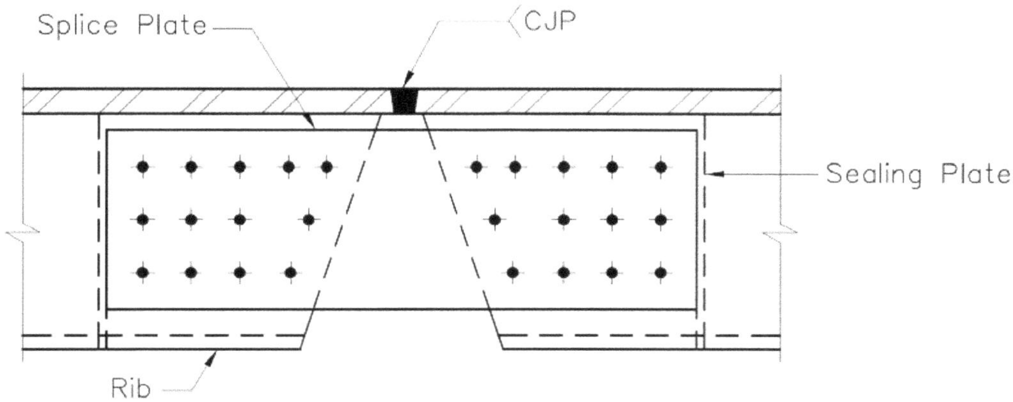

Figure 6-4 Bolted Rib Splice and CJP Deck Splice with the Backing Bar Removed showing the rib sealing plates. The Rib has been Cut Back to Provide Sufficient Access

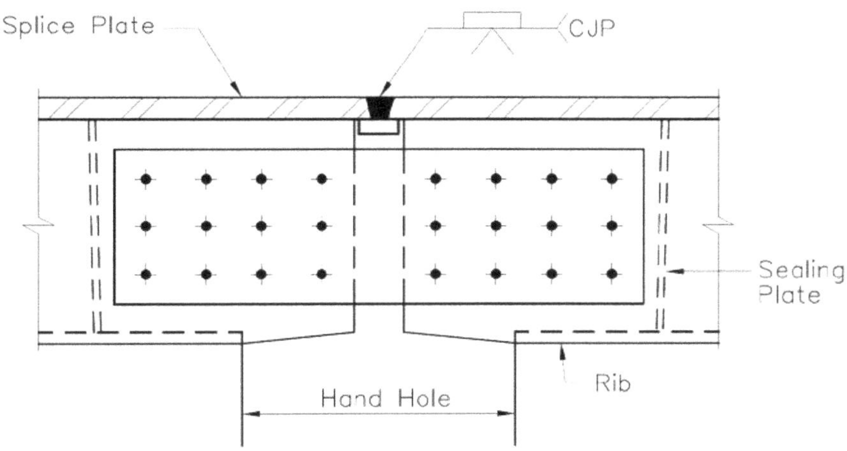

Figure 6-5 Bolted Rib Splice and CJP Deck Splice with the Backing Bar Remaining showing the rib sealing plates. The Lower Portion of the Rib has been Cut Back to Provide Sufficient Access

The termination of the rib on each side of the joint has an abrupt transition. To the extent that part of the stress in the upper part of the rib will go into the deck plate, there will be a stress concentration at that termination. Keeping the bolted splice close to the deck plate and locating the splice at points of low moment make the detail viable. But some practitioners prefer that the splice not be so close to the RD, lest the stiffness be increased such that more moment (transverse) is attracted to the joint with more opening stress in the weld.

The gap between the rib plate terminations should be kept to the minimum required by fabrication or field work (approximately 63 mm to 76 mm [2 ½ inches to 3 inches]). Also, flexural and punching shear effects occur at each rib termination. To alleviate these effects, some practitioners design a 13mm (1.5in) end return of the fillet weld as illustrated in Figure 6-6, or alternately place a CJP weld over the 13mm (1.5in) length.

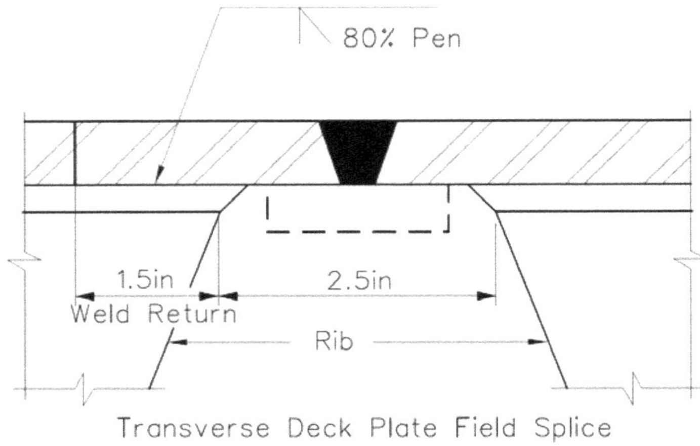

Figure 6-6 Detail of the Rib-to-Deck Weld Return at the CJP Field Deck Splice

Floorbeam Splices

FB splices are typically welded when performed in the shop. Field splices are typically bolted. When they are bolted, attention must be paid not to increase the out-of-plane stiffness of the FB close to the rib cut-out. To accomplish this, the width and thickness of the splice plate must be minimized and the distance from the splice plate edge to the rib cut-out must be maximized, as shown in Figure 6-7.

The FB must also have a rathole to accommodate a backing bar for the longitudinal deck plate splice weld, if the longitudinal deck plate splice is done in the field. The stress range in the deck plate at that point is usually low enough to be satisfy the Category E detail allowable stress range. The joint is shown in Figure 6-7.

The comments in the previous section regarding the opening for the backing bar in the transverse deck plate joint detail is also valid for the rat-hole dimension for the backing bar for the longitudinal splices shown in Figure 6-7. These details must resist wheel pressure at discontinuities. Therefore, typical rathole details specified in other codes, which have rathole length of 120 mm (4 3/4 inch), will likely result in deficiencies. In the event that a welded FB splice is needed or desired, the rat-hole detail as shown in Figure 6-8 could allow for a smooth cut surface where the weld termination is removed.

The 64mm (2 1/2 in.) opening is typically required in most cases, but fabricators and contractors will seek approval of greater space for greater ease of work. The effects that would results from such choices should be weighed before design comprises are made.

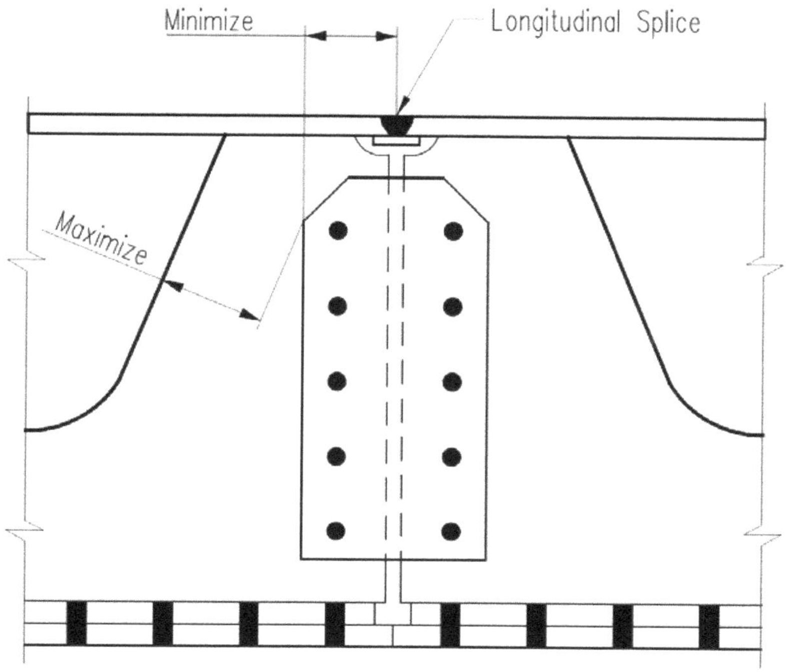

Figure 6-7 Transverse Floorbeam Web Bolted Splice and Longitudinal CJP Deck Splice Minimizing the distance from the Deck Splice and Maximizing the Distance from the Rib to the Bolted Splice Plate

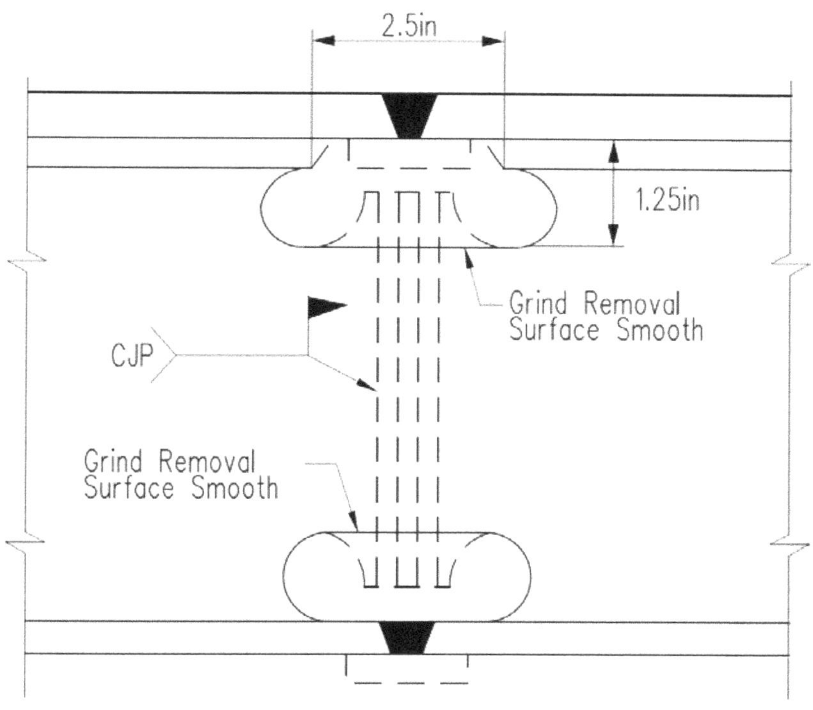

Figure 6-8 Transverse Floorbeam Web Welded Splice and Longitudinal CJP Deck Splice with Grinding Limits

It is noted that the rib splices shown in Figure 6-4 through Figure 6-6 are the standard used in American practice. They require the use of a sealing plate, adjacent to the bolted splice to prevent condensation inside the rib. Many European practitioners prefer the detail illustrated in Figure 6-9 where the rib is mad continuous by using a small rib connector and CJP welding.

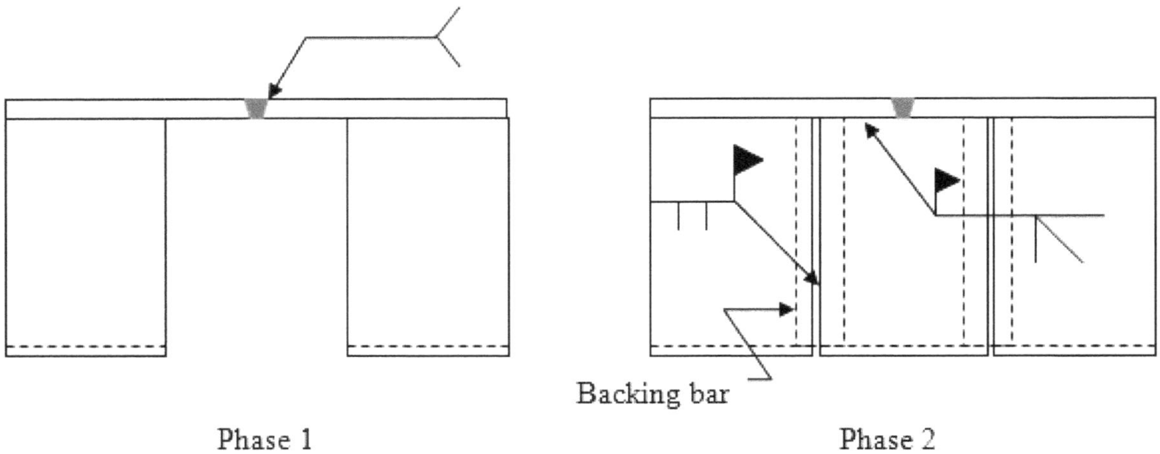

Figure 6-9 Preferred Rib Splice Procedure in Europe showing the Welding of the Deck Plate and then the Application of the Welded Short Rib Splice with Backing Bars

In the completed detail the, weld with the backing bars left in place is defined in the AASHTO LRFD Specifications as a Category D. A category has not yet been defined for the detail at the termination of the backing bar at the deck plate, were rib transverse bending develops. is the detail proposed by Eurocode (ECS, 1992). A V-butt joint, however, is an improved detail. Practice shows that the square butt weld is more prone to defects and requires larger root gaps than a V-joint. Conversely, the V-butt weld is more expensive to prepare.

6.8. REDECKING DETAILS

Differences between OSDs in new bridges and in deck replacements stem from the availability of space above the primary FB flange of the existing bridge floor framing verses that of a new bridge. These differences may lead to different choices of rib and FBs with or without the cut-out.

In terms of practicability, OSDs can span a maximum of 7.62 m (25 ft) within the typical available space over the sub-framing. But often the space between the deck surface and the FB is limited. Shallower depth ribs can be used if supported over closely spaced crossbeams so that they can be accommodated within the available space, with minor profile grade increases, if any.

In many viaduct structures, the deck spans transversely between stringers which in turn span between 15.2 m and 21.3 m (50 and 70 ft), approximately. Supplemental crossbeams can be installed to support an OSD which must bypass the primary FB at the bent. An example of such a scheme is shown in Figure 6-10. In this example, the ribs are shallow enough to bypass the primary FB's top flange and must have supplemental crossbeams spaced sufficiently close to permit shallow ribs to work effectively. The deck plate is yoked to the stringers by virtue of inverted “T”, and the crossbeams are connected so that they are sufficiently flexible to allow for rotation of the rib without damage to the crossbeam. This type of design is said to be longitudinally constrained. But it could be floating transversely (i.e. allowed to deform) unless this causes structural problems requiring restraint in that direction as well.

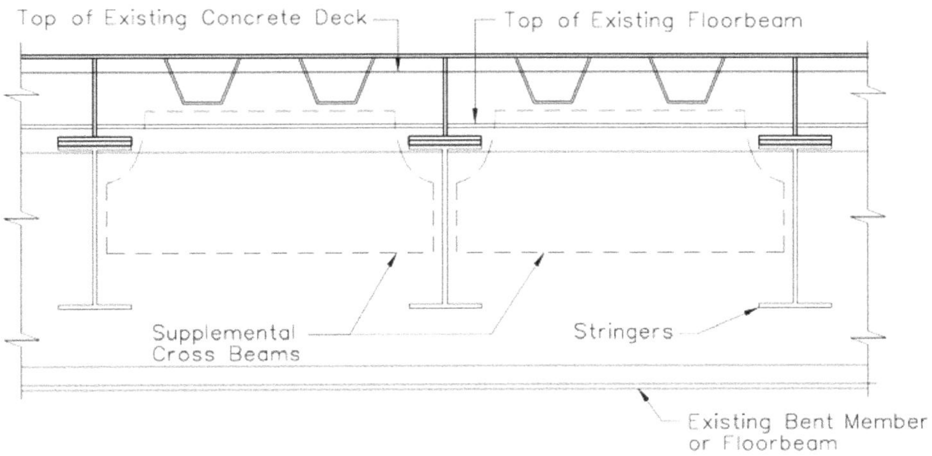

Figure 6-10 Redecking Example of a Floorbeam and Stringer System with an Orthotropic Deck. The Retrofitted Deck Bypasses the Existing Floorbeams

In cases where the stringers rest on top of the FB flange and the spacing of the FB does not exceed 7.62 m (25 ft), the deck can be rigidly attached to the existing framing in both directions, as illustrated in Figure 6-11.

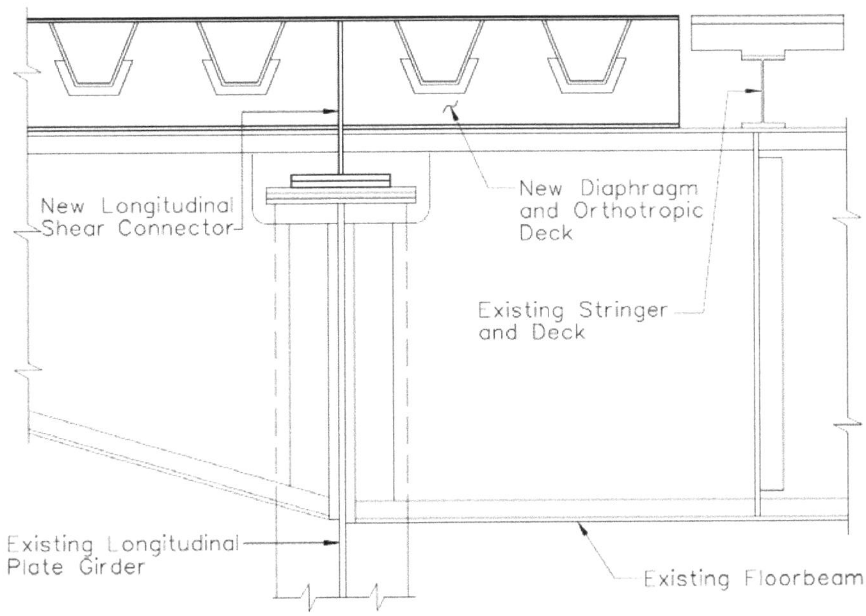

Figure 6-11 Redecking Example of a Floorbeam and Stringer System with an Orthotropic Deck where the Existing Floorbeam is Attached to the new Orthotropic Floorbeam (Diaphragm)

Figure 6-11 shows that the FB constrains the deck transversely and the longitudinal shear connector constrains the deck longitudinally. The longitudinal connector is a "hard spot" and may require a thickening of the deck plate and/or CJP welding as the adjacent ribs are comparably much "softer" (are actually carried by the hard spot) and cannot share load to eliminate high transverse deck plate stresses.

In cases where the required space is available over the bridge FBs, but their spacing is too far for the OSD to span (exceeds 7.62 m [25 ft]), intermediate crossbeams can be installed over the stringers as illustrated in Figure 6-12.

An example of this type of design is the proposed redecking of the Verazzano Narrows Bridge, where stringers span approximately 15.1 m (49 ft - 51/2 inches). Here, OSD diaphragms are being spaced at 3.77 m (12 ft - 41/2 inches) to support an OSD with a 305 mm (12 inch) closed rib, intended to replace a filled grid deck that runs longitudinally over crossbeams, spaced shy of 1.52 m (5 ft) on center and resting on the stringers. This represents a major reduction of crossbeam weight as well as deck weight.

The type of OSD shown in Figure 6-12 is restrained transversely. But it could be restrained longitudinally if connected to fixed stringers. This depends on constructability and other design issues.

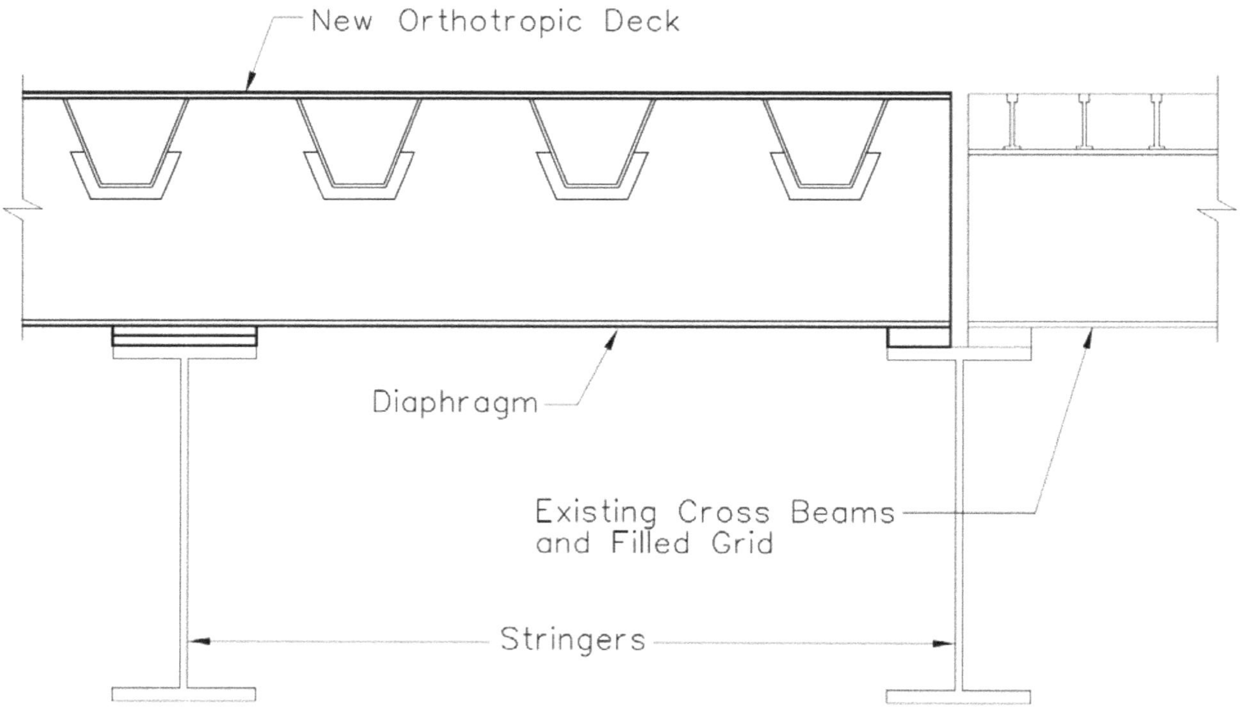

Figure 6-12 Redecking Example of a Beam (Stringer) and Cross Beam System with an Orthotropic Deck where the Floorbeam (Diaphragm) is connected to the Longitudinal Stringers

These examples illustrate the issue of available depth. The other main issue is whether the OSD can be rigidly integrated to the existing floor in both directions, as in Figure 6-11 or, for practical purposes, should be designed with inherent flexibilities. It is noted that a project where the deck was designed to float in both directions did not perform well leading to subfloor failures. Remedial measures proved difficult and thus this situation should be avoided where possible.

When the deck requires diaphragms or sub-floorbeams over stringers (Figure 6-12), the diaphragm is typically less flexible than what would be prescribed for new designs. Regardless, the diaphragms must be able to rotate with the rib without great distress either at the web-to-flange fillet, or at the connection to the stringers. As described in **System 4**(see Chapter 2). This motion is driven by the rotation of the rib under the passage of wheels. Additionally, other forces can cause a similar our of plane distortion in the diaphragm. For Example, the transverse movement of the ribs that results from live load acceleration or breaking forces can impart a rotation at the FB to stringer connection and must also be limited.

6.9. BARRIER DETAILS

The design of roadway barriers changed drastically in 1998 when AASHTO LRFD proposed new design criteria for barriers. They indicated that the forces of trucks impacting roadway barriers were much greater than had previously been presumed. The new AASHTO regulations, for the states that would adopt them, required also that the barrier be tested using prototypes on actual substrates, deck, slab, or structure that would apply, with a standard truck for the level of traffic carried by the highway in question, and code-specified direction of impact. Some bridge authorities prefer to use the specified design forces without conducting prototype tests.

Barriers installed on OSD bridges have been made of both concrete and steel; however, steel is the more common material used since it is the least weight and connections are more easily made. It is the responsibility of the owner to determine which test level (TL) is applicable to the bridge site, and typically they will have standard shapes that have been tested and approved for use. However, the barrier and its attachment to the OSD must be shown to satisfy crash test requirements or provide improved capacity as compared to an existing crash-tested and approved design. It is the attachment that often requires new design and approval. When a minor detail is changed on a railing system that has already been tested and approved, engineering judgment and analysis should be used when determining the need for additional crash testing (AASHTO LRFD). If additional testing is deemed necessary, a simplified pendulum test may be sufficient in lieu of a full scale crash test.

It has been shown that for OSDs of major bridges, which often have to sustain a TL-4 loading, the edge of the OSD may have to be reinforced lest it sustain permanent deformation of the deck plate edge and nearby ribs. This depends on the longitudinal span between floor supports of the deck and spacing of intermediate barrier supports between major deck supports, such that the impact load is well distributed longitudinally to two or more restraints. Connections may employ a system of thru-bolts to the OSD, or by welding to the surface.

The forces on barriers that are impacted include a transverse force and a longitudinally-applied force at the top of the barrier. Typically, the transverse force is the more critical and more difficult to manage. One rational way to manage these forces is to:

- Provide adequate distance between ribs resisting barrier forces such that the forces on the ribs are sufficiently low to be resisted without yielding of the material.
- Make the ribs under the barrier sufficiently strong that they would not yield under barrier impact.
- Use a combination of both of the above approaches.

These methods are diagrammatically illustrated in Figure 6-13:

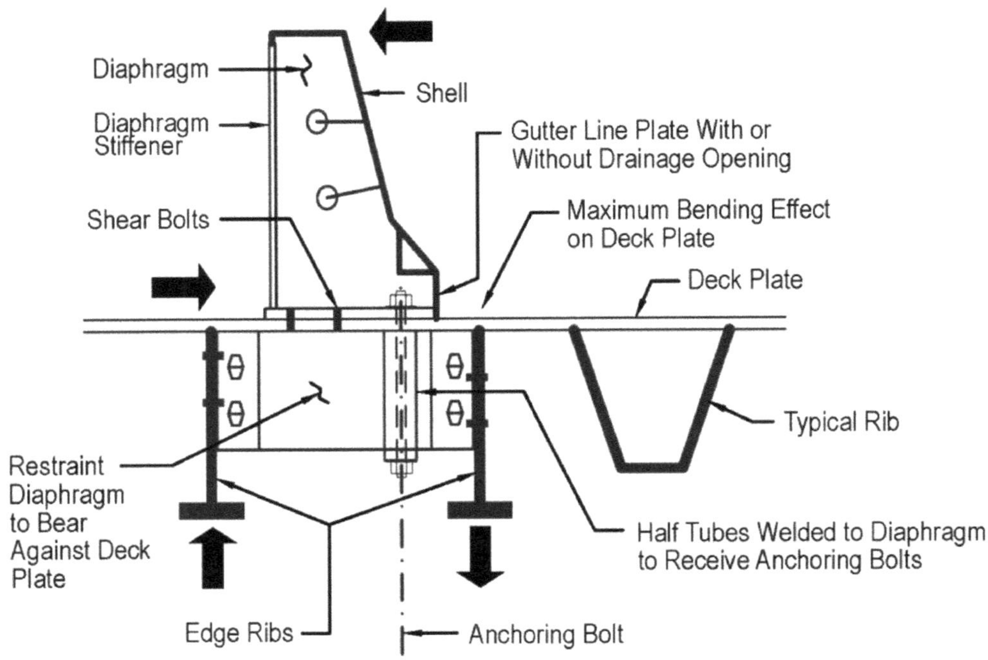

Figure 6-13 Schematic OSD Barrier Connection Detail Showing Supplemental Edge Ribs, Restraint Diaphragm, Shear Bolts, and Anchor Bolts to Resist Crash Loading

The illustration is only a schematic of what has been used in actual projects. For instance, on the San Francisco Oakland Bay Bridge, the inboard edge rib is actually a truss within the closed box girder (Figure 6-14). The ribs on this bridge span 5 m (16.4 ft), but the truss anchoring the barrier has verticals at 2.5 m (8.2 ft). Also, the anchoring bolts straddle the truss top flat plate stiffener, instead of the FB, but restraint FBs are more frequently spaced to transfer the moment. Many design variations are possible and are left to the ingenuity of the engineer and the specific problems they encounter.

It is noted that Caltrans has tested a steel barrier with specifics similar to that of the San Francisco Oakland Bay Bridge. The design performed satisfactorily. This suggests that when forces specified in the AASHTO LRFD Specifications are used to design a barrier and substrate system, they will likely result in an acceptable design.

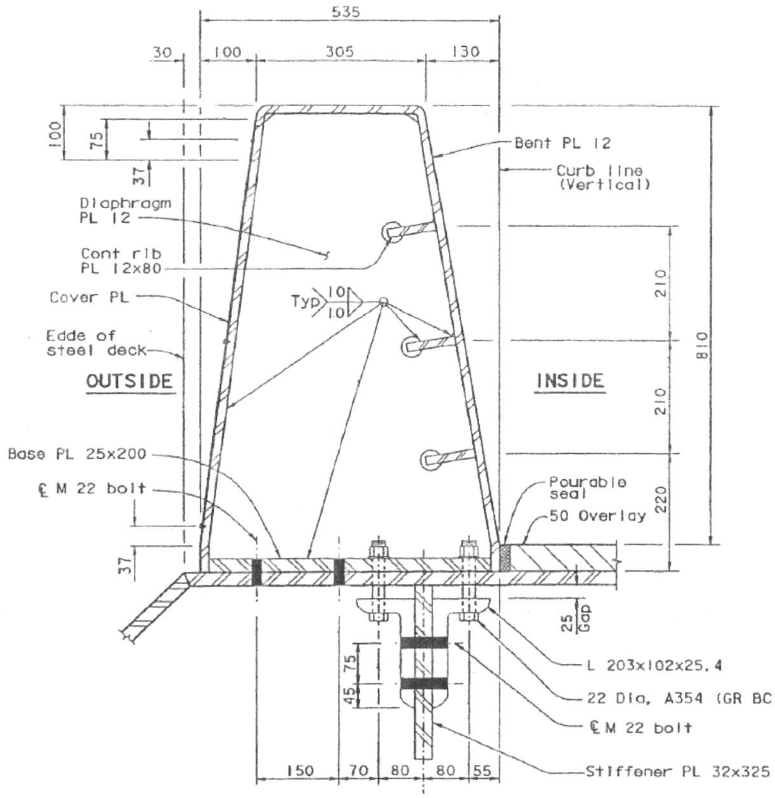

Figure 6-14 San Francisco Oakland Bay Bridge Barrier Detail showing the Barrier Stiffening and Anchorage System

7. CONSTRUCTION

This chapter provides basic fabrication, welding, and erection guidelines for orthotropic steel deck (OSD) bridge components, illustrated by photos of shop and field practices. Discussions highlight the controls that are necessary to give reasonable assurances that the final product is of acceptable quality. Emphasis is placed on automation to minimize the chances of mistakes. It is important that the engineer, fabricator, and erector work together to control the number and complexity of the welds, which in turn will decrease the cost of fabrication, and erection. This chapter is also intended to assist in development of language that can be used for the development of a project special provision.

7.1. FABRICATION DOCUMENTS

7.1.1. Fabrication Specifications

It is customary to address special fabrication needs in project special provisions. This is also the case for OSDs. Most bridge owner specifications in the United States require steel bridges to be made to the combination of AASHTO, AWS D1.5 Bridge Welding Code, owner standard specifications, and/or any project special provisions. Such specifications cover basic fabrication processes, such as cutting, bending, welding, and heating. Generally, D1.5 and owner specifications are well-suited to I-girder bridges and (to a lesser extent) to other bridge types, such as tub girder bridges, arch bridges, trusses, and orthotropic bridges.

Special provision items for consideration include:

- A requirement for a welding quality control plan from the fabricator that addresses items considered important to the engineer (items that are identified in the contract plans).
- Welding procedure qualification and associated production control – the D1.5 method is suitable for most orthotropic steel deck (OSD) welding, but the special provisions must address qualification of rib-to-deck plate weld (RD) welding procedures.
- Mock-up requirements, if any.
- Non-destructive testing. D1.5 covers technique, but the special provision should cover inspection frequencies and any special acceptance criteria.
- Dimensional and distortion flatness tolerances.
- Pressure test for closed ribs.
- Assembly requirements and associated procedures, if any.
- Ultrasonic testing requirements for RD welding depth of penetration verification and identification of any "hot cracking".
- Transportation requirements, particularly sea transportation requirements.

It is recommended that owners provide specific performance requirements for items they address in the project special provisions.

Further, owners should apply tolerances judiciously, with consideration of fabrication method and performance demand. For example, deck plate flatness is important for bridge performance (ride quality), so a tolerance should be specified; however, rib flatness is not as critical, especially considering that fabrication methods will not readily result in a rib flatness issue.

7.1.2. Certifications and Qualifications

AISC maintains a program for steel fabricator certification, including a specific bridge fabricator certification. Though AISC does not have a specific program for OSD bridges, the bridge certification including fracture critical endorsement, should be required as a minimum. Most bridge owner standard specifications already require this certification.

Beyond the AISC certification, the fabricator needs experience with advanced bridges. Experience with OSD bridges is advantageous but not necessary. A fabricator with extensive experience in advanced bridge structures such as trusses, arches, and signature spans will be readily able to adapt to OSD work.

In some countries outside of the United States, it is customary to require the fabricator to achieve ISO 9001 certification.

7.1.3.Fabrication Plan

As is customary on many bridge projects, requiring a fabrication plan is recommended. The plan should be submitted for approval and address the fabricator's means of accomplishing the job, including:

- Control lines
- Welding procedures
- Nondestructive evaluation (NDE)
- Heat straightening
- Distortion control
- Assembly
- Loading
- Shipping
- Quality plan

Review and approval prior to fabrication will help ensure everyone approves the procedure before the project starts. This can be particularly important on international teams where the customary practices and usual specification requirements that the fabricator is used to may be different from owner expectations.

7.1.4. Drawings

The fabricator's shop floor operates from shop drawings. Most modern shops increasingly rely on electronic data and sketches documents. For example, where previously the shop would read a shop drawing to lay out a cutting pattern with soap-stone, straightedge, and tape, today a computer numerically controlled (CNC) equipment operator may simply load the appropriate program, verify the pattern by viewing a sketch on a monitor, and then proceed with cutting without ever looking at a traditional shop drawing or marking steel. Though some fabricators still rely on traditional methods, it is preferred that a fabricator be capable of fabricating an OSD bridge with CNC driven equipment.

It is customary in the United States for the fabricator's shop drawings to be reviewed by the owner for approval prior to fabrication. The scope and depth of this review varies from owner to owner and engineer to engineer. For OSDs, the engineer and contractor should take advantage of this opportunity to ensure that the fabricator will execute fabrication as the designer and owner originally intended. As with all bridge projects, it is important to handle the review process expeditiously to keep a project on schedule. Experience has shown that the use of "electronic" shop drawings is greatly encouraged to help achieve this expedition.

7.2. FABRICATION SEQUENCE

For OSD bridges, many fabrication operations are similar to those used in fabrication for other bridge types, but deck plate and rib fabrication is unique. The fabrication sequence is basically the same for box girder and deck panel systems. Components are fabricated from steel plate, except that diaphragms and FBs may be made from split W sections.

Steel is ordered as shown on the design and detail drawings. ASTM A-6 is the standard governing specification for steel plate quality; it is referenced for quality by structural steel plate specifications such as ASTM A709. The fabricator may consider ordering the plate to tighter flatness tolerances than the ASTM A-6 tolerances, but this could prove cost-prohibitive.

When steel is received it is inspected, verified against material test reports, and stored. Generally the fabricator will want to use the steel as quickly as possible. If the steel must be stored for an extended period, it is to be suitably protected to minimize remediation later. The fabricators may pre-blast the steel to remove loose mill scale, rust, and other deleterious material to facilitate welding. Alternately, the fabricator will clean local weld zones closer to welding time.

The steel will then be cut to size and marked. As previously noted, it is common for both of these operations to be conducted by CNC equipment, particularly for advanced structures such as OSD bridges. Cutting is by plasma or oxyfuel. Oxyfuel is more conventional and is suitable for all structural steel thicknesses. Plasma is faster and produces a cleaner cut, but is only efficient up to about two to 64mm (2 ½ in.), depending on the system. Cutting is likely to be over water to catch smoke and reduce noise, in particular for plasma.

Marking will most likely be zinc but could also be plasma. Like the cutting tool path, the marking paths are programmed with the CAD/CAM system. The fabricator uses the marking to lay out locations for members, tack welds, support components, and rib edges.

The ribs are usually formed on a large press-break to the specified shape and length. The ribs can also be rolled into shape, but this practice is less often employed because it is more difficult to achieve suitable tolerances with this method. Depending on the number of dies that must be used, rolled ribs can vary considerably. However, either method is considered acceptable when the proper tolerances are observed. It is preferred to form the ribs full length, if the press has the capability. Alternately, shorter length ribs can be formed and spliced by welding with a ground-flush CJP joint without difficulty. In such cases, ultrasonic testing can readily be used to verify quality. Diaphragms and end plates are welded to the inside of the ribs after welding, but before being fit to the deck plates.

The plain deck plates are welded into larger pieces and moved to the rib fit-up station, where the ribs are both aligned and the RD weld joint gap is minimized before being tack welded to the deck plates.

The rib-deck plate subassemblies are moved to the welding station where the partial penetration RD welds are made. Placement of the ribs on the deck plate must be controlled such the ribs align with the openings that are cut into the FBs. There, subassemblies can be preloaded before welding so that the panels come out relatively flat and straight after welding. This will reduce the amount of heat straightening required. After welding, the deck panels are heat straightened to the specified tolerances, if necessary.

The subassemblies are moved to the final assembly area to be assembled into the final shipping pieces. At this time the beam support structure is welded to the deck plate assemblies and all shop welds and bolted connections are checked for alignment. The shipping pieces are placed into a lay-down assembly. Any attachments that have dimensions that are critical for alignment from one assembly to the other are finalized at this time. The field weld joints are fit-up for alignment and are also checked for weld joint tolerances at this time. The rib bolted splice joints are also checked for fit during assembly.

In summary, an overall schematic of the fabrication process is presented in Figure 7-1.

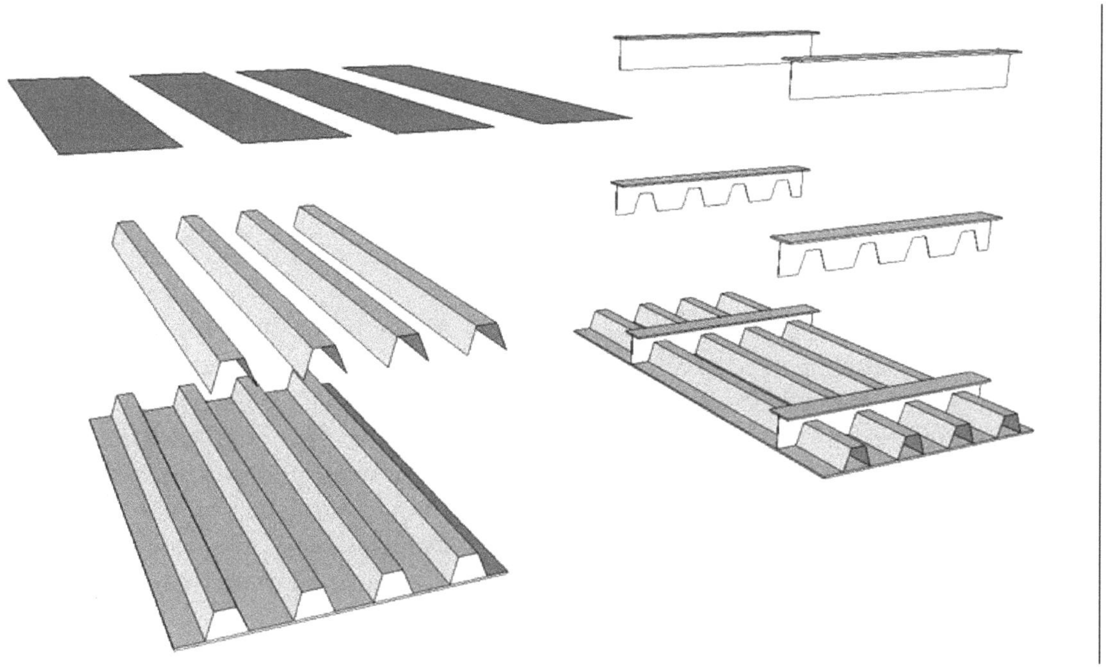

Figure 7-1 Fabrication Sequence of Orthotropic Deck Panels including Forming of the Ribs, Attachment of the Ribs to the Deck Plate, Fabrication of the Floorbeams, and finally Attachment of the Floorbeams to the Ribs and Deck Plate

7.3. FABRICATION PROCESS

OSD fabrication uses many of the same processes as those used in fabrication of more common bridge types, such plate girder bridges, arches, and truss, including the following:

- Material control (use of MTRs)
- Marking
- Cutting
- Fitting
- Welding
- Drilling
- Assembly
- Painting
- Shipping

However, there are some inherent differences in OSD bridges that affect fabrication processes. This includes:

- Thinner plates mean that more coil steel could be used. For example, it is most likely that ribs will be made from coiled steel.
- Thinner plates mean that more welding distortion may result. This in turn indicates more heat correction will be necessary and also motivates the use of cooler welding processes, such as GMAW instead of SAW.
- Plasma cutting being very prevalent. Plasma is a faster, cleaner, and more consistent means of cutting steel. However, present technology faces a practical limit of 2 inches in thickness which is satisfied by most OSD plate materials.
- Water spray may be used during cutting to minimize distortions.

Proper methods of maintaining flatness as OSDs are fabricated is required

7.3.1. Distortion Control

Generally, distortion control for OSD bridges is the same as for other steel bridges, with the important exception of achieving ride-quality flatness in the deck panels (see Section 5.2.2). Flatness in the panels may also be needed where the OSD serves as a compression flange in a box girder, to provide structural stability. Geometry control is also important to provide alignment for field splices, which are often welded. Because the deck plates have a significant amount of welding, and especially because this welding is only accomplished on one side of the deck plate, the deck plate distorts from welding. The fabricator will pre-bend deck plate subassemblies to minimize distortion prior to welding and then use heat correction to ensure tolerances are met.

There are four significant issues to address for distortion control:

1. Amount and type of preloading to be used.
2. Placement of the tack welds toward the face of the weld joint or in the root. Both approaches raise issues that must be addressed.

3. What combination of preloading before welding and heat straightening after welding should be selected to flatten the deck panels.
4. Risks associated with the development of hot cracks during welding.

There are also four basic approaches that can be used to achieve acceptable deck plate flatness after welding:

1. Preload the panel so it comes out flat after welding.
2. Preload the panel to the extent necessary so the panel needs only a limited amount of heat straightening.
3. Use no preloading and use only use heat straightened to flatten the panel after welding.
4. Counterbalance with heat during the weld process.

All methods have limitations and risks based on the results of past OSD projects that should be seriously considered when selecting and approving the methods and technologies used for making the tack and RD welds. In practice, method No.2 is the most prevalent because it is reasonable to assume fabricators will have to do some heat straightening, even if method No. 1 is used. That is, it is almost impossible to predict the actual amount of preloading that will result in the OSD panel being flat after welding. See Figure 7-2 and Figure 7-3 for examples of preloading and pre-bending to facilitate fabrication. Each preloading method is described below. Figure 7-3 shows pre-bending in the transverse direction, but it is equally appropriate and possible to pre-bend in the longitudinal direction (or both directions), depending on the circumstances.

The reluctance to allow for heat straightening in the panel fabrication stems from speculation that the process will degrade the fatigue resistance of the welded connections. However, the research into this issue by Sim and Uang (2007) was inconclusive. The work by Connor et. al (2008) found that damage and repair cycles in steel bridge girders did not have any appreciable effect on the fatigue life.

Figure 7-2 Preloading is Used to Help Control Distortion by applying Weights to the Inverted OSD

Figure 7-3 Pre-bending by Clamping the OSD to a Curved Assembly Bed is used to Offset Deck Panel Distortion due to Welding

Method No 1: Preloading the panel to achieve flatness after welding with no heat straightening required

This method locks stress into the panel during both fit-up and welding and therefore is the riskiest. One of the problems associated with this is that the ribs are fit to the panels by force and tack welded in place, locking that force in the tack welds. After fit-up, the assembly is moved to the welding station for RD welding and preloads are applied to the extent necessary so the panel will be flat after welding. Preloading may be applied using temporary weights that stay on the panel during welding or by the use of wedges, shims, and jacks that stay in place during welding. Both ways add additional stress to the tack welds. Another method for reducing the total stress in the tack weld is to apply heat to the outside of the rib during welding. This counters the weld shrinkage stresses.

In Europe, panels are preloaded to a pre-weld configuration before tacking the ribs to the deck plate. They also apply heat to the outside of the rib, as described above, to lower stress in the tack welds during welding with good results.

Method No. 2: Preloading the panel so it requires a limited about of heat straightening after welding

The preloading can be less if heat straightening is used after welding. Lower preloading results in lower residual stresses and less demand on tack welds. Heat straightening the panel after welding will place residual stresses in the entire RD and not just in the tack welds. The preloading methods can be the same as those used in Method No. 1, only with less magnitude.

Method No. 3: No preloading of the panel before the welds are made

This method creates the least risk to the panels during welding because the only stresses placed on the tack weld are those holding the rib in place during fit-up and tack welding. This same stress will also be present for the other methods because the method used for closing the root gap to acceptable limits is the same for all distortion control methods. Using this method to straighten the panel distributes the stresses throughout the panel because no preloading has been used.

Method No. 4 Counterbalance with heat during the weld process

This method equalizes the effects of the weld with an equal and opposite heat input. The thermal expansion of the preheated components pre-forms the ODS such that the deck cools to the proper shape. Although this method, when properly applied, can reduce or help to eliminate locked-in stresses from the cooling process, controlling the heat input to control local distortions and maintain the proper root gap is difficult. As such, preloading is often still required for this method.

To achieve the proper proportions in the final, completed panel, some fabricators have used deck plates with extra (sacrificial) material past the nominal dimensions of the assembled piece. This extra material is then trimmed off after welding to set the final geometry of each completed panel.

7.3.2. Tack Welds

Tack welds can become a critical fabrication step for OSDs. When proper procedures are observed, they are a vital tool that greatly aid proper fabrication. However, two things can happen with respect to the tack welds during preloading and welding. First, some of the tack welds may crack from thermal and preloading stresses created when the RD weld is made. Second, the rib may move minutely when the tack weld is remelted, creating a serious potential for hot cracking. When the tack weld is placed in the weld joint (root or face area), the extent to which it is remelted during welding, and how the preloading is applied to the section both play an important role in the seriousness of the risks. For example, if the tack weld is in the front of the weld joint, it will be remelted 100 percent during welding. Conversely, if the tack weld is placed in the back of the weld joint, it may not totally remelt. Furthermore for this condition, any lack-of-fusion or other discontinuity that exists in the tack weld may not be removed because the tack weld is not totally remelted. Another concern is that tack welds that crack during welding and are not remelted will have cracks buried in the root of the weld once finished. In either case care must be taken to make sure that the tack weld area is sound after the RD has been made.

7.3.3. Splices

As with any bridge project, effective achievement of splices for field-bolted and field-welded connections is essential for a good OSD project. The technologies and methods used by fabricators will vary depending upon the fabricator's equipment, experience, and skills. For example, the fabricator may use traditional ream-assembly or CNC drilling, drill templates, and/or fit-up fixtures (Figure 7-4) to build connections. Methods alternate to ream-assembly may preclude the need for shop assembly; if so, the fabricator should demonstrate this through check assemblies.

Figure 7-4 Fit-up Fixtures such as the One Shown for a Box Girder are Used To Ensure Alignment without Assembly

7.3.4. Internal Bulkheads

The internal bulkhead is welded inside the rib before the rib is positioned and welded to the deck plate. Afterwards, the FBs are positioned over the ribs but without any possibility of visual check of alignment between the internal bulkhead and the FB web with the deck plate in place. Experience indicates that considerable misalignments occur from time to time. In the United States, panel lengths of 18.3m (60 ft) have been used on redecking projects. This suggests that shrinkage caused by RD welding can be a source of misalignment between the diaphragm and the bulkhead. Misalignment is a defect that introduces relatively high additional secondary stresses and may have a large influence on the fatigue life of the detail and proper procedures are necessary to eliminate it.

7.4. WELDING

AWS D1.5 addresses most procedure and workmanship requirements. However, any missing considerations or requirements should be addressed in the contract special provisions. As required by D1.5, no welding should be conducted without an approved procedure.

Preproduction weld trials should be conducted to develop procedures for special applications and to demonstrate fabricator proficiency. This specifically needs to include RD welding, but should also include any non-standard joints or processes. As described below, the project specifications should address requirements for RD welds. (D1.5 already provides a path for other non-standard processes.) The trials should demonstrate that the fabricator's process will consistently achieve the required value (such as minimum penetration) in fabrication.

Under the preproduction trials, the fabricator should develop the procedure that is desired and will then conduct the D1.5 Procedure Qualified Record (PQR) testing for use in developing the final actual welding procedure specification (WPS).

7.4.1. Rib-to-Floorbeam (RF)

The various types of details used for this joint are discussed in detail in Section 5.7.5 and Section 6.5. This detail require substantial effort in terms of manual welding and grinding As fabricators in the U.S. have typically executed this weld by manual methods. However, it is evident that there is potential economy for robotics to be used here due to the repetition involved. Projects of larger size may justify the investment into robotics.

Where no cut-out is used and the rib is welded all-around to the FB, particular attention should be paid to the weld at the bottom portion of the rib as this is where cracking is most probable. Where a cut-out is used with a smooth termination, current practice is to fabricate this weld as a CJP ground flush with the rib (Figure 6-3). A fillet weld is substantially less time consuming, but the wrap-around or start/stop must be executed with care and checked for quality. In both cases, the quality of the weld in this transition region is important as this is the most probable location for cracking when a cut-out is used.

In all cases, specifications for the welds follow those set forth by D1.5. The project special provisions need to indicate, by reference to a standard, the geometrical condition of the finished surface and, if required, the desired hardness, depending on the type of system selected for this connection.

7.4.2. Rib-to-Deck Plate (RD)

It is not possible to achieve a conventional complete penetration weld where closed ribs are joined to the deck plate because there is no weld access to the back of the joint from inside the rib. Therefore, it is customary to join ribs to deck plates with a one-sided PJP weld as discussed in earlier chapters. The fabricator will accomplish welding in the means that is best suited to their equipment, but will most likely be some form of mechanized process (for example, see Figure 7-5).

Figure 7-5 Fabricators Automate where Possible to Improve Productivity and Quality such as Using the Gantry as shown which Allows Three Ribs o be Welded at One Time

7.4.2.1.RD Welding Procedure

The fabricator will develop a procedure that best accomplishes the performance demand of the weld, including joint design (size and angle of bevel, size of root face), welding process for production and tack welding, electrical parameters, travel speed, consumables, and number of passes. It is important to give the fabricator flexibility in accomplishing the RD weld because many factors can play into the effectiveness of this welding. Example of a flexible procedure includes:

- The fabricator should be allowed to select a preferred welding process. This will likely be SAW or GMAW. Due to the high heat input associated with SAW, joint bevel will be smaller; in fact, thinner ribs may require no bevel preparation, though care must be taken not to melt-through or to drive the weld too deep into the base metal in an effort to melt off the corner of the rib plate. Conversely, GMAW is cooler and would likely require a bevel in order to achieve proper penetration. However, since it is a cooler process, GMAW may result in less distortion and therefore may require less heat straightening.
- The fabricator should be allowed to prepare the RD joint on the rib plate as necessary to accomplish the required penetration and also ensure a suitable profile. Hotter processes (like SAW) have more penetration and therefore can tolerate a larger root face area, more root gap, and less beveling than cooler processes (like GMAW).
- The bevel will impact the quality of the weld on the backside. If the fabricator uses a bevel that is too large and does not provide sufficient root face area, the weld will readily burn through or melt through behind the rib wall.

- The procedure must identify essential variables and associated limits that must be respected by the welder to ensure weld quality. One of the primary reasons for not achieving repeatability during welding is failure to use the original welding parameters, including wire position and angle.

7.4.2.2. Procedure Development

The fabricator will achieve proper RD welds by developing a procedure that will achieve the performance demand, including penetration and profile requirements. Because successful welding is a function of the rib thickness and rib slope, it may be necessary to redevelop a procedure even if the fabricator has conducted OSD work before. See Figure 7-6 through Figure 7-8. The fabricator should:

- Demonstrate that the procedure will be effective by conducting trials and examining macroetches.
- Set parameters and then conduct D1.5 PQR testing in accordance with the appropriate Clauses in the AWS D1.5 Code.
- Write the final procedure and submit for engineer approval, if PQR testing is successful.

Proper procedure development and implementation is critical. Achieving the proper RD depends on a number of factors. First is wire placement, including:

- Angle of the wire transverse to the joint.
- Lead angle (drag).
- Distance that the wire projects into the weld from the face of the rib.

The second factor is oversized wire diameter which can result in:

- Poor weld profile, including overlap, undercut, and oversized reinforcing.
- Lack of fusion because the fabricator may speed up to correct profile issues.
- Poor width-to-depth ratio and therefore center bead cracking.
- Excessive penetration and melt-through.

The third factor is undersized wire diameter which can result in:

- Inadequate weld size.
- Increased passes, which in turn increases distortion.
- Undersize reinforcing.
- Overlap due to slower travel speed in attempts to include weld size.
- Undercut.

Figure 7-6 The Fabricator will Develop a Procedure for Rib-to-Deck Welding as the Worker in the Photo is Testing

Figure 7-7 Macro-Etches Like the One Shown Here are Used to Ensure Weld Penetration is Suitable, as well as Other Weld Requirements, Such As Width-To-Depth Ratio and Profile

Figure 7-8 Wire Angles and Diameter are Critical for Suitable Rib-to-deck Welding. The Picture shows a Demonstration Piece with the Wire Size and Angle Prepared

7.4.2.3. Proficiency Testing

In the course of developing the welding procedure, the fabricator must demonstrate the following proficiencies for accomplishing the RD partial penetration welds:

- Required penetration (for PJP joints).
- Melt-through does not exceed specifications.
- No blow-through.
- Tack weld remelting (consistent with procedure requirements).
- Welding both sides of the rib at the same time.
- Mechanized or fully automatic welding.
- Producing a suitable weld profile per AWS D1.5.

Depending on the specifications, it can take a fabricator a number of months to develop the proper welding procedure on their first OSD project. From that time forward, the development process should be lessened and will depend on how much the RD weld configuration changes.

With respect to the current specifications, the AASHTO LRFD Specifications prior to the revisions accepted in 2011 required a one-sided PJP with 80 percent target penetration. The revised current requirement, however, allows a minimum penetration of 70 percent for the weld, providing the fabricator a target of 80 percent, but not penalizing for failure to meet the requirements exactly at every location. Other codes are slightly different. For example, the Eurocode (ECS) and Japanese code (JRA) allow 70 percent penetration. Further, other codes also allow simple fillet welding, though this likely does not satisfy fatigue design requirements. More discussion on the updated requirements is presented in Chapter 6.

7.4.2.4. Pre-production Weld Trials

Preproduction weld trials are an important part of any OSD project. The cost impact associated with repairing production welds is magnified well beyond what is seen on other forms of bridge construction because of the amount of welding on an OSD bridge. On any bridge, the number of weld defects will be directly proportional to the total amount of deposited weld metal, and since the length of welds on an OSD are greater than what is seen on other bridge types, the chance of having more weld defects is significant. Identifying and eliminating the problems discovered during the preproduction weld trials will have a major impact on reducing the cost of fabrication.

Full Scale Weld Mock-Ups

OSD fabrication requires full-scale weld mock-up trials (see Figure 7-6, not to be confused with Level 1 Design). These trials will verify that the work can be performed as described in the Contract Specifications. That is, problems occur during fabrication that cannot be anticipated and the use of full scale mock-ups allows the contractor to find and eliminate these problems so they do not occur during production. For RD welding, full-scale mock-ups demonstrates that the fabricator's method of welding works with his method of panel assembly and distortion control.

Start of Shift Production Weld Macroetches

Start of shift production weld samples are used to verify that the welding is completed the same way that was approved in the preproduction weld trials. This is required to be done daily at the start of each shift on the RD partial penetration welds by each welder. When making the orthotropic RD welds, it is important that the welding operators demonstrate that they have the automated welding machines and guidance systems adjusted properly on a test panel before the start of production welding. Hence, when production welds are done, there is reasonable assurance that they will comply with the quality requirements spelled out in the Contract Specifications. Taking macro-etches of rib extensions on a regular basis is an effective way to verify that the welding is being completed in accordance with the approved welding and fabrication procedures.

7.4.2.5. Fit-Up

Good, consistent fit-up is essential for consistent weld quality; this includes weld penetration and weld profile. Jigs may be used to help accurately position the ribs to the deck plate (Figure 7-9). In order to achieve proper penetration, the root face fit-up gap is relatively tight. Experience shows that 0.5 mm (0.020 inch) is necessary and provides satisfactory results. 0.5 mm (0.020 inch) is also the maximum allowed by AASHTO LRFD. The fabricator determines this gap as part of the procedure development. The actual root face gap that is used for production welding is established during the procedure development tests. Note that experience has shown that larger gaps have been shown to create excessive melt-through and blow-through.

Figure 7-9 Fabricators use Jigs Such as the One Shown to Help Achieve Proper Alignment in the Ribs

7.4.2.6. Tack Welding

Good tack welds are particularly important in RD welds because they must be large enough to hold the rib in place against pre-restraint (if any) and the advancing production welding equipment, but they must also be minimized to avoid compromising the weld profile.

It is desirable, though not necessary, to entirely remelt the tack weld with production welding. The more the tacks remelt, the less likely they are to compromise weld profile and the less likely they are to trap slag or cause lack of fusion at the root (Figure 7-10). Tack remelting can be facilitated by grinding tacks.

Further, as mentioned in Section 7.3.2, tacks that do not remelt may result in unique problems. Tack welds inherently do not have complete fusion at the start and stop. Further, depending on the technique and proficiency of the welder, there may also be lack of fusion along the length of the tack. **Complete remelt of the tack is the most desirable condition.** The likelihood of remelting the tack depends on many factors, but especially the process used for the cover pass(es). SAW is much more likely to remelt a tack (and thereby eliminate tack weld lack-of-fusion) weld completely than cooler processes, such as GMAW and FCAW.

Figure 7-10 The Tack Welds Under this Cover Pass Were Re-melted, Yet Their Location is Still Obvious (This Condition is Acceptable because Profile Meets Code). The Larger Tack Welds Can Influence the Weld Profile

Tack welds are placed manually by a hand-held welding gun, which makes it difficult to place the weld exactly in the root of the RD partial penetration weld. Grinding tack welds is necessary if the cover pass is not intended to remelt the tack, because otherwise travel speed would need to be unreasonably increased. At such an increased rate, it will be difficult for the PQR to pass. However, thicker tack welds must be ground to a thickness that is compatible with the overall welding parameters.

In procedure development, the fabricator should demonstrate the suitability of tack welding process by taking macroetches through the tack welds. If the tack welds are remelting entirely, no further consideration is necessary. But if the tacks do not remelt entirely, further macro-etch examination must demonstrate that lack of fusion does not occur along the sides and at the starts and stops. The likelihood of remelt is a function of many things, but in particular the thickness of the tack and the process used for the cover pass. A SAW cover pass is much more likely to remelt the tack than a GMAW cover pass.

Where the tack weld is located within the weld joint is important. Panels are preloaded or prebent so they are as flat as possible after welding, and while this practice helps to reduce heat straightening, it puts more strain and therefore demand on the tack welds. This increases the likelihood that tacks will crack as RD welding progresses along the length of the panel, as well as the associated risk that ribs will move and cause defects.

7.4.2.7. Penetration, Melt-Through, and Blow-Through

Generally PJPs are avoided in bridge design and construction because, depending on the joint configuration, associated stiffness, and the applied stress, PJPs can be a fatigue concern. In fact, the RD weld is an exception to the provision of AWS D1.5 not allowing PJP welds at T-joints subjected to tension in the root. This is why the penetration, melt-through, and root gap must be carefully controlled in production.

Even with stringent project special provisions, prequalification of the welding procedure, verification tests, and quality controls in production, melt-through will occur from time-to-time and has been used as a reason for rejecting a given weld process or panel. However, as discussed in Chapter 6, it may not necessarily be detrimental to the weld performance (Sim and Uang, 2007). Hence, it is reasonable to set a tolerance on this parameter as well. Note that a distinction should be made between melt-through and blow-through. In melt-though, a small amount of weld material oozes into the backside during the welding process. With blow-through, the weld material spatters. Both of these conditions may create sites of potential crack initiation and should be limited. Furthermore, it has been demonstrated that with proper welding procedure, blow-through can be avoided. Hence, **blow-through should not be tolerated**As discussed in Section 6.4, a moderate amount of melt-through is permissible. See Figure 7-11 for an example of melt-through without blow-through.

Figure 7-11 Melt-Through as Shown in this Picture is a Sign that the Joint Landing May Be Too Small

From a practical point of view, melt-through and blow-through can only be verified in production monitoring since it is not possible pragmatically (based on current technology) to examine the inside of the ribs once the welding is complete. Ultrasonic testing cannot establish the presence or lack of melt-through or blow-through.

In order to establish and demonstrate a proper procedure, a welding procedure should be determined with a mock up prior to any fabrication. The mock up is intended to be of an appropriate size and scale relative to the complexity of the new construction and is specified by the engineer. When completed, the mock up is cut for macro-etching of the weld. The macroetching need only demonstrate the extents of the weld showing the penetration, size, and throat, as shown in Figure 7-12. Additionally, in order to verify consistency of the weld and complete remelting of the tack weld, 20 percent of these etchings must be taken at a tack weld locations.

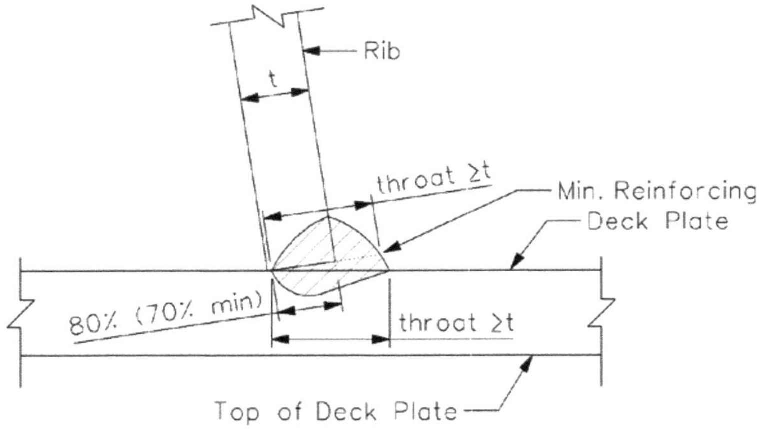

Figure 7-12 Rib-to-Deck (RD) Weld Detail showing the Proper final Dimensions for the Weld Relative to the Rib Plate Thickness

7.4.2.8. Production Monitoring

RD welding must be monitored over the course of the project to ensure proper weld quality is maintained, particularly for weld penetration. The best way to track it is to require periodic macroetches and ultrasonic testing. As it is a destructive process, macroetches are conducted on pieces that will not actually be part of the bridge. It is important, however, that the test pieces represent that actual work adequately. Mock-up RD weld specimens can be made, or an occasional actual production rib can be made longer than required (and welded to a run-off tab extended beyond the edge of the actual piece).

Ultrasonic penetration testing (UT) should be conducted throughout the fabrication process for each job. The UT quality control procedures can be reduced over time using statistical process control and an appropriate confidence interval. Such reduction is prudent since UT testing is time consuming and conducting more testing than is necessary causes unnecessary delays and cost. Additionally, note that ultrasonic testing is accurate to about 10 percent for depth verification, which translates into about 1 mm (0.04 inches). Hence, a testing procedure can be established by verification of the depth check from UT to macroetches.

7.4.2.9. Weld Procedure Qualification (PQR) and Specifications (WPS)

PQR and supporting WPS are required to give the engineer and owner reasonable assurances that the fabricated welds will perform as intended. The contractor is responsible to perform PQR testing as described in the AASHTO/AWS D1.5 Bridge Welding Code to establish a basis for the WPSs that will be used in production welding. Additional performance testing will be specified by the engineer to give assurances that non-standard weld joints, such as the RD welds, will be acceptable and comply with the standards in the Contract Specifications.

7.4.3. Field Welding

Field welding is useful and often essential for OSD construction. There are negative perceptions about the suitability of field welding for bridge construction. These are based on instances where improper procedures were either specified or implemented. Field welding has been dismissed out-of-hand in the past, but it has been routinely with confidence. For example, in Texas and Georgia, girder flanges and webs are commonly welded together in the field (in lieu of a bolted splice). The rail industry also frequently uses field welds with much success. The keys for achieving good field welding are to provide a suitable environment, including protection from wind, and to conduct sample radiographic testing (RT) or UT for full penetration welds. Generally UT is more prevalent than RT in OSD fabrication and has been proven to be suitable.

It is most common to join OSD plates by full penetration field welding, particularly because bolted connections interfere with the deck surfacing. To help ensure quality and facilitate production, these full penetration field welds are accomplished primarily from the top side of the deck. A steel backing bar may be used. The backing bars should be attached in the field to facilitate fit-up and should be attached in the throat, as shown in Figure 7-13. Ceramic backing bars have been used as an alternate to steel internationally, but are not as common in the United States

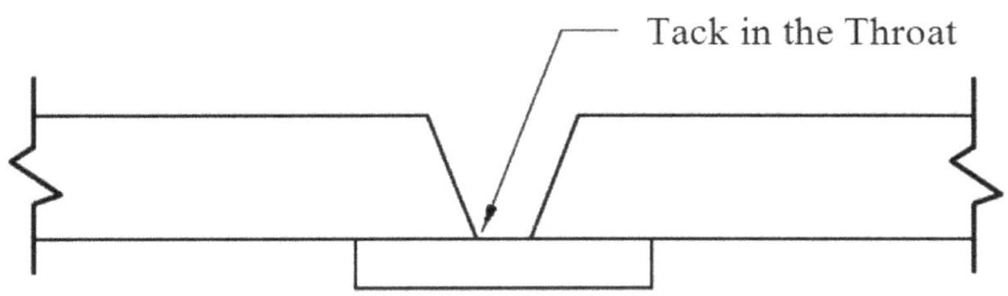

Figure 7-13 Typical Deck Plate Field Weld

Deck welding should be accomplished with an automatic process to the most extent possible. SAW is well suited to welding in the flat position required for deck welding from the top side, though other traditional bridge processes are also acceptable, as shown in Figure 7-14. However, SAW can only be used from the top. If a backup bar is used, all welding can be done with SAW from the top, but often the back-up bar must be removed. Alternately, a double-bevel joint can be used, with an alternate process used to weld the back-side overhead. For non-flat position welds like this, FCAW is the process most preferred by field welders, though some welders prefer to use SMAW to get better penetration.

In order to avoid melt-through when placing the flat SAW, the contractor may need a small semiautomatic first pass after back-gouging the weld to sound weld metal. This may be necessary to prevent the first SAW pass from melting through the weld joint. If a backup bar is used and removed, the operator should be qualified, including a qualification test to demonstrate proficiency.

In redecking projects, longitudinal splices between OSD panels are welded while traffic is maintained on nearby deck work already erected. It is necessary to yoke the decks at intermediate FBs and at temporary connections to ensure that no relative displacement occurs during the production of such welds.

Figure 7-14 Field Welding of the Transverse Deck Plate Joint during Construction

7.5. SHIPPING AND HANDLING

The transportation method selected for an OSD depends on whether it is a new structure or a redecking project. Transportation to the bridge site of new box girders typically occurs by waterway, if the boxes are deeper than overpasses allow. Long overland spans require special attention. On redecking projects, OSD widths depend on construction staging coordinated with transportation routes. Excessive movements of elements during transport can create unnecessary damage to the decks. Shipping across oceans requires special attention. Furthermore, if part of the wearing surface is shop installed, damage prevention practices are necessary.

Salt deposits and brine also cause concerns related to corrosion of OSD panels. Thus, salt spray should be rinsed off from all surfaces exposed during transport to sufficiently remove all brine or salt deposits. Alternate methods for washing the panel can be specified by the engineer if additional requirements are necessary.

7.6. ERECTION

Erection of OSDs is similar to other steel bridge structure erection. In recent experiences with major bridge redecking, OSDs were laid down as fast as 122 m (400 ft) of lane per night (without field welding). This was achieved with prior removal of the existing deck within a permanently closed lane and with necessary preparation of the existing FB. The desirability to place decks at night and to allow peak traffic to flow without any lane closures have been seriously considered by bridge authorities where major bridges were being redecked. Because of nightly mobilizations, Nondestructive Evaluation (NDE) time, and deck removal time, the time for deck installation and welding was limited, in order to reopen the lane again during peak traffic. On long-span bridges, deck replacements using OSDs have been conducted with closed lanes for optimal cost efficiency and shorter duration of construction. For short spans where detours are permitted to adjacent streets, OSDs offer superior construction erection time and quick return to normalcy.

Fit-up attachments such as alignment lugs are required for aligning large field sections (Figure 7-15). Such hardware requires temporary welds. Negative perceptions are common about temporary welds, but if they are accomplished and removed properly, they do not create a condition any worse than the adjacent field weld. Additional information on erection can be found in an American Bridge publication by Callaghan (2010).

Figure 7-15 Temporary Fit-Up Lugs are Necessary for OD Construction. Tensioning Rods Bridge Two Sections to Pull Together and Align the Adjacent Panels

OSDs are being used more and more as components of wide box girders in which the width tolerances in individual panels and poor control of longitudinal deck plate joints may cause geometric fabrication "errors" in total width, which create difficulties in abutting adjacent box elements.

7.7. INSPECTION AND TESTING

Tests and inspections are necessary to give reasonable assurances that the quality of the product is in conformance with the contract documents. Some of this has been touched upon in the previous sections. This section (and the chapter in general) are based on the premise that the quality of a product is controlled by the specifications and not by NDE inspections after the work is done. Fabrication/erection inspections and tests shall be performed as necessary prior to assembly, during assembly, during welding, and after welding, to ensure that materials and workmanship meet the requirements of the contract documents. Inspections and supplemental tests must be specified in clear and concise terms so there is no misunderstanding regarding what is required. NDE procedures should be developed for the welds being tested.

Due to the critical nature of the proper fabrication of the RD weld, a special UT procedure should be developed for monitoring the depth of penetration. Experience has shown that it may take up to 50 test trials on mock-up weldments to develop the proper procedure. UT will not give a precise penetration value; rather, it will read the depth to + / - 5 percent, but coupled with production macroetch testing, this will be enough to ensure adequate penetration. More importantly, UT will discover the presence of any hidden hot cracking that may occur, which has proven to be very costly if left undetected. Retrofits of miles of welds in the field are vastly more expensive than catching it during production.

The phased array UT technique may be suitable for checking penetration depth to be used to determine the depth-of-penetration of the RD. Based on other applications of phased array, the technique looks promising for the RD depth check. However, its use is not yet customary, so a suitable procedure must be developed and verified. Further, research has not been conducted to demonstrate suitability for checking lack of fusion, so it may be prudent to be specific about the UT method for this application. A detailed UT testing procedure should be required and the procedure needs to demonstrate that the desired accuracy is achieved on a consistent basis.

Continual production monitoring of penetration depth is prudent. This can start at a high rate of monitoring 15 to 25 percent of the welds and then taper off over time as confidence improves, but it should not taper off to zero.

All CJP field welds should be inspected 100 percent by UT.

8. INSPECTION, EVALUATION, AND REPAIR

To satisfy basic criteria for long-term safety and serviceability, all bridge structures require maintenance. Orthotropic steel deck (OSD) bridges, like all other types of bridges, require routine inspection, evaluation, and sometimes rehabilitation. New OSD designs are expected to provide problem-free performance with appropriate maintenance. It is noted and has been mention in previous chapters that many of the service problems that have occurred in the past are the result of inadequate construction quality control and/or lack of experience, which cannot necessarily be eliminated by good design. Recognizing that all structures are subject to varying levels of deterioration over the design life of the structure, this chapter provides recommended methods for maintaining and evaluating OSD bridges, including inspection, rehabilitation strategies, retrofits, and load rating.

This chapter does contain some information previously presented in other chapters. It is repeated here, in lesser detail, with the understanding that this chapter may serve as a stand-alone reference for bridge inspectors.

8.1. INSPECTION

Inspecting orthotropic bridges presents unique challenges as compared to other more common bridge types. This is mostly due to the inherent configuration of the OSD, which is comprised of a thin steel plate stiffened by a large number of closely spaced longitudinal ribs welded to the underside of the plate at right angles to the transverse floorbeams (FBs). Additionally, OSDs typically are constructed with significantly more weld per square foot. In Chapter 0, Figure 1-1 depicts typical OSD bridge schematics with open ribs and closed ribs, respectively, identifying the major components and representative rib stiffener types. As shown in these schematics, the steel deck plate acts as the flange for the ribs, FBs, and the main girders. The advantages of this system are well covered in the previous chapters.

Comparing the schematics in Figure 1-1, the open rib stiffeners have a tighter spacing between the ribs and between the FBs. Comparatively, the inherent stability and increased stiffness of the closed rib stiffeners result in wider spacing between the ribs and between the FBs. Regardless of type, the ribs typically maintain continuity through openings cut into the webs of the FBs.

From the inspector's perspective, an advantage of the open rib configuration is simplified access to the welded joints. All welds can be visually inspected within arm's reach, subject to acceptable access for the inspector. Conversely, the closed ribs limit an inspector's ability to visually inspect the root of the welds located within the closed shape of the stiffener. But this is no different from attempting to inspect the root of a fillet weld in a plate girder flange to web weld. Additionally, as has been demonstrated through research, when the rib-to-deck weld (RD) of the closed rib has been properly designed and fabricated, root cracking will not control. Should root cracking be suspected due to cracking in weld throat or in deck plate, there are NDE methods available to inspect the weld. UT methods have been developed to detect fatigue cracks in the longitudinal welds. The UT evaluations can supplement the inspection of orthotropic bridges with closed ribs, if root cracking is of concern. As there is high redundancy with numerous longitudinal ribs 100 percent inspection is not required. That is, UT evaluations can be incorporated by selective sampling regions of the weld in an overall inspection program.

Guidelines for inspecting Orthotropic Bridges are detailed in Publication No. FHWA NHI 03-001, Bridge Inspector's Reference Manual (Ryan et al 2006). In particular:

- Section 5, Inspection and Evaluation of Bridge Decks, Topic 5.3 Steel Decks.
- Section 8, Inspection and Evaluation of Common Steel Superstructures, Topic 8.5 Steel Box Girders.

The Manual for Bridge Evaluation (AASHTO, 2008) also covers Inspection of Orthotropic Decks in Section 4: Inspection, Subsection 4.8.4.3 Steel Decks.

8.1.1. Biennial Inspections

In accordance with the National Bridge Inspection Standards (NBIS), for a Routine Inspection, each bridge should be inspected at regular intervals not to exceed 24 months (biennially). Due to the large number of welded connection details, it is prudent that a sampling of welds of representative orthotropic details (number/location to be determined as part of an overall inspection program) receive inspections every 24 months. These predetermined details are then monitored over time to ascertain whether the detail is exhibiting any fatigue cracking. Historically speaking, fatigue cracking has been problematic for certain orthotropic details. As detailed in Chapter 4, many of these problems arose from the trial and error nature of the development of the orthotropic detail, as well as poor fabrication. As the technology continues to evolve it is important for the inspector to understand which details are most susceptible. These details include the RD, the rib-to-floorbeam (RF), rib-to-deck weld at the Floorbeam (RDF), the Floorbeam-to-deck weld, and the deck-to-girder weld. Figure 8-1 through Figure 8-3 show illustrative examples of potential crack sites in details associated with OSDs.

As with all structures, attention should be paid to details located in regions of tensile stresses (such as over piers in continuous girders, or cantilever deck overhangs), or where located in the wheel paths. For OSDs, this obviously includes the entire deck portion of the structure.

OSD bridges, due to their relative newness in the United States and their fatigue characteristics, are considered an advanced bridge type. As such, they could be considered as complex bridges according to the NBIS regulations. As demonstrated in this chapter, however, OSD bridges should not be labeled complex for inspection purposes. However, should an owner choose to define their OSD bridge as complex, the NBIS requires identification of specialized inspection procedures, and additional inspector training and experience to inspect complex bridges. Regardless of complexity designation, just as with inspection of other bridge types, the inspector must be properly trained and familiar with OSD bridges. Although not classified as fracture critical in most cases, it is recommended that OSD bridges be inspected by inspectors who have successfully completed the FHWA/NHI Training Course - Fracture-Critical Member Inspection Techniques for Steel Bridges. This ensures that the inspectors have been exposed to the kinds of details that will be encountered on OSDs. Additionally, customized inspection forms can greatly facilitate reporting findings in a systematic manner.

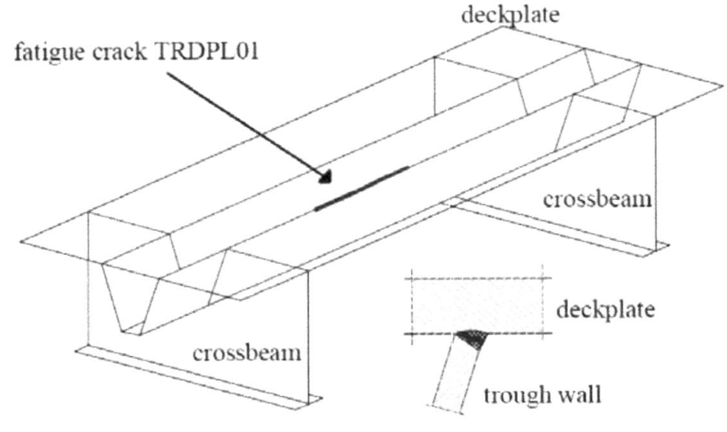

Figure 8-1 Cracks Identified in the Longitudinal Rib-to-Deck Weld shown Growing through the Rib-to-Deck (RD) Weld (de Jong, 2006)

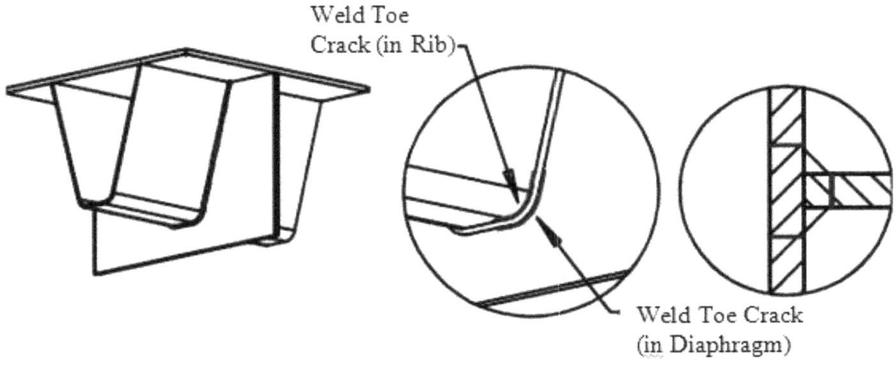

Figure 8-2 Fatigue Cracks Identified in the Rib-to-Floorbeam (RF) Connection at the Base of the Trapezoidal Rib. The Rib is Continuous, Passing through Floorbeam (de Jong, 2006)

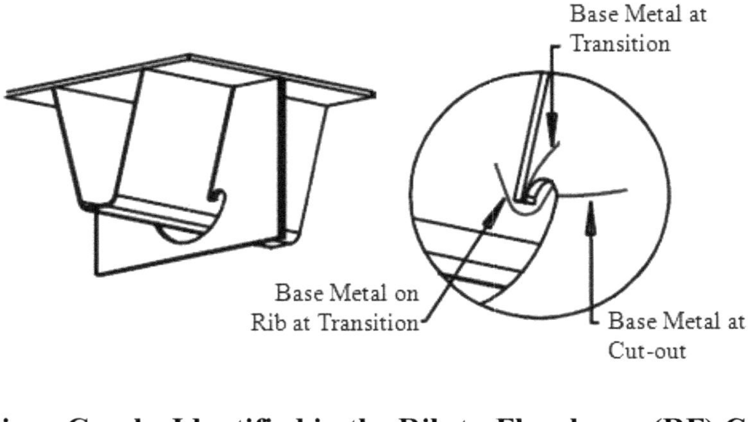

Figure 8-3 Fatigue Cracks Identified in the Rib-to-Floorbeam (RF) Connection at the Cut-out Transition. The Rib is Continuous with the smooth transition Cut-out (de Jong, 2006)

With respect to the wearing surface, asphalt, epoxy asphalt, and polymer concrete wearing surfaces placed over orthotropic steel plate decking have shown tendencies to de-bond and/or rut, depending on the material, climate, and flexibility of the supporting steel deck structure. The potential for wearing surface de-bonding along the topside of the plate and wearing surface cracking can be evaluated by inspectors walking on the wearing surface in work zones provided by appropriate traffic control. Typical wearing surfaces for OSDs are asphalt overlays (25 mm to 76 mm [1 inch to 3 inch] thickness) and polymer concrete paving surfaces (19 mm to 51 mm [3/4 inch to 2 inch] thickness).

A loss of bond can occur in the wearing surface if the top surface of the steel plate was not properly prepared (blast cleaning, corrosion-protective painting, and bonding layer application). Evidence of de-bonding includes pavement shoving, raveling, rutting, and potholes (findings observed from the top of deck). Topside cracking of the wearing surface, also observed from the top of deck, typically occurs due to excessive local OSD flexibility manifested with a thick asphalt wearing surface over a thin steel deck plate. In addition to visual inspection (see Figure 8-4), other NDE testing methods typically used include hammer sounding, chain dragging, delamination detection machinery, acoustic wave sonic/ultrasonic velocity measurements, ground-penetrating radar, impact-echo testing, and infrared thermography. Once detected in the wearing surface, destructive methods are used to confirm deck plate cracking (see Figure 8-5).

Figure 8-4 Visual Observations Indicating Deck Plate Crack as Evidenced by the Overlay Deterioration (de Jong, 2006)

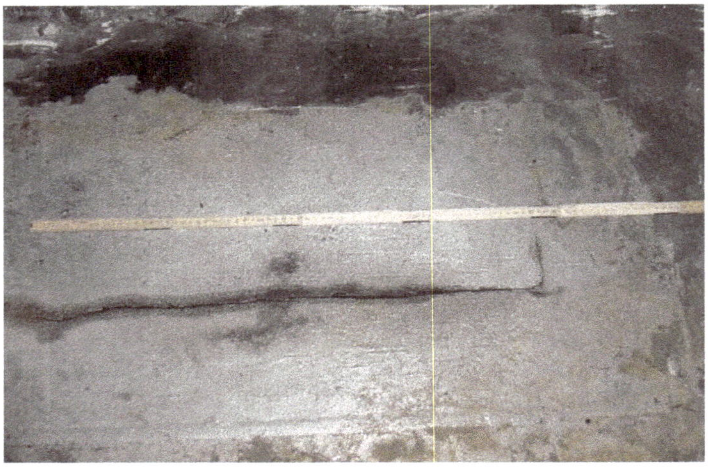

Figure 8-5 The Deck Plate Crack is Verified after Removal of Asphalt Layer (de Jong, 2006)

8.1.2. Inspection Access

Access to the underside of the OSD for visual inspection of the RD and the RF is required. Very often, the OSD is part of a closed steel box girder (Figure 8-6), which facilitates the inspection of the rib stiffeners on the interior of the box girder by allowing the inspector to enter through a manhole in the web and climb through the girder from one end to the other. On other bridge configurations, access equipment/vehicular lifts are required to place the inspector in position to evaluate the rib stiffener details. Even on box girder bridges, access equipment/vehicular lifts may be required to inspect the exterior of the box girder. It is critical that the inspector have knowledge about where the fatigue-prone details exist, so these areas are examined more carefully. Common steel defects of concern when inspecting OSDs include: bent, buckled or damaged members; corrosion; section loss; and fatigue and other stress-related cracks. In addition to visual inspection, other NDE testing methods employed may include: dye penetrant, magnetic particle, UT (phased array and straight line), radiographic testing, pulse eddy current, and acoustic emissions testing.

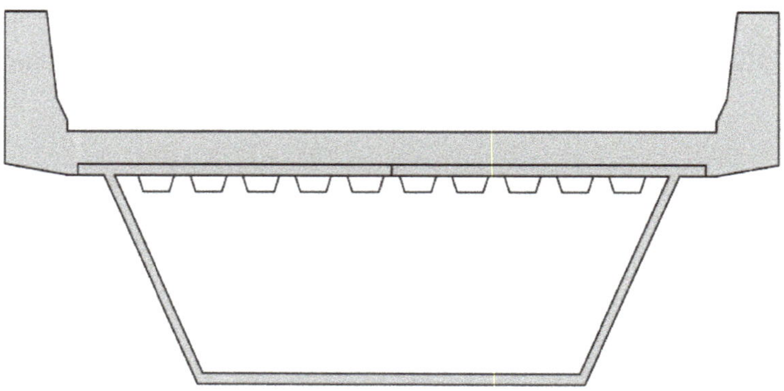

Figure 8-6 Conceptual Sketch of a Box Girder Cross-section with Orthotropic Steel (Ryan, 2006)

The two primary methods of gaining access to an orthotropic bridge are access equipment and vehicular (aerial) lifts. Access equipment includes ladders, rigging, and scaffolds. Typical vehicular lifts are manlifts, bucket trucks, and under-bridge inspection vehicles. Usually, a vehicular lift will be less time-consuming than access equipment. However, the time savings must be offset by the higher costs associated with operating vehicular lifts. The purpose of either method is to position the inspector within arm's reach of bridge components to facilitate a hands-on inspection. This is of particular importance for the fatigue-prone connection details.

In assessing the timesaving effectiveness of vehicular lifts, the following questions should be answered:

- What type of lift is available?
- How much of the bridge can be inspected using the lift?
- How much of the bridge can be inspected from one set-up?
- How much time does it take to inspect at each set-up?
- How much time does it take to move from one set-up to the next?
- Does the lift require an operator and/or driver?
- Will using the lift require special traffic control?

Inspection time and vehicular lift costs can then be compared to the time and costs associated with using conventional access equipment.

8.1.3. Biennial Inspection Report

A written report shall be prepared for each biennial bridge inspection performed by certified bridge safety inspection personnel. The report requirements are described in the applicable specifications adopted by each state department of transportation (or equivalent entity) and various other transportation agencies (turnpike commissions, thruway authorities, and port authorities, etc.). In addition to the required information delineated in the appropriate specification, the report should also include the following information about the OSD system:

- Separate listings of the ratings applied to each component.
- Photographs of deficiencies.
- Reasons for ratings lower than five.
- Recommended action items.
- Reasons for recommending an in-depth/NDE inspection (if applicable).

There are two major rating guideline systems currently used throughout the country (FHWA's Bridge Inspector's Reference Manual):

- FHWA's Recording and Coding Guide for the Structural Inventory and Appraisal of the Nation's Bridges is used for the National Bridge Inventory (NBI) component rating method:
    - A 1-digit code for Item 58 on the Federal Structure Inventory and Appraisal (SI&A) sheet indicates the overall condition of the deck.
    - A 1-digit code for Item 59 on the SI&A sheet indicates the overall condition of the superstructure.

- Condition Rating from 0 to 9, where 9 is the best possible rating.
- Use both the current and previous inspection findings to determine the rating.

- The AASHTO Guide for Commonly Recognized (CoRe) Structural Elements is used for the element level condition state assessment method:
  - The AASHTO CoRe elements for an **Orthotropic Steel Deck** are:

Element No.	Description
030	Steel Deck – Corrugated / Orthotropic

  - The unit quantity for Element 030 Steel Deck is "Each" and the entire element must be placed in one of the five available conditions based solely on the top surface condition. Condition state 1 is the best possible rating. The total deck surface area is needed in order to calculate a percent deterioration and fit into a given condition state description. Some bridge owners use the total area (m^2 or ft^2). When total area is used, the total area must be assigned to one of the five available condition states depending on the extent and severity of deterioration.
  - For connections of steel decks showing rust packing between steel plates, use the "Pack Rust" Smart Flag, Element No. 357, with one of the four available condition states. The unit quantity for Element Smart Flags is "Each" and the entire element must be placed into one condition state.
  - The AASHTO CoRe elements for a ***Steel Box Girder*** (if applicable) are:

Element No.	Description
101	Unpainted Closed Web/Box Girder
102	Painted Closed Web/Box Girder

  - The unit quantity for the steel box girder is "meters or feet" and the total length must be distributed among the four available condition states for "unpainted" and five available condition states for "painted" structures, depending on the extent and severity of deterioration. For both cases, Condition state 1 is the best possible rating.
  - For steel box girder damage due to fatigue, use the "Steel Fatigue" Smart Flag, Element No. 356, with one of the three available condition states.
  - For steel box girder rust, use the "Pack Rust" Smart Flag, Element No. 357, with one of the four available condition states.
  - For steel box girder damage due to traffic impact, use the "Traffic Impact" Smart Flag, Element No. 362, with one of the three available condition states.
  - For steel box girders with section loss, use the "Section Loss" Smart Flag, Element No. 363, with one of the four available condition states.

8.1.4. Element Level (Pontis) Manual Excerpts

An example of a state's CoRe element level field inspection manual that incorporates OSDs as a Deck Element and steel box girders as a Superstructure Element is the Michigan Department of Transportation (MDOT) **Pontis Bridge Inspection Manual** (2007). The MDOT Pontis Manual includes applicable Smart Flag pages for steel fatigue, pack rust, traffic impact, and section loss, with several condition state photographs.

8.2. LOAD RATING

This section outlines the approach to load rating of orthotropic bridges by evaluation of applicable limit states using the LRFR Methodology consistent with the 2008 AASHTO Manual for Bridge Evaluation (MBE).

Load rating of orthotropic bridges varies depending on the type of application. Simple decks that only serve to transfer local wheel loads for another superstructure system need not be **routinely** evaluated for load capacity. Fatigue, and not strength, limit states control most of the design of the orthotropic components. **However, there exists the potential for development of localized section loss that may control the strength rating. Also, a specified permit vehicle with a large axle load may be heavy enough to control.** In these cases, detailed load rating should be carried out. Additionally, when the orthotropic panel is used as an integral component of a superstructure system, such as a box girder flange or cable-stayed bridge deck, it must be loaded as a part of that steel bridge member.

The rating of orthotropic long-span bridges, movable bridges, and other complex bridges may involve additional considerations and loadings not specifically addressed in MBE Section 6 and the rating procedures should be augmented with additional evaluation criteria where required. This section extends the LRFR provisions with additional criteria for orthotropic bridges and defines the limit states and performance criteria for load rating of design loading, legal loads, and permit loads.

Consistent with the MBE, the following general equation should be used in determining the load rating of each component and connection in the orthotropic panel:

$$\mathbf{RF} = \frac{\mathbf{C_r} - \gamma_{\mathbf{DC}} \cdot \mathbf{DC} - \gamma_{\mathbf{DW}} \cdot \mathbf{DW} - \gamma_{\mathbf{P}} \cdot \mathbf{P}}{\gamma_{\mathbf{LL}} \cdot (\mathbf{LL+IM})} \qquad (8\text{-}1)$$

where:

RF	=	Rating Factor
C_r	=	Capacity
γ	=	LRFD load factor
DC	=	Dead load effect due to structural components and utilities
DW	=	Dead load effect due to wearing surface and utilities
P	=	Permanent load other than dead load (where any post-tensioning is used, this is generally ignored for OSD)
LL	=	Live load effect
IM	=	Dynamic load allowance

8.2.1. Live Loads for Evaluation

Bridge evaluations are performed for varied purposes using different live-load models and evaluation criteria. Three load-rating procedures that are consistent with the load and resistance factor philosophy have been provided in the MBE for the load capacity evaluation of in-service bridges. Each procedure is geared to a specific live-load model with specially-calibrated load factors aimed at maintaining a uniform and acceptable level of reliability in all evaluations. The load-rating procedures are structured to be performed in a sequential manner, as needed, starting with the design load rating:

- Design load rating (first level evaluation).
- Legal load rating (second level evaluation).
- Permit load rating (third level evaluation).

In the LRFR process, the HL-93 Design load rating serves as a screening process to identify bridges that should be load rated for legal loads. This was based on the longitudinal analysis of member load effects for the HL-93 loading.

Global effects, as a result of the OSD participating as an integral part of the bridge superstructure, and local effects from the application of the wheel loads should be superimposed to estimate the extreme force effect in the panel. Different live load components govern the global and local effects. Local effects in OSDs are governed by wheel loads, not truck loads. In effect, the screening level evaluation will then consider the design truck or design tandem wheel loads. The magnitude of wheel loads specified in the LRFD Specifications are 71 kN (16 kips) for the HL-93 design truck and 55.6 kN (12.5 kips) for the HL-93 design tandem. The refined HL-93 truck for orthotropic decks is shown in Figure 5-1 and should be used when evaluating the local effects.

In 2005, AASHTO adopted new legal loads for single-unit Specialized Hauling Vehicles (SHV) that include closely spaced multiple axles on short wheelbase trucks to model the new generation of hauling trucks. These vehicles are represented by the new Notional Rating Load (NRL) model, which is an 8-axle truck weighing 356 kN (80 kips), as shown in Figure 8-7. For framing members, the HL-93 will envelope the NRL load effects. It is uncertain if this will always be the case for OSDs. The design load rating in LRFR may need to consider both the HL-93 load model and the NRL load model to serve as an adequate screening level for OSDs. The impact of closely spaced multiple axles on OSD behavior needs to be further investigated.

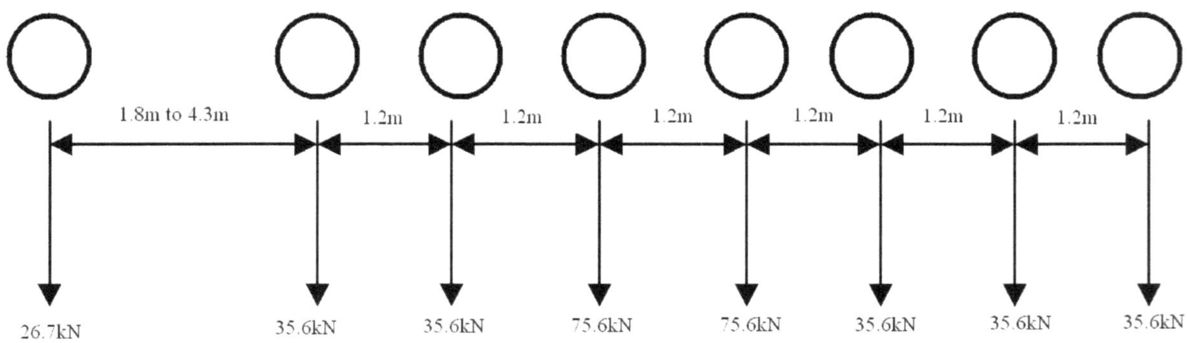

Figure 8-7 Notional Rating Load for Single Unit SHVs

The NRL is a suitable rating load model for states that comply with Federal weight laws and Federal Bridge Formula B (AASHTO LRFR). States that allow exclusion vehicles usually have legal loads that will exceed the Formula B limits for axle groups. It would be important to consider these state legal loads in the load-rating process. Checking of OSDs for permit loads should consider the specific axle configuration and axle loads as given in the permit application.

8.2.1.1. Application of Wheel Loads

Local response of OSDs is sensitive to how the wheel loads are applied. As discussed in Chapter 5, the simplified, but unrealistic, approach of using concentrated wheel loads should not be used for OSDs. The wheel load should be distributed over the tire contract area yielding a uniformly distributed pressure to be applied to the OSD contact surface. Furthermore, the notional 71 kN (16 kip) wheel loads of the HL-93 Design Truck should be more accurately modeled as two closely spaced 35.5kN (8 kip) wheels four feet apart to more accurately reflect a modern tractor-trailer with tandem rear axles (see Figure 5-1).

For most orthotropic component evaluations, it is the single truck loading that will control. Multiple presence of live load can typically be ignored, except for the evaluation of FBs with longer (4.6m [15ft] +) spans.

8.2.1.2. Load Paths and Load Distribution

OSD evaluation requires analysis of load effects that are primarily in two orthogonal directions for the ribs and FBs and involves localized distortions at the FBs. The design of OSDs is often governed by fatigue mainly where the ribs interact with the FBs. Analyzing the effects that occur at the intersections of the ribs and FB and how these effects impact the long-term performance of OSDs is an important aspect of fatigue life evaluation. However, due to the highly redundant nature of orthotropic bridges, this is of lesser importance than a strength evaluation.

As described in Chapter 5, the required level of analysis can vary depending on the test data available to the engineer and the limit states under investigation. Strength and Service limit states, and certain fatigue details, can be addressed by simplified 1-D and 2-D analysis methods. However, certain details require refined 3-D analysis methods for evaluation of Fatigue limit states. These same analysis requirements apply to evaluation of orthotropic bridges. (See Chapter 5 for more information.)

8.2.1.3. Live Load Factors for Evaluation

LRFD calibration reports a target LRFD reliability index (β) of 3.5. The LRFD design criteria based on this index are derived for a severe traffic-loading case (including the presence of 5000 ADTT). The LRFR procedures adopt a reduced target reliability index of approximately 2.5 calibrated to past AASHTO operating level load rating. Sets of load factors targeted to specific types of live loads (design, legal, and permits) are calibrated to achieve a certain level of reliability under various combinations of loads.

The magnitudes of the load factors in a combination reflect the uncertainty of the loads and the probability of the simultaneous occurrence of the loads represented in the combination. These load factors are calibrated for force effects due to trucks, not individual wheel loads. There are key differences between truck multiple presence and axle multiple presence. There will be a greater probability of side-by-side axle events than side-by-side truck events, due to the fact that each truck has two or more axles. However, OSD components are mostly controlled by the single truck loading. LRFD has specified a Multiple Presence Factor (MPF) of 1.2 to account for the possibility of a heavy single truck.

The LRFR live load factors for legal loads provided in the MBE account for the multiple-presence of two heavy trucks side-by-side on a multi-lane bridge. The live load factors for permits were derived to account for the possibility of simultaneous presence of non-permit heavy trucks on the bridge when the permit vehicle crosses the span. Their appropriateness for OSD evaluation is uncertain. The issue of design load calibration for LRFD deck design using recent traffic data was a focus of NCHRP Project 12-76. Rigorous calibration of load and resistance factors for deck design requires the availability of statistical data beyond live loads. LRFD did not specifically address deck components in the calibration.

As discussed in Chapter 5, the HL-93 design truck axle weight of 142.5kN (32 kips) is slightly less than the 75-year extrapolation value calculated by Nowak (2008) for decks. However, it is believed that the live load factor of 1.75 is somewhat conservative considering the lesser variations found in typical truck axle load spectrum. Until future research addresses the LRFR live load calibration for deck components, the current factors are assumed to be acceptable for load rating of OSDs.

8.2.1.4. Dynamic Load Allowance

Static stresses are amplified by dynamic load allowances to estimate dynamic stresses. The dynamic load allowances (IM) of the LRFD Specifications are 0.33 for Strength and Service limit states and 0.15 for Fatigue limit states. The specified values of IM consider surface roughness to be the predominant characteristic that is not apparent to the designer but can be easily verified in the field by the inspector. Further, they assume the potential surface roughness associated with potholes in typical reinforced concrete decks that are not anticipated in OSDs, depending on whether the wearing surface is thick or thin. The MBE provides a reduced dynamic load allowance for load rating primarily as a function of pavement surface conditions. The dynamic load allowances specified in the LRFD Specifications may be reduced for OSDs considering the nature of the wearing surfaces. However, as mentioned in Section 5.5.3, special consideration for the need for higher impact factors in the regions of expansion joints or other details that may result in amplified loads may be considered by the engineer.

8.2.2. Resistance and Resistance Modifiers

LRFR load capacity C in the load rating equation is given as:

$$\mathbf{C} = \varphi_c \varphi_s \varphi \mathbf{R}_n \qquad (8\text{-}2)$$

where,

φ_c	=	Condition factor
φ_s	=	System factor
φ	=	LRFD resistance factor

LRFD Design provisions of Chapter 5 of this Manual and other sources referenced are to be used as applicable for determining the nominal resistance R_n of existing OSDs. Resistance factors, φ, for the Strength limit state, shall be taken as specified in LRFD. Nominal resistance shall be based on the findings of the most recent field inspection, including consideration of effects of deterioration or cracking. A deteriorated structure may behave differently than the structure as originally designed and different failure modes may govern its load capacity. The condition factor (φ_c) should be applied to the orthotropic bridge in a manner consistent with evaluation of any typical steel bridge component.

Load modifiers (η) relating to ductility, redundancy, and operational importance contained in the AASHTO LRFD Bridge Design Specifications are not included in the general load rating equation. In load rating, ductility is considered in conjunction with redundancy and incorporated in the system factor φ_s. When the orthotropic panel is used as a floating deck system, the ribs should be evaluated with φ_s = 1.0 due to their high local redundancy and the FBs should be evaluated with φ_s = 0.85. When the orthotropic panel is an integral component of a global superstructure system, φs should be determined based on the global bridge system.

8.2.3.Limit States

Generally, the Strength condition is the primary basis for LRFR evaluation of bridge members. The focus of serviceability checks in evaluation is to identify and control live load effects that could potentially damage the bridge structure, and impair its serviceability and service life. Serviceability checks are necessary even though the live load may have been determined to be safe at the Strength limit state. Consequently, serviceability considerations in evaluation are aimed at avoiding or minimizing bridge damage due to live loads by placing limits on service load stresses under normal use and controlling permanent inelastic deformations under authorized or unauthorized overloads. In LRFR, the serviceability checks are generally given as optional, as there may be a high-cost penalty for imposing certain non-Strength related limit states in evaluation compared to design where the cost impact may be negligible. Service limit states can be used for evaluation of OSDs to protect the integrity of the wearing surface. Fatigue need not be considered for routine evaluation.

8.2.3.1.Strength Limit State

Strength limit states maintain the safe load-carrying capacity governed by material properties, such as yield strength and/or geometric properties, including loss of stability. Global and/or local geometry may govern stability considerations. For the case of a floating OSD, the ribs and FBs, including connections, should be independently evaluated for shear and flexural strength from local demands. When the orthotropic panel is used as an integral component of a superstructure, it must also be evaluated for stability by local and panel buckling, and must consider both local and global demands.

The Strength limit states, including live load and dead load as the primary loads in the combinations, govern the evaluation of OSDs. In the MBE, these load combinations are grouped as Strength I and Strength II limit states. Strength limit states must be satisfied for both buckling and yielding. The Strength I load combination is applied in conjunction with the HL-93 design load model and legal load models for rating and posting, while the Strength II load combination is applied for checking overweight permits during the permit review process.

The design-load rating at the Strength I limit state assesses the performance of existing OSDs using the LRFD design loading (HL-93) and design standards as defined in Chapter 5 of this Manual. In LRFR, the design load rating of bridges may be performed at the same design level (Inventory level) reliability adopted for new bridges by the AASHTO LRFD Bridge Design Specifications, or at a second lower-level reliability comparable to the Operating level reliability inherent in past load-rating practice. The need for two levels of reliability for screening were based on meeting the needs of states that comply with Federal weight laws and states that allow exclusion vehicles. For OSDs, a single screening level consistent with the design level reliability may be adequate.

Bridges that do not have sufficient capacity under the design load rating are load rated for legal loads to establish the need for load posting or strengthening, also at the Strength I limit state, using specially calibrated live load factors for Federal and state legal loads. The LRFR Posting methodology in the MBE addresses weight limits for truck loading and not axle loading. This is a reliability-based method that aims to maintain the same reliability target for posted bridges. The posting equation has a built-in truck overload cushion for low-rated bridges. The application of this methodology to the posting of decks governed by axle loads has not been thoroughly established and needs to be investigated.

MBE Section 6A.4.5 provides procedures for checking bridges to determine the load effects induced by the overweight permit loads and their capacity to safely carry these overloads. This is done at the Strength II limit state. Permit live load factors are selected based on the permit type, loading condition, and site traffic data. Permit live load factors were specially calibrated considering likely multiple presence probabilities for Routine and Special permits under various crossing scenarios using live load distribution factors. It is uncertain how these live load factors will apply to deck evaluations governed by axle loads.

8.2.3.2.Service Limit State

Service limit states are applied to maintain the service life of the deck. In LRFR, Service II load combination check is provided for the control of permanent deflection in steel girder bridges. However, this will never control in evaluation orthotropic steel components since Strength limit state checks are already based on the yielding limit. The Service I live load deflection check is not usually carried out for LRFR evaluation of bridge members. However, it is recommended that this combination be applied to protect the integrity of the wearing surface on the OSD. The owner will likely consider it unacceptable if the wearing surface is allowed to fail prematurely.

OSD wearing surfaces have performed relatively poorly in some instances, attributable to excessive deck flexibility, improper materials, and construction-related problems. Chapter 5 specifies that for OSDs, the Service I limit state must be satisfied for overall deflection limits for the deck plate (span/300) and the ribs (span/1000) and relative deflection of adjacent ribs (0.10 inches). These deflection limits are intended to prevent premature deterioration of the wearing surface, but are based on limited research. Service I limit state check for deflection should be performed during the normal course of load rating OSDs. The results will be very helpful in identifying decks that may be susceptible to premature deterioration of wearing surfaces and could guide inspections and future maintenance programs. However, these limits cannot be applied in a strict sense, and do not allow for meaningful calculation of rating factors or loads for posting.

In the special case of a thin polymer concrete wearing surface, there are measurable material performance criteria that may be used for detailed load rating calculations with the Service I limit state. Since there has been no LRFD or LRFR research conducted on design of these materials, appropriate load and resistance factors will have to be selected at the discretion of the engineer based on material test data.

Two aspects of Service I check will need further discussion and study:

- What deflection limits are appropriate for the older inventory of OSDs that are known to be more flexible than modern designs? Are the design deflection limits too stringent for existing bridges?
- Should the Service I check be mandatory for setting load restrictions such as posting or for use in permit reviews?

A mandatory requirement for deflection checks could have operational consequences that need to be considered. Most OSDs are on major bridges that serve as key transportation links, which may make it harder to justify service restrictions based only on subjective serviceability considerations. Ultimately, the owner must decide on the course of action, whether to load post the bridge or take some other corrective action to achieve the necessary performance.

8.2.3.3. Fatigue Limit State

The MBE does not currently employ the Fatigue limit state for the load rating of steel bridge structures, and orthotropic bridges are no different in this regard unless the owner desires to restrict traffic to increase the service life. It is more likely that an existing orthotropic bridge will require a fatigue evaluation of remaining life for maintenance or rehabilitation planning. Section 7 of the MBE gives fatigue evaluation procedures to address load-induced fatigue in components from in-plane stresses, using the nominal stress S-N curve approach. As noted in previous chapters, these Category details do not specifically apply to OSD fatigue issues that are more distortion induced.

Fatigue life evaluation of existing orthotropic bridges can be highly intricate. It requires refined Level 3 or Level 1 analysis, as described in Chapter 5, and deep understanding of the structural fatigue science. Evaluation requires more detailed application of live load and more accurate structural modeling than is necessary for design. Stresses at fatigue-prone details are very sensitive to the precise transverse position of the wheel patch load, and maximum stress cycles are often the result of two different trucks in slightly different positions causing a reversal. For design, this is typically addressed by conservatively assuming the truck wheel loads to be applied at the most critical location for all cycles and stresses are simply kept below the threshold (infinite life design). For evaluation, these simplifying assumptions will lead to excessively conservative results. The live load applied must represent the effective truck rather than the maximum truck for finite life calculation and must account for the spectrum of transverse positions and multiple truck events. Another complication is in accounting for the stiffening effect from the wearing surface, which is dependent on the temperature variations. For accurate results, this may require Monte-Carlo simulation for rational transverse positioning of load and stress cycle counting or performing a site-specific assessment by monitoring stresses at critical details. That is, **it is unlikely thathere will ever be an effective method for measuring the exact wheel loads and respective wheel positions to compile an accurate stress history for each orthotropic detail**

8.3. REHABILITATION STRATEGIES

Orthotropic bridges designed according to the guidelines provided in this Manual are expected to produce a relatively trouble-free performance. However, there are many existing bridges that were designed prior to the full understanding of behavior gained by research and development in recent years. Additionally, it is possible that even well-designed bridges may fail by displacements and stresses that arise from other non-design issues. For example, bridges experience operational issues like joint lock-up, or construction-related problems like hot cracking in welds. The two most common problems found in orthotropic bridges that require rehabilitation are wearing surface failure and metal fatigue cracking.

Wearing Surface Failure

Wearing surface failure is most often the result of an overly-thin, flexible steel deck plate design, or due to lack of quality control during construction. These circumstances will typically call for simple replacement of the surface with a properly applied system that is designed to accommodate the increased deflections. Stiffening the steel deck plate to improve surface performance has shown to be impractical. Some surface failures are attributed to localized cracking in the deck plating or from water penetration and freeze-thaw cycling that can be mitigated. These situations can be addressed by special retrofit and repair strategies. Additional data on maintenance and repair of wearing surfaces is covered in Chapter 9.

It is also noted that a different approach to rehabilitation may be required, depending on whether the wearing surface is thick or thin (see Chapter 9 for description). Thick wearing surfaces that have delaminated will result in deeper potholes, and will leave moderate levels of debris on the roadway. Although it may not be a safety concern, this may impact the traveling public in terms of driving comfort. Thin wearing surfaces will not have this problem, and could allow for more time to plan and execute the necessary repairs.

Metal Fatigue Cracking

OSDs are highly redundant structures and many details are displacement loaded. Fatigue cracking eventually proceeds at a very slow pace, except for the possibility of crack propagation in the deck plate. Here, field experience and research have shown that several crack promulgating phenomena interact to drive cracking. Depending on the size and importance of the bridge, monitoring by visual inspection must be evaluated verses the cost of replacing given deck elements. Since there is often no life-safety concern with many fatigue problems, the options considered may be as follows:

- Do nothing
- Monitor
- Retrofit bad details
- Remove and replace bad details
- Full bridge replacement

The strategy adopted should be based on life-cycle cost analysis of the various options. Very often, orthotropic bridges are major structures for which decisions are dominated by traffic control considerations much more than the basic cost of executing the structural repair.

The fist step in development of any fatigue rehabilitation strategy is to determine the cause and extent of the problem. Orthotropic bridges are highly repetitive in terms of details, and there is the potential that any problem will also be found throughout the entire bridge. Construction-related problems will likely manifest within the first few years of the service life, while fatigue-related problems will not likely be observed until a much later time. In-depth inspection is recommended to assess the root cause of any problem.

Redecking Applications

OSDs used in redecking application are particularly sensitive to fatigue since the interactions with the supporting structure are often complex and difficult to assess. In floating decks, where subfloor structure resting on stringers was introduced to support the deck, the likely source of fatigue problems in the subfloor, or strap plates connecting the ribs to it, is the excessive relative displacement between deck and subfloor. This problem is solved either by increasing the flexibility of the ligaments, or by introducing stiff ligaments that prevent relative movement.

Strap Plates' Phenomena

Strap plates, as shown in Figure 8-8, are susceptible to uplift effects, especially if adjacent to a roadway joint. The problem is most often the result of prying action or poor initial installation.

During installation, if the OSD was forced onto the subfloor beam by placing a weight on the deck, experience indicates that bolt failure is almost guaranteed. In such a case, a necessary, but not necessarily sufficient remedy is to shim the space between strap and subfloor when the deck is free. Prying action effects must be checked regardless.

High stresses at the toe of the RD are obtained when the bolt is far from the weld. Unfortunately, due to lack of access, strap plate welds to the rib are not field repairable. Where open ribs may be repairable by bolting, as shown in Figure 8-8, welds in strap plates of closed ribs cannot be repaired for lack of access.

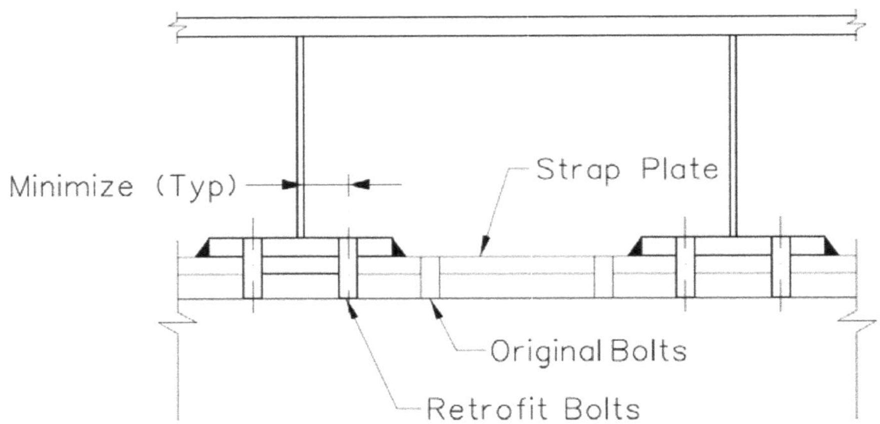

Figure 8-8 Retrofit Connection using Strap Plates to connect the ODS to the Existing Structure

Subfloor Structure

Subfloor structure problems surface in jointed decks, which are typically allowed to float longitudinally, and the subfloor connections to the stringers must take the longitudinal impulses of trucks entering and exiting the panel (i.e., the deck structure between joints). Usually transverse resistance is sufficient to handle effects arising from change of lane.

Panel lengths can be in the order of floor beam spacing such as 18.3 m (60 ft), or girder span length up to 61 m (200 ft), and possibly more. The engineer needs to investigate the more effective strategies of displacement prevention vs. increasing flexibility. Alternately, increasing the length of a short panel by joining two or three of them increases the resistance to deformation of the ligaments accordingly and could be sufficient to abate the distress permanently or reduce it to a manageable level.

When the transverse flexibility of the sub floor is high, causing top deck displacements relative to the floor beam, effects could be created and could cause crack propagation of strap plate welds. In such cases, displacement prevention by adding FBs usually provides satisfactory solutions. The ability to perform this in the field has been questioned, but despite difficulties, it is deemed possible if the space is available.

8.4. FATIGUE RETROFIT

Although there has been considerable research focused on the fatigue performance of OSDs as a result of the in-service cracking that has been observed, there is much less information in the literature specifically focused on repair and retrofit strategies. Furthermore, although much of the research focused on new design is intended to be used for general applications, retrofit strategies in the literature are, for the most part, applicable to a specific bridge and/or geometric configuration. In other words, they are often specific repairs to a specific shortcoming in a given design. Nevertheless, important information can be gleaned from the literature and by applying fatigue retrofit strategies commonly used for other bridge components, such as hole-drilling or adding bolted splices. Such retrofit concepts have a proven track record and can often be applied to a multitude of details, many of which can be found on a typical OSD.

One major point to keep in mind with respect to cracking in OSDs is related to the consequence of cracking. Contrary to the effects a fatigue crack may have in the tension flange of fracture critical bridge, fatigue cracking in OSDs is, in most cases, simply a maintenance or serviceability issue. This is primarily due to the inherent substantial level of redundancy in these deck systems (i.e., cracking in one member will not compromise the strength of the deck system). However, if left unchecked, the potential for continued crack propagation exists and retrofit or repair is a prudent choice. Nevertheless, the inherent redundancy is often advantageous, as many types of cracks will not need to be retrofitted in an emergency type response.

In this section, a few common problem areas will be reviewed along with retrofit/repair strategies that have either been used in the field or would appear to be effective based on previous experience. The reader is encouraged to obtain the cited references for further detail as the information provided herein is intended to only provide an overall description of the retrofit. Although references are provided in this manual, the reader is cautioned that they are not necessarily endorsed by this manual. It is the reader's responsibility to analyze any retrofit strategy and verify the effectiveness, robustness, and applicability of a given retrofit prior to selection.

Lastly, regardless of the strategy employed, it is highly recommended that field measured strains and displacements should also be obtained to ensure the effectiveness of the retrofit. OSDs on larger bridges contain hundreds if not thousands of duplicate details. Assuming that cracking has been observed at one such detail, the number of potential retrofit locations can quickly become significant (potentially hundreds or thousands of locations). Obviously, it would be advantageous to accurately determine that a given retrofit is going to perform as intended prior to implementing it on a wide scale. Hence it is recommended that a few prototype retrofits be installed, instrumented, and monitored for a period of time in order to determine the effectiveness of the retrofit and identify if any improvements or modifications are required.

Deck Plate Cracking

When the deck plate is relatively thin, failures can occur all along the rib stem going through the deck plate, instead of the RD. This condition has been a persistent problem in Europe, where deck plates were originally designed to be thin (10 mm to 12 mm [3/8 inch to 15/32 inch]). Where, for a long time, these cracks were back-gouged and welded before replacement of the asphalt, the current solution, after the cracks are repaired, is to place shear studs on the deck plate and a layer of reinforced concrete, which considerably increases the composite transverse rigidity of the deck and its resistance in flexure. Although analyses indicate promising results, the duration of this fix relative to deterioration of concrete, which may or may not be protected by asphalt, has yet to be determined by actual experience.

In emergency repairs, it is acceptable to weld or bolt plates over the area where deck plate cracking has occurred. Such retrofits must be considered temporary and used if there is concern regarding the integrity of the deck or if there is some other safety concern (e.g., a piece of deck falling underneath the bridge or a piece of steel puncturing a motorist's tire).

More permanent solutions that have been used include cutting a section of the thin deck plate out and installing a thicker plate in its place. Obviously, this may require shortening the height of the ribs if a much thicker plate is used to minimize the mismatch in the height at the driving surface. Special consideration must be given to the quality of all repair welds to ensure the repair will perform as desired. Depending on the orientation and position of the crack, it may be feasible to simply arc-gouge the crack out to the crack tips, and place a CJP weld. Obviously, the success of any weld repair is heavily dependent on the quality of the weld, as well as the driving mechanism that produced the crack in the first place. For example, if the cracks are the result of the deck plate being too thin, simply repairing the crack with a CJP weld, cracking could be expected to reinitiate in the future, as the cause of the problem has not been corrected (Kelly and Dexter 1997).

Other stiffening strategies include using a thin layer of high-performance concrete that is reinforced with a steel mesh or other similar reinforcing. Bond to the steel deck plate is essential to ensure a composite system is achieved and the greatest increase in stiffness. In some cases, very short shear studs have been used to ensure adequate bond between the concrete and steel deck plate. Such an approach was used on the McNaughton Bridge rehabilitation located in Pekin, Illinois. An overly flexible deck plate led to cracking of the overlay, which in turn led to corrosion of the deck plate.

One interesting repair strategy that has been proposed, apparently with success, involves filling the ribs beneath the wheel paths with a special epoxy that bonds to the deck plate and rib wall. The material possesses enough stiffness to ensure it provides vertical support to the deck plate (between the rib walls), thereby reducing bending stresses. It is reported that the decrease in deck plate stresses can be reduced by a factor of two or three, although no specifics regarding durability, laboratory data, or strain measurements are given (Boersma, 2003). It would seem that the actual effectiveness of this approach, especially the long-term environmental durability issues, would need to be verified before implementing on a large-scale.

It has also been proposed to simply bond an additional steel deck plate to the existing deck plate using an appropriate adhesive. According to Boersma, field tests suggest this is a viable option, though no specific data are presented in the paper (Freitas, 2009).

Depending on FB and deck proportions, it is possible that cracking through the plate will occur in the deck plate due to stress concentrations at the RDF. In this case, remedial measures, such as back-gouging and welding, will likely have a short duration. Rib replacement with a bulkhead may provide a satisfactory solution.

Rib Cracking (Weld Toe in Rib)

Many RF failures are caused by wheel passage over the rib or ribs nearest to the wheel mean position. Hence, for every traffic lane, there are two problem areas (possibly four) out of six ribs per lane. An assessment of the costs of monitoring vs. that of replacing two ribs or the entire truck lane with improved details should be part of the engineering study necessary for a recommendation.

When the failure occurs in the rib material at the end of the FB cut-out transition, the crack will extend longitudinally (horizontally along the rib length) to several inches. Drilling a hole at the crack termination will arrest the crack, because once the crack has reached a given value, the effective displacement no longer causes the crack to grow. This alternative solution to replacement requires monitoring. Caulking and monitoring are part of this solution.

Rib Splice Cracking

Fatigue cracking of the welded transverse splices in the longitudinal ribs has been known to occur at field splices. Although bolted field splices seem to generally perform quite well, field welded splices have not enjoyed such success. The reasons for the cracking are not much different than typically noted for other field welds. These included poor quality and the lack of complete fusion or penetration. Lack of penetration can't be avoided in some of the details used to make the joint. In most cases, a large plate (serving as a backing bar) is fitted and welded inside one of the ribs. The next rib is then slid over the backing plate and a weld placed within the joint. Cracks have been known to grow out of the lack of fusions zones, as well as the weld toe.

Repairs have included rewelding the joint or replacing the splice with a bolted joint. Regardless, it is important to isolate the crack in the rib from entering the deck plate if possible. Hence, it would be advisable to drill out the crack tip to reduce the likelihood of the crack propagating further. A recommended strategy would be to design and install the common bolted rib splice if possible. This effectively eliminates or bypasses the existing welded splice. In order to install and access the bolts, a hand-hole will be required. In new design, this is typically placed in the bottom flange of the rib. The addition of the hand hole does not really decrease the capacity of the rib if sufficient splice plates are used. Then, using relatively thick splice plates that are sufficiently long to accommodate the effects of shear lag, an effective bolted splice can be introduced.

Fatigue cracking will likely reinitiate if repair welds are used, unless the quality of the repair welds are greater than originally placed. Lastly, the bolted connection will likely have a greater fatigue resistance than the welded repair. Hence, repair welds are discouraged unless complete joint penetration and excellent weld quality can be assured.

Rib-to-Floorbeam Weld Cracking

As discussed in other sections of this manual, there are a number of modes of fatigue cracking at the RF joint that may occur. Considering the many variations of this detail, the source of cracking is also quite varied. As with most cracks, hole-drilling often provides a useful and simple method of repair. Such strategies would be very useful in cracks in the wall of the rib as one-sided access is all that is likely to be available. Cracking in the FB or web plate can also be effectively controlled with hole-drilling in some cases. Of course, for hole-drilling to be effective, the driving forces that caused the initial crack must be less than those required to cause reinitiation within the drilled hole. In some cases, a pretensioned high-strength bolt, which introduces beneficial compressive stresses, should be inserted in the hole to further improve the fatigue resistance of the retrofit. Depending on the amount of rib rotation, hole drilling may not always be an effective strategy for the same reasons hole drilling is not always effective when used to mitigate out-of-plane distortion cracks in the web of plate girder (i.e., the demand is too high).

Although simple hole-drilling may be effective, other more sophisticated retrofit strategies have been developed. (Even if another retrofit strategy is being implemented, a hole should always be placed at the crack tip to blunt the crack.) Kolstein (2008) illustrates several concepts and strategies that appear to be effective. These included bolting doubler plates to the web or FB plate in the region of the RF, either with or without the cut-out. This retrofit, used in combination with hole-drilling, will likely be effective in mitigating fatigue cracks that have grown into the FB or web plate.

Another interesting repair is applicable where the weld between the RF or web plate is cracking. It can also be effective where a cut-out has or has not been used. The retrofit essentially consists of a welded one piece "jacket" that fits around the rib and includes a portion of a FB plate. The two components (the jacket on the web and the plate that is in the same plane as the FB or web plate) are welded together before installation. Once fabricated, the piece is bolted to the rib and FB or web plate and effectively bypasses the original connection. The detail would be installed on both faces of the FB or web plate. Schematics of the detail are presented in Kolstein (2008).

Failure of the FB may be due to the presence of an internal rib bulkhead, causing high in-plane stresses in the web. Cracking occurring at the toe, or root, of the FB means that the weakest link is impacted by VQ/I effects. This condition requires replacement of the web with a thicker one. Investigation would be required as to whether this could create unacceptable out-of-plane bending. In any case, it would be an expensive endeavor. It could be studied whether better transition would be beneficial. If the expected life is of short duration, or the cracking is expected to proceed at a faster pace than manageable, redesign and replacement are necessary.

Root cracking is likely to be caused by a large gap of unwelded material. It could be improved or completely resolved by gouging and back-welding with larger welds, as long as the FB web is thick enough to prevent toe cracking.

9. WEARING SURFACES

9.1. INTRODUCTION TO WEARING SURFACES

Wearing surfaces serve several important purposes. Primarily, they provide a skid resistant surface with good ride quality. They also provide corrosion protection to the deck plate, level out deck plate irregularities, and potentially contribute to increased fatigue life of the deck plate. The latter benefit arises from a resulting reduction in stress levels in the steel plate as compared to a bare surface. Desirable characteristics include light-weight, stability, and stiffness over a wide temperature range, long-term durability, good bond to the deck plate, resistance to shoving/rutting, crack-resistance, superior fatigue response, ease of application, and maintenance (Rigdon, 1990). As one selects a wearing surface system for an orthotropic steel deck (OSD), it is important to understand the mechanical behavior of the wearing surface system from a materials perspective, as well as from a structural perspective. This dual perspective is important to understand, as most surfacing materials are rigidly bonded to the steel deck plate and act as a composite system. Table 9-1 summarizes desirable characteristics in a wearing surface system for use on OSDs. Wearing surface materials are applied as complete systems, comprising some of the following components:

- waterproofing membrane that also provides corrosion protection.
- bond coat facilitating composite action of the wearing surface with the deck plate.
- leveling layer that evens out deck welds and irregularities.
- wearing surface layer that may include embedded surface aggregates to enhance skid resistance.

Table 9-1 Desirable Characteristics in a Wearing Surface System

Desirable characteristics	Remarks
Skid resistance	Provide a safe, skid-resistant surface with polish-resisting aggregates for the life of the surfacing.
Ride quality	Provide smooth riding.
Bond strength	Have high bond strength to the steel deck to provide composite action between the surfacing and the steel deck, to reduce fatigue stresses in the steel deck and in the surfacing, and to resist de-laminations from shear stress due to flexure and from thermal effects.
Tensile cracking	Provide fatigue resistance against pavement cracking under millions of loaded truck-wheel loads.
Deformation	Be resistant to shoving, rutting, and raveling due to millions of wheel passages, braking, lane-changing forces, and temperature extremes.
Durability	Be resistant to environmental influences, such as sunlight, oxidation, and temperature changes, and be impervious to salt water and fuel and oil contamination from traffic.
Water proofing/Corrosion protection	Be impervious to the passage of water through the surfacing. In addition to being impervious, provide a corrosion-resisting coating to protect the steel plate.
Fatigue life	Meeting or exceeding the design life.
Ease of repair and maintenance	Easy-to-repair in case of mechanical or fire damage.
Ease of construction	Can use existing construction equipment, mixing plants, and crews for the installation.
Economy	Offer competitive life-cycle costs.

9.2. POPULAR CLASSIFICATION AND SUCCESSFUL APPLICATION OF WEARING SURFACE SYSTEMS

Proper selection, placement, and maintenance practices of surfacing systems should form an integral part of OSD design (Gopalaratnam et al., 1993). Many types of classifications can be used for wearing surface systems, including:

- Primary surfacing materials such as polymers concrete, asphalt concrete, Portland cement concrete, and various combinations of these materials.
- Wearing surface thickness from thin ranging from 9 mm to 15 mm (3/8 inch to 19/32 inch), and thick ranging from 30 mm to 60 mm (1 3/16 inch to 2 3/8 inch).
- Reinforced versus unreinforced systems.

Some surfacing systems that use a combination of bituminous and epoxy resin (polymer) as binder (e.g. epoxy-asphalts) defy such generalizations. Similarly, composite systems using a Portland cement concrete layer, anchored using shear studs to increase deck rigidity and topped with asphalt ride surfaces, also make unique classification difficult. Another classification of surfacing system used in published literature is based on whether the binder used is "thermoplastic" or "thermosetting" (Seim and Ingham, 2004). Thermoplastic materials soften, deform, and melt with heat. Surfacing using such binders can rut and shove when subjected to a combination of high deck temperatures (60°–70° C) and wheel loads (like some asphalt-based wearing surface systems described in Section 9.2.1). Regardless of the classification used, it is important to understand how different wearing surface systems perform under a combination of mechanical and thermal loads due to composite action with the steel deck plate.

This section presents a brief review of the relevant mechanical and physical properties of the two predominant types of wearing surfacing systems: (1) bituminous surfacing systems, including mastic asphalts, latex-modified asphalts, and reinforced asphalt systems; and (2) polymer surfacing systems, including epoxy-resins, methacrylates, and polyurethanes. Although not mandatory, many bituminous surfacing materials used on OSD bridges are 50 mm (2 inches) or greater in thickness, while most polymer surfacing materials are 15 mm (19/32 inch) or less in thickness. Since epoxy-asphalt systems are typically 50 mm (2 inches) or thicker, for the purposes of discussions in this chapter, this type of surfacing is included along with bituminous surfacing. The above generalizations are useful to understand influence of the important parameters that affect performance of surfacing materials and OSDs, even if they defy easy or unique classification.

Table 9-2includes summary information on different wearing surface systems successfully used in select bridges in the United States and abroad. Remarks regarding their performance while in service are also provided. While not exhaustive, the list demonstrates the diverse options in terms of wearing surface types and thicknesses, deck plate thicknesses, and differing conditions of environmental exposure and vehicular loading patterns.

Table 9-2 Examples of Successful Wearing Surface Systems

Bridge Name	Location	Deck Thickness	Wearing Surface System Remarks
Ben Franklin	Philadelphia, PA	16 mm (5/8 in.)	32 mm (1 ¼ inch) thick epoxy asphalt layer topped with a 32 mm (1 ¼ inch) thick asphalt wearing surface.
Bronx Crossing Span of Triboro (Robert Kennedy)	New York, NY	16 mm (5/8 in.)	Polymer bond coating, 9-mm (3/8 inch) polymer slurry with wear surface embedded with basalt aggregates for traction. System was used on factory-fabricated steel panels when the span was redecked in 1999.
Bronx Whitestone	New York, NY	16 mm (5/8 in.)	Three layers of polymer concrete using the multi-layer broom and seed method. Total wearing surface thickness of 9 mm (3/8 inch) was built up in prefabricated steel deck panels. Deck was sealed using polymer used for the surface. Surface in use since 2006.
Coronado Bay	San Diego, CA	10 mm (13/32 in.)	Inorganic zinc waterproofing, epoxy asphalt bond coat, 50 mm (2 inch) epoxy asphalt surface placed in two 25 mm (1 inch) lifts with epoxy asphalt bond coat between lifts. Less than 4% voids. Resurfaced in 1993.
East River Suspension Span of Triboro (Robert Kennedy)	New York, NY	16 mm (5/8 in.)	Three layers of polymer concrete using the multi-layer broom and seed method. Total wearing surface thickness of 9 mm (3/8 inch) was built up in prefabricated steel deck panels. Surface in use since 2003.
Faro	Denmark	12 mm (15/32 in.)	4 mm (1/8 inch) Special mastic rubber-resin adhesive bond coat, 25 mm (1 inch) binder layer of mastic asphalt, and 30 mm (1 3/16 inch) Wearing course of mastic asphalt. Wearing surface has been performing well since it was installed in 1998.
Freemont	Portland, OR	16 mm (5/8 in.)	Epoxy asphalt bond coat, 64 mm (2 ½ inch) epoxy asphalt placed in 2 lifts, epoxy asphalt bond coat between lifts, compacted to <4% voids.
Golden Gate	San Francisco, CA	16 mm (5/8 in.)	
Great Belt East	Korsor, Denmark	12 mm (15/32 in.)	4 mm (1/8 inch) Special mastic rubber-resin adhesive bond coat, 25 mm (1 inch) binder layer of mastic asphalt, and 30 mm (1 3/16 inch) wearing course of mastic asphalt. Wearing surface has been performing well since it was installed in 1998.
Hagenstein Bridges	Netherlands	10-14 mm (13/32-9/16 in.)	Reinforced high performance concrete wearing surface, 60 mm (2 3/8 inch) thickness. Bonded using 4-mm epoxy layer with embedded aggregates. Installed in 2005.
Luling	New Orleans, LA	11 mm (7/16 in.)	Deck thickness is both 11 mm and 12 mm (7/16 inch and 15/32 inch) at different locations.
Moerdijk Bridge	Netherlands	10-14 mm (13/32-9/16 in.)	Reinforced high performance concrete wearing surface, 50 mm (2 inch) thickness. Bonded using an epoxy layer with embedded aggregates. Installed in 2005.
Poplar Street	St. Louis, MO	14 mm (9/16 in.)	Inorganic zinc waterproofing coating, polymer bond coating, 12 mm (15/32 inch) polymer slurry with surface embedded with basalt aggregates for traction. Surface used first in 1992. Same surface was replaced in 2006.
Port Mann	Vancouver, BC	11 mm (7/16 in.)	Polyamide cured red lead epoxy, coal tar epoxy with 6 mm (1/4 inch) stone chips bond coat, 16 mm (5/8 inch) sand asphalt, and 35 mm (1 3/8 inch) asphalt.
Queensway	Long Beach, CA	13 mm (1/2 in.)	Inorganic zinc waterproofing, epoxy asphalt bond coat, 50 mm (2 inch) epoxy asphalt surface placed in two 25 mm (1 inch) lifts with epoxy asphalt bond coat between lifts. Less than 4% voids.
San Mateo	San Mateo, CA	14 mm (9/16 in.)	Inorganic zinc waterproofing, epoxy asphalt bond coat, 50 mm (2 inch) epoxy asphalt surface placed in two 25 mm (1 inch) lifts with epoxy asphalt bond coat between lifts. Less than 4% voids. Resurfacing scheduled within 5 years.
Shonan Ohashi Bridge	Tokyo, Japan	12 mm (15/32 in.)	Hybrid reinforced (steel fibers and CFRP mat) ultra rapid hardening cement concrete. 70 mm (2 ¾ inch) layer bonded using steel studs and epoxy layer. In service since 2005.
Throgs Neck	New York, NY	25 mm (1 in.)	

Figure 9-1 shows schematics and details of the component layers in two typical thick (asphalt-based [a, b]) and two thin (polymer concrete [c, d]) wearing surface systems. The 60 mm (2 3/8 inch) thick mastic asphalt systems, as seen in Figure 9-1a, have been successfully used in Europe (Great Belt East Bridge in Korsor, Denmark). The 50 mm (2 inch) thick epoxy asphalt wearing surface systems in Figure 9-1b have been used successfully in many orthotropic bridges in California (Coronado Bay Bridge, San Mateo Bridge, and Golden Gate Bridge). The 12 mm (15/32 inch) polymer concrete wearing surface system in Figure 9-1c, a polymer concrete slurry system) has been successfully used on the Poplar Street Bridge in St. Louis, Missouri. The 9 mm to 10 mm (11/32 to 13/32 inch) polymer concrete wearing surface system in Figure 9-1d is a multi-layer polymer concrete using the broom and seed method of placement (painting or brushing on a clean substrate and casting sand on the resin) and has been used more recently in a few New York Bridges (East River Suspension Span of Triboro [now Robert F. Kennedy] Bridge and Bronx-Whitestone Bridge).

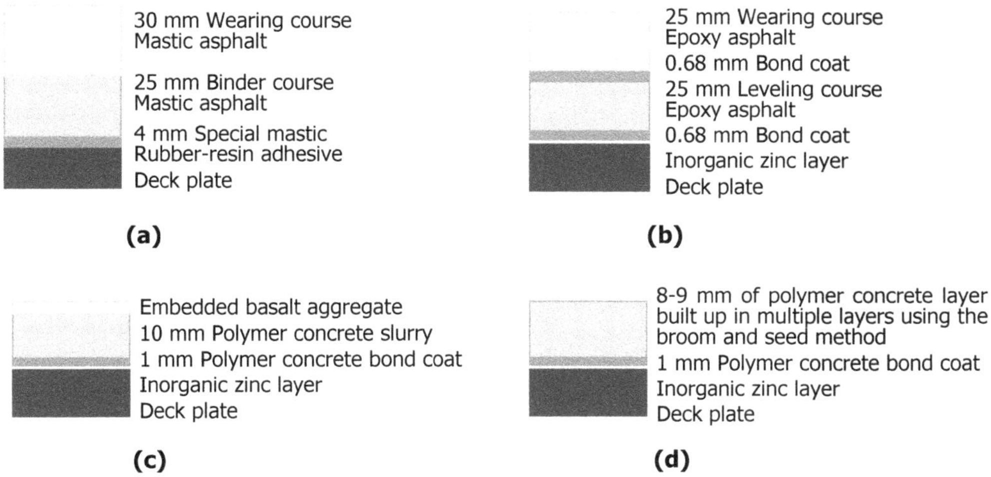

Figure 9-1 Typical Component Layers in (a) Mastic Asphalt, (b) Epoxy Asphalt, (c) Polymer Concrete (Slurry Method of Placement), and (d) Multi-Layer Polymer Concrete (Broom And Seed Method Of Placement)

9.2.1. Bituminous Surfacing Systems

These types of surfacing systems include materials where the primary binder is asphalt-based. The surfacing systems, which are typically thick (50 mm [2 inches] or greater) and applied while hot, have performed well on relatively rigid decks plates. Stiffness characteristics of such surfacing systems are influenced very significantly by temperature, as discussed later. Limited results are also available to show strain-rate sensitivity of the elastic modulus. At cold temperatures, the contribution of the surfacing to the flexural stiffness of the composite deck is significant due to the reasons described earlier. Among the thermoplastic surfacing systems, Gussasphalt, a poured asphalt developed in Germany, uses low penetration-grade bitumen. This is a very hot and dense void-less mix requiring special procedures and equipment for placement. Such surfaces were popular in Germany and Japan in the 1960s. Mastic asphalts, popular in the 1980s and 1990s in Europe and Japan as surfacing for major suspension bridges, use higher binder content than the poured asphalt. This type of surface also requires special paving equipment for placement. Neither of these types of surfacing has been used in major OSDs in the United States. Conversely, a hot-mix asphalt surfacing comprising epoxy resin and bituminous hardener, resulting in a thermosetting binder surfacing popularly called "epoxy-asphalt," has been used in several major bridges in the United States, including several bridges in California. It has been used more recently in bridges in China and South Korea. There has been only limited success with conventional asphalt or asphalt modified with latex or other similar additives as surfacing for OSDs. A good survey of surfacing materials used worldwide can be found in Touran and Okereke (1991), Gopalaratnam et al. (1993), Hulsey et al. (1999), and Wolchuk (2002), even if each of these lists are not exhaustive by themselves. Rutting, shoving, and tensile cracking are among the common types of failures observed in some thick bituminous surfacing systems placed on OSDs.

9.2.2. Polymer Surfacing Systems

Two types of polymer surfacing systems include: slurry systems where fillers are integrally mixed with polymers and placed on the steel deck plate, with additional aggregates embedded on the surface to provide the requisite traction; and surfacing systems built-up in layers with alternate layers of neat polymers and surface embedded aggregates. Polymer surfacing systems are typically thin (10 mm to 15 mm) and applied after cold mixing. Such thin surfacing materials do not contribute very much to the deck stiffness, even at colder temperatures, and hence do not contribute to reduction in steel plate stresses when subjected to dynamic wheel loads. On the plus side, they contribute to a reduction in dead loads on the deck and hence may offer unique opportunities to increase fatigue life (by reducing inertial loads) of older bridges needing replacement surfaces. They also offer opportunities for prefabrication of composite deck panels (Weidlinger Associates, Inc., 2002) (Wolchuk, 2006) for new bridge decks (factory-produced steel deck panels with pre-placed wearing surface) with better quality control of the placement surface (clean and dry shot-blasted white metal surface with prescribed roughness), and environmental conditions (dust-free, dry, and optimum placement temperature). Most of the polymer surfacing materials used on OSDs are of the thermosetting type. These include epoxy resins, polyurethanes, and methacrylates. Tensile cracking, loss of surface aggregates, and delamination of the surfacing from the steel deck plate are common types of failures observed in some thin polymeric surfacing systems.

9.2.3. Concrete Surfacing Systems

Concrete overlays and wearing surfaces have been used for bridge decks, including steel grid decks, for a long time, however only in the last decade have concrete and reinforced concrete-wearing surface systems been researched systematically for use on thin OSDs (Cao 1998, De Jong and Kolstein (2004), Walter et al. (2007), Buitelaar and Braam (2008), Kolstein and Slidrecht (2008), Braam et al. (2008), and Kondo, Goto and Iwashita (2008). Several OSDs in the Netherlands were retrofitted with reinforced high performance concrete (RHPC) overlays from 2003 to2008 (Buitelaar and Braam 2008). The Caland Bridge was used as a pilot project for the implementation of the RHPC wearing system in 2003. Several applications of the RHPC wearing surface system have been successfully completed since, including the Moerdijk bridge and the Hagenstein bridges. The RHPC wearing surface system in both these bridges replaced an older 60 mm (2 3/8 inch) mastic asphalt wearing surface. The RHPC wearing surface system typically comprises a 4 mm (5/32 inch) bonding layer of epoxy with embedded aggregates to provide mechanical bonding with the 60 mm (2 3/8 inch) layer of reinforced high performance concrete. A minimum concrete compressive strength of 50 MPa (7.25 ksi) was required before the RHPC wearing surface was opened to traffic on the Hagenstein Bridge. The reinforcement consisted of three layers of reinforcement rolls (8 mm [5/16 inch] bars at 50 mm (2 inch) spacing (Buitelaar and Braam, 2008). Later wearing surfaces used less reinforcement to cut down on reinforcement congestion and mitigate the difficulty in placing a "zero" slump HPC mix. All RHPC wearing surfaces were placed using a temporary hemispherical tent to control the placement and curing environment (moisture and temperature protection).

A 70 mm (2 ¾ inch) thick Steel Fiber Reinforced Concrete (SFRC) wearing surface was used for the Shonan Ohashi Bridge near Tokyo, Japan, in 2005, replacing a 70 mm (2 3/4 inch) asphalt wearing surface. The objective in this case was to provide stiffness to a 12 mm (15/32 inch) deck plate, which had experienced fatigue cracking in the deck plate. Bonding of the wearing surface to the deck was achieved through welded shear studs and a layer of epoxy waterproofing (Kodama et al., 2008). The SFRC also uses a mat of carbon fiber reinforcement (100 mm x 100 mm [4 inches x 4 inches]). The matrix of the wearing surface comprises ultra-rapid hardening cement concrete (with a design strength of 24 MPa [3.5 ksi] in 3 hours).

These recent concrete wearing surfaces, including conventional and hybrid reinforcement, offer good opportunities, particularly for retrofit and upgrade projects where deck stiffening is desired.

9.3. TEMPERATURE-DEPENDENT PROPERTIES OF WEARING SURFACES

The temperature dependence of wearing surface properties, particularly the elastic modulus, plays a significant role in the stress magnitudes in both the wearing surface and deck plate, and hence their fatigue life as well. It is important to understand how different surfacing materials behave at various temperature ranges and at different loading rates in order to engineer an integral surfacing-deck plate composite system (Kolstein and Wardiner, 1997, Wolchuk, 2002).

Figure 9-2 shows typical variations in elastic moduli of surfacing materials as a function of test temperature, as compiled by Wolchuk (2002) from numerous investigations. The surfacing materials in all cases were subjected to flexural tensile stresses. Results are back-calculated for some surfacing materials from tests on composite specimens comprising steel-plate and wearing surfaces. Also, since test geometries, controlled thermal environment, and loading rates (static and dynamic test frequencies of 1 Hz – 30 Hz) are not readily comparable among the different referenced studies, the actual magnitudes of the modulus may not be directly comparable. Despite these differences, observations of the relative performance of the different surfacing materials provide some useful information. Asphalt surfacing exhibits a more significant influence of temperature on the elastic modulus compared to polymer surfacing. The elastic modular ratio, n ($E_{steel}/E_{wearing\ surface}$), of steel to surfacing can vary by more than an order of magnitude in the range of temperatures normally expected on bridge decks. At the hot temperatures, there is little difference in elastic moduli of the different wearing surface materials. At the cold temperatures, asphalt surfacing can contribute significantly to composite stiffness of the deck system, both due to a larger elastic modulus and due to the typically thicker section used for such materials.

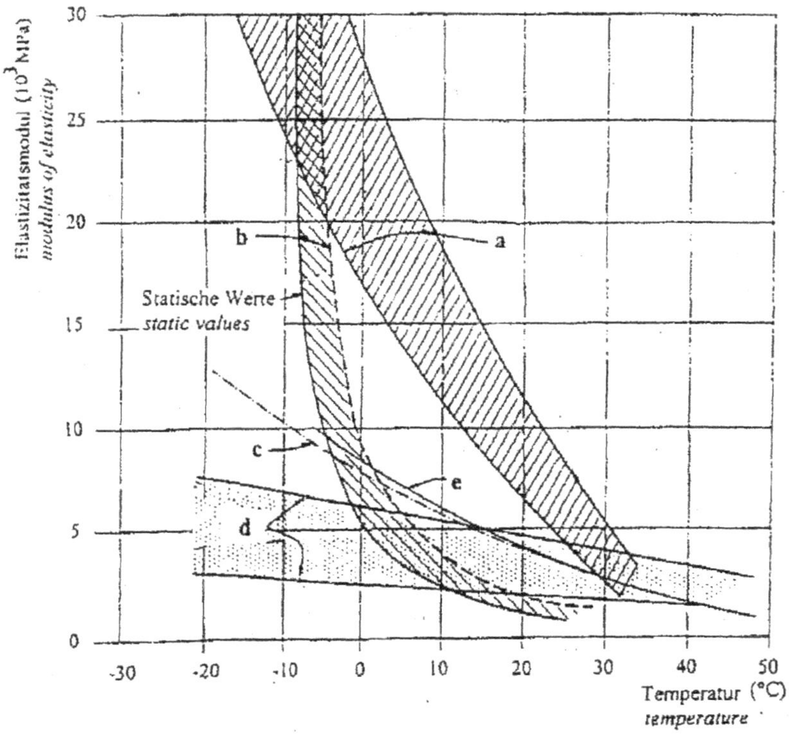

Figure 9-2 Elastic Moduli of Surfacing from Tests of Composite Specimens with Surfacing in Tension (Wolchuk, 2002)

It should be noted that almost all of the surfacing materials typically used as wearing surfaces for OSDs exhibit visco-elastic behavior at room temperature (both viscous, i.e. time-dependent, and elastic, i.e. instantaneous linear deformation on load application and recovery on unloading). At the cold temperatures, their response is largely elastic. Additionally, several of these materials also exhibit modulus values that are different in compression and tension (Cao, 1998). These characteristics (temperature-dependent modulus, visco-elastic behavior exhibiting rate-sensitive modulus, differences in modulus values in compression and tension) have yet to be systematically quantified and understood, make design of wearing surface systems a semi-empirical undertaking at the present time.

9.4. BASIC MECHANICS AND GENERAL BEHAVIOR

OSDs are known for their flexibility relative to other types of commonly-used bridge decks. While the stresses in surfacing material in the longitudinal (along the traffic) direction are globally compressive in nature, tensile stresses manifest in regions that experience negative moments due to bending in the longitudinal direction (e.g. over bridge supports). Tensile stresses in the surfacing, however, are predominant due to local transverse (transverse to the traffic) direction bending. These stresses in the wearing surface are located immediately over the trapezoidal ribs of the OSD in the vicinity of wheel load application. As illustrated in Figure 9-3, they are attributed to a combination of effects, including (a) transverse bending of the wearing surface-deck plate composite with no significant relative displacement between adjacent trapezoidal stiffening ribs, and (b) additional stresses due to relative displacements between trapezoidal ribs and associated shear effects. The first effect is more likely dominant in the vicinity of the transverse FBs, where there is little relative vertical displacement between adjacent legs of the trapezoidal stiffeners (Figure 9-3a). A combination of the two effects is more likely midway between two adjacent FBs (Figure 9-3b). The second effect is more pronounced where stiffness differences between the web of the main girder and the adjacent trapezoidal rib is significant (Figure 9-3c). Cracks in the wearing surface can develop longitudinally immediately over the stiffeners (both main girder web as well as trapezoidal ribs).

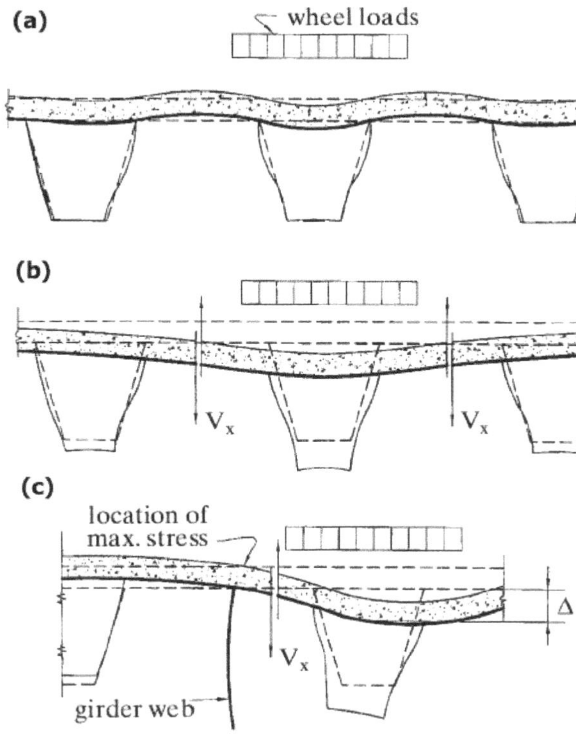

Figure 9-3 Transverse Bending of the Wearing Surface-Deck Plate Composite Showing Flexural and Additional Shear Effects Depending on Relative Vertical Displacement Between Stiffeners. (a) Closer to the Floorbeams, (b) Midway Between Floorbeams, and (c) In the Vicinity of the Main Girder Web (Wolchuk, 2002)

Stress analysis of the wearing surface is complicated due to location-dependent stiffness variations in the deck plate, elastic and thermal mismatch of "wearing surface–deck plate," visco-elastic behavior of the wearing surface, and dynamic/fatigue loading typical in such applications. Despite these complications, it is possible to consider reasonable idealizations for analytical purposes to provide estimates of stress magnitudes for practical design. Semi-empirical approaches to establishing stress magnitudes in the wearing surface have been used (Hartnagel et al., 1991) (Hartnagel, 1993) (Cao, 1998).

Modeling using elaborate 3-D finite element approaches has also been used with custom geometries for the decks under consideration (Hulsey et al., 1999) (Seim and Ingham, (2004). Given the need for numerous idealizations in the material and numerical models and complexities of the various boundary conditions and loading parameters involved, the accuracy of stresses from a more involved 3-D finite element investigation may not normally be warranted. A simplistic mechanics model of composite action in the wearing surface-deck plate system using elastic behavior and a multi-span continuous beam analogy can be very effective in establishing stresses in the wearing surface due to wheel loads, as discussed next. The basic model, popularly termed "finite strip model" is not new or unique to analysis of stresses in wearing surfaces of OSDs, but has been used extensively in analysis of one-way bending in reinforced concrete slabs.

9.4.1. Transverse Bending and Relative Deck Displacements

Analysis of stresses in the surfacing material from transverse bending can be accomplished using a composite continuous beam-analogy model with adequate consideration of deck-plate thickness effect, as well as relative settlement of supports (between trapezoidal ribs or between rib and the main girder web).

This section will address the complexities of stress computations in the surfacing material while incorporating the surfacing-deck-plate composite action due to local transverse bending action in the vicinity of the wheel loads. Additional effect of shear due to relative deformations between the trapezoidal ribs will also be presented and analyzed.

Stresses in the wearing surface due to transverse bending (assuming idealized 'one-way" action, Figure 9-4), can be estimated using the "finite strip" approach, as illustrated in the schematics shown in Figure 9-5. Modeling the wearing surface-deck plate system as composite multi-span continuous beam captures, the localized transverse bending effects reasonably accurately (Rigdon, 1990) (Rigdon et al., 1991) (Gopalaratnam et al., 1993). The number of spans used to compute stresses does not affect the computed stress magnitudes significantly, as long as edge effects from supports too close to the loading patch are eliminated by using between five to seven spans as verified by Rigdon (1990). Shear and moment envelopes can be established for desired loading per AASHTO LRFD Bridge Design Specifications by placing load patches to get maximum positive and negative moments at critical locations (over longitudinal stiffeners), as illustrated in Figure 9-5a and Figure 9-5b. Applicable load factors also need to be applied to account for impact, multiple presence factor, and live load factor.

The supports in Figure 9-5 represent stiffener locations in the OSD. It is recommended that such analysis of stresses in wearing surface also consider cases that include yielding supports to model permissible differential deflection between trapezoidal ribs, per AASHTO LRFD Bridge Design Specifications (Figure 9-5c).

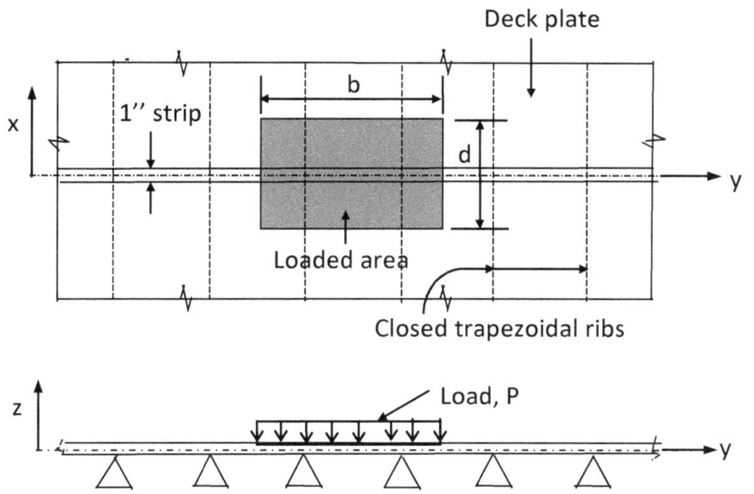

Figure 9-4 Idealized Multi-Span Continuous Beam Model of the Wearing Surface-Deck Plate Composite Used to Analyze Stresses in the Wearing Surface Due to Localized Transverse Bending (Gopalaratnam Et Al., 1993)

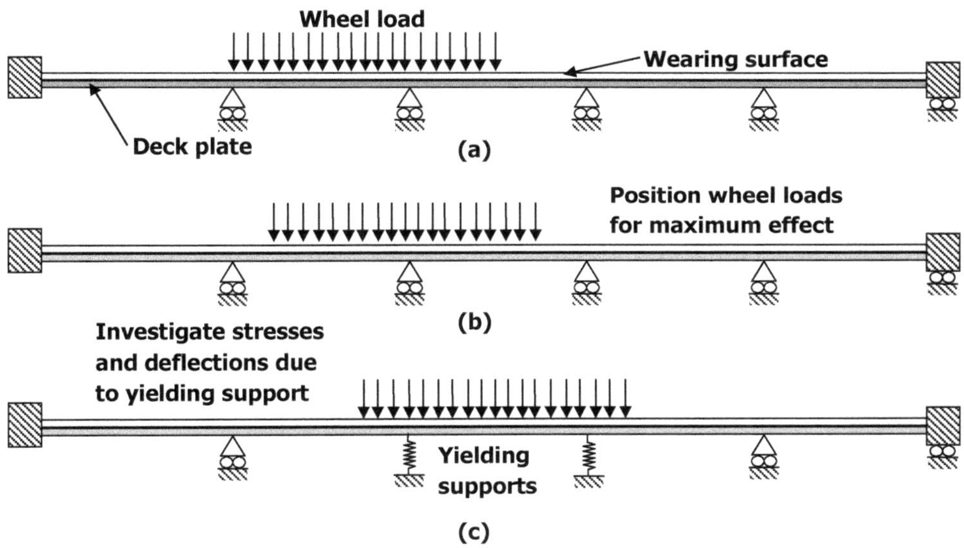

Figure 9-5 Finite Strip Models for Computing Stresses in the Wearing Surface with Different Loading Conditions to Generate Maximum Moment and Shear Force Envelopes: (a) Maximum Positive Moment Midway between Stiffeners; (b) Maximum Negative Moment at Stiffeners; (c) Stresses in the Wearing Surface Due to Combination of Loads and Yielding of Supports (Differential Movement between Stiffeners)

9.4.2. Surfacing Stresses and the Effects of Deck Plate and Wearing surfaces Thickness

Once shear force and bending moments envelopes are established based on the prescribed loading using finite strip analysis, as described earlier, it is possible to compute:

- shear stresses: at the interface between the wearing surface system; the steel deck plate;
- tensile stresses: in the wearing surface using elastic analysis of the wearing surface; steel deck plate composite (using transformed section analysis).

These stresses, it should be noted, account only for static elastic behavior at the prescribed temperature. These stress magnitudes will depend on temperature, as described earlier, and location-specific temperature extremes are to be used to investigate expected stress ranges due to daily and seasonal temperature excursions.

In addition to static wheel loads and composite action with the deck plate, thermal loads, and elastic mismatch between the deck plate and the surfacing material, one also needs to use engineering judgment to allow for fatigue effects that should be incorporated in the design of wearing surface systems. Presently this can only be accomplished in a semi-empirical manner by conducting flexural fatigue tests of composite specimens, comprising custom wearing surface systems applied to steel plate of the same thickness as the deck plate. Results from such flexural fatigue tests conducted under prescribed temperature conditions, together with service traffic data (average daily traffic, percent truck traffic, associated load information) allow one to establish approximate fatigue life of the wearing surface. Additionally, engineering judgment should be applied considering the effects of ageing on the material may negatively affect the life.

Even while it is not easy to estimate fatigue life of a wearing surface system for OSDs, the importance of this characteristic of a wearing surface system cannot be overemphasized. Typically, if it is estimated that trucks will produce about half a million heavy-wheel loads a year, or 10 to 15 million fatigue cycles within the surfacing service life of 20 to 30 years. This is the fatigue and structural demand on the surfacing. The bridge designer or owner should select a target service life for the surfacing, just as a target service life of the bridge is selected at the start of the bridge design process. The next step is to make bridge-specific calculations of the demands that traffic will impose on the bridge, using the best estimate of future traffic to find the number of fatigue cycles the wearing surfacing will experience during the selected service life of the surfacing. This is the fatigue demand that the surfacing must resist.

When a separate waterproofing membrane or bonding layer is used on the steel deck plate, it has been observed that composite interaction between the wearing surface and the steel deck plate is significantly reduced. This is largely because the bonding layer is relatively compliant due to use of a low modulus material or a thick layer, or both. In such cases, one can expect the wearing surface to be subjected to smaller strains compared to when no bonding layer (or a stiff bonding layer) is used. Zinc primer or coatings of polymeric resins have often been used to protect the deck plate from corrosion before the wearing surface is placed on the deck. In addition to potential protection against corrosion of the deck plate, the advantage of using such a layer includes reduced tensile cracking in the wearing surface. The obvious disadvantage is a less stiff composite system where the wearing surface and the deck plate resist bending independently. In extreme cases, this soft layer can also lead to potential delamination of the wearing surface from the steel deck. Figure 9-6 illustrates experimentally-measured strains along the depth of the wearing surface, bonding layer, and steel deck plate highlighting the nonlinear strain distribution and "shear lag" effect. It is possible to compute stress distributions in the composite system using simple mechanics, knowing the elastic and thermal properties of the deck plate, the bonding layer, and the wearing surface based on the applied bending moment.

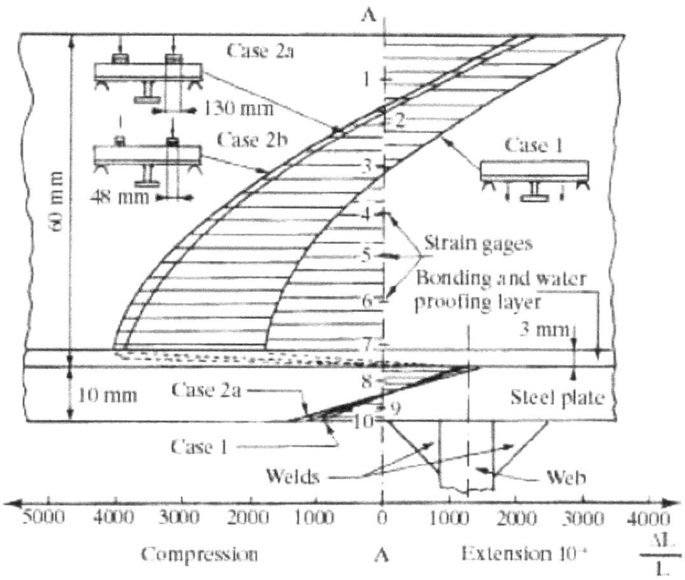

Figure 9-6 Experimentally Measured Strains along the Depth of the Wearing Surface in Thick Mastic Asphalt for Different Loading Cases Exhibiting Nonlinear Strain Distribution (Hemeau Et Al., 1981)

The influence of wearing surface stiffness on the stresses in the deck plate has been well documented both experimentally (Hauge, 1998) and analytically (Wolchuk, 2002) (Gopalaratnam, 2009). Figure 9-7 shows analytically-computed variations in steel deck plate stress and stress in the wearing surface as a function of the modular ratio ($n = E_{steel}/E_{wearing\ surface}$) for thin (10 mm [3/8 inch]) and thick (50 mm [2 inch]) wearing surfaces. Figure 9-7a is for a steel deck plate of 14 mm (9/16 inch) thickness (based on the plot in Wolchuk, 2002), whereas Figure 9-7b is for a steel deck plate of 20 mm (25/32 inch) thickness. Both plots are made for an assumed negative moment (causing tension in the wearing surface) magnitude of 4,000 N-mm/mm and are plotted to the same scale. In each case, wearing surface stresses are plotted on the left axis while the steel deck plate stresses are plotted on the right axis. The elastic modulus of the wearing surface is temperature dependent, as observed earlier. At hot temperatures, the wearing surface contributes very little to the stiffness of the composite (steel stress of 122 MPa (17.7 ksi), approximately comparable to that without any wearing surface for the 14 mm (9/16 inch) deck plate and of 60 MPa (8.7 ksi) for the 20 mm (25/32 inch) deck plate), whereas at cold temperatures, the contribution of the wearing surface to the composite stiffness is significant enough to influence the stresses in the deck plate. This influence, as observed from Figure 9-7, is more significant for the thick wearing surface (l_{ws} = 50 mm [2 inches]) than for the thin wearing surface (l_{ws} = 10 mm [3/8 inches]). The reduced steel stress is likely to result in improved fatigue life of the deck plate. Figure 9-7 also highlights the effect of cold temperature on stresses in the wearing surface. The larger stresses expected in the wearing surface at the cold temperatures are likely to increase the potential for tensile cracking. This is particularly significant for the thin surfacing. To mitigate this potential for cracking at cold temperatures, use of low-modulus surfacing is recommended. A multi-layer build-up of polymer surfacing discussed earlier or use of less aggregates and more binder in the slurry type systems will result in low-modulus thin wearing surfaces.

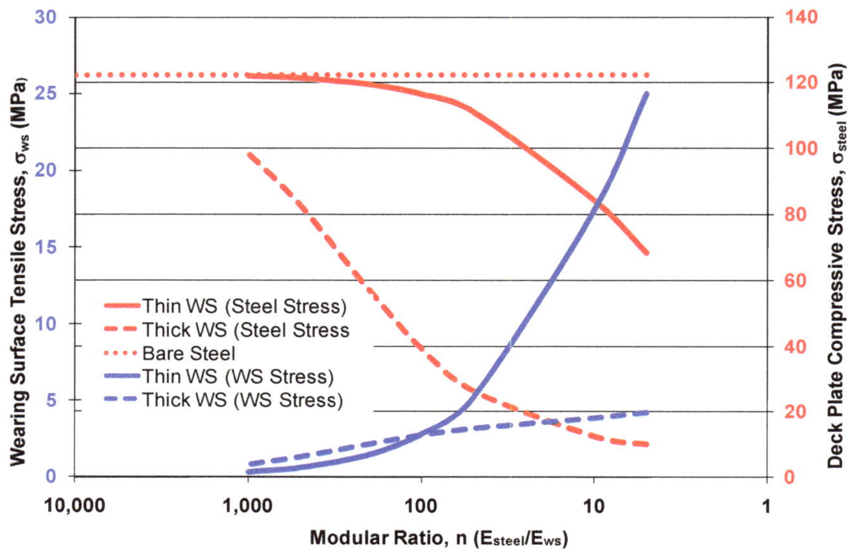

(a)

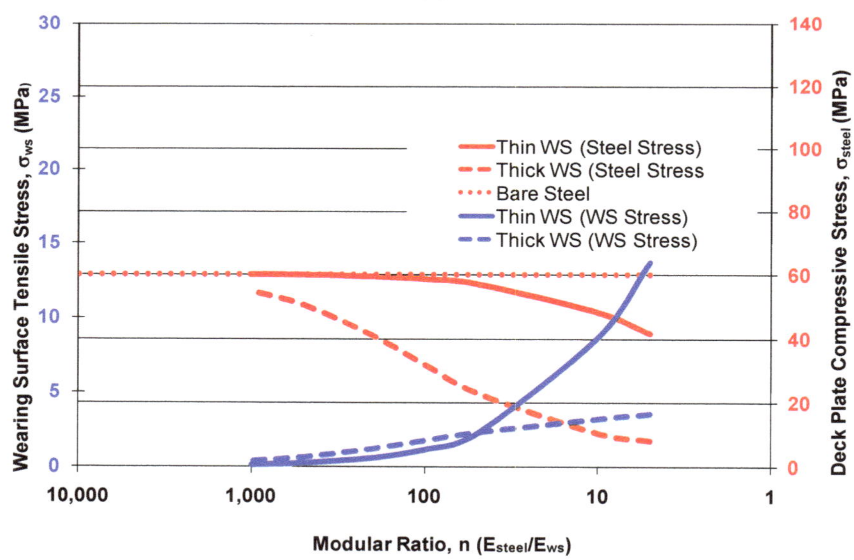

(b)

Figure 9-7 Stresses in Thick and Thin Wearing Surfaces due to Variations in the Elastic Modular Ratio, n, for (a) Thin Deck Plate (14 Mm [9/16 Inches]), and (b) Thick Deck Plate (20 Mm [25/32 Inches])

A comparison of steel and wearing surface stress magnitudes for the 14 mm (9/16 inch) thick deck plate and 20 mm (25/32 inch) thick deck plate clearly shows the benefits of using thicker deck plates. When life-cycle costs are considered, it is likely that the benefit of longer fatigue lives for both the deck plate as well as the wearing surface will outweigh the increased initial investment. It is also interesting to note that when modular ratios are higher than 50, wearing surface stresses are not very sensitive to either the wearing surface thickness or deck plate thickness. Summer deck temperature extremes are likely to result in such modular ratios.

9.4.3. Limit States in the Different Types of Surfacing Systems

The limit states (failure modes) observed in surfacing systems depend on the type and thickness of the wearing surface. Even while generalizations may not be always be applicable in all cases, generally thin polymeric concrete wearing surfaces often fail due to tensile cracking, delamination due to local shear at the interface with the deck plate, or debonding of large chunks of the surfacing material. Generally, thick asphalt wearing surfaces often fail due to shoving and rutting resulting from inelastic deformations from braking/lane-changing loads on the surface. Tensile cracking has also been observed in thick surfacing.

9.5. DESIGN AND DETAILING

The design of wearing surface systems to-date has largely been semi-empirical and based in part on field and laboratory testing of wearing surface systems, testing of wearing surface-deck-plate composite specimens, and idealized analysis of plate bending. Composite action, bi-axial fatigue loading, influence of thermal and elastic mismatch between the surfacing and the deck plate, and visco-elastic behavior of the surfacing material complicate the analysis of stresses in the wearing surface. Despite these complexities, it is possible to approximate ranges of stresses in wearing surface systems to facilitate design that withstands specified service loads and estimate approximate fatigue life. **Experience indicates that wearingurfaces have performed well when thicker (5/8"+) deck plates are used.**

9.6. TESTING OF WEARING SURFACES

Engineering design of wearing surfaces to date has relied very heavily on customized testing. This has been necessary because, as noted earlier, the performance of wearing surfaces depends on thickness of the steel deck plate, type and thickness of wearing surface, typical temperature ranges likely to be experienced by the wearing surface, steel plate composite, temperature-dependent elastic modulus of the wearing surface, and the location-specific fatigue loading characteristics of the bridge deck (a function of strain magnitudes from specified wheel loads and local traffic pattern, which is a function of volume of truck traffic compared to total traffic). These parameters influence the overall performance of the wearing surface steel-plate composite system in an inelastic and coupled manner making simplified generalized analysis nearly impossible. Despite the complexities in allowing generalizations, wearing surface testing can still be effectively used to (1) evaluate the relative performance of different wearing surface systems subjected to a prescribed set of test parameters for a particular orthotropic bridge, (2) accept minimum specifications of a wearing surface system in simplified fundamental tests to ensure proper design performance, and (3) monitor service performance allowing timely maintenance with a view to improve ride quality and enhance service life. Three of the commonly used tests described below have been employed for one or more of the above purposes. In addition, full-scale laboratory testing of composite bridge deck panels (Hague, Denmark) and experimental wearing surface test sections on OSDs subjected to service loads and temperatures (Gopalaratnam et al.) have also been successfully employed to evaluate the performance of different wearing surface systems. While practical, cost considerations may preclude such full-scale testing in many projects. When possible, they are likely to simulate the most realistic testing conditions to evaluate performance of wearing surface systems.

9.6.1. Flexural Testing

Small-scale laboratory flexural testing has often been used to simulate tensile stresses in the wearing surface over trapezoidal rib stiffeners and over the web of the main structural girder. This test can also be designed to reproduce the interfacial bond stress between the wearing surface and the steel deck plate. Figure 9-8 shows a typical loading geometry used for evaluation of different types of wearing surfaces using a composite specimen. Steel plate thickness to be used for fabricating the composite specimen must be identical to that used for the bridge deck. Steel tabs welded to the plate at its two ends simulate the stiffening provided by the trapezoidal ribs in the bridge deck. These tabs are also used to support the composite specimen, as shown in Figure 9-8, using four rollers, two on each side of the specimen. The loading span used is to be identical to rib spacing on the bridge deck (a 13 inch span used in the illustration represents 13 inch trapezoidal rib spacing on the Poplar Street Bridge, Gopalaratnam et al., 1993). The wearing surface is to be placed on the specimen steel plate in a manner identical to that expected to be used on the bridge deck. The thickness and composition of the wearing surface should be identical to that planned for application on the bridge. If corrosion protection coating or waterproofing membranes are to be used on the deck, then similar coatings/membranes should also be used while fabricating the laboratory specimens. A nominal width of the specimen of 4 inch is used here to allow for representative wearing surface characteristics to be captured. This dimension is a compromise between observing edge effects from using too small a width, and requiring significantly larger loads while using larger widths.

The test configuration in Figure 9-8 allows convenient inspection of the wearing surface for cracking and delaminations. Parameters typically monitored include load applied, specimen deflection at midspan (on both sides along the width), end-slip between wearing surface and the steel plate at either end along the specimen length, and electrical continuity of the wire glued on the top of the wearing surface. Load-deflection response in static tests or specimen stiffness and its degradation in fatigue tests (computed as ratio of load range versus deflection range) are typically recorded in real-time during the test. Cracking can be monitored either through a drop in the composite stiffness or via electrical continuity of thin wires glued on the top of the wearing surface. Delamination can be monitored using transducers to measure relative slip between the wearing surface and the steel plate (Gopalaratnam et al., 1993).

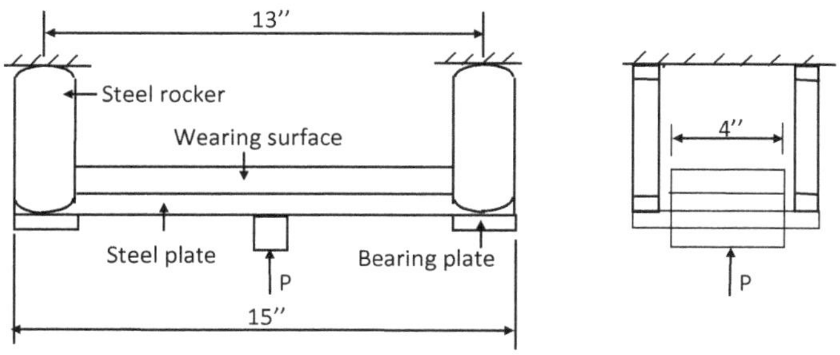

Figure 9-8 Recommended Laboratory Test Configuration for the Wearing-Surface – Steel Plate Composite Specimens Subjected Simultaneously to Prescribed Temperature History and Flexural Fatigue Loading (Gopalaratnam et al., 1993)

Loading can be static (linear ramp loading) or fatigue with specified upper and lower limit loads (sinusoidal loading at a prescribed frequency) depending on the specific test requirements. Often it is convenient to use servo-controlled electro-hydraulic test equipment for flexural testing of wearing surface steel-plate composite specimens. However, it may also be possible to adapt other conventional machines (non-servo-controlled or electro-mechanical testing machines) to conduct such tests. Static tests can be used to calculate static elastic modulus of the wearing surface as a function of test temperature from the stiffness characteristics of the composite flexural test. Results from such tests can also be used to determine cracking strain or debonding strains as a function of test temperature. Fatigue tests can be used to measure dynamic modulus of the wearing surface as a function of test temperature. These tests can also be used to establish overall fatigue performance and to estimate service life under prescribed test conditions.

It is convenient to test multiple composite specimens in parallel at one time using displacement control (to simulate equivalent load control, by using a stiff distributor beam and soft load cells to measure individual specimen loads) if replication is desired to facilitate testing of a statistically significant number of specimens for the fatigue test program (see Figure 9-9, Rigdon et al., 1991). Simulating Load control while displacement control is used can be ensured by using soft load cells in series with each specimen. This allows cracking of one specimen to not influence loads on the other specimens.

While testing one specimen at a time, load control is often used to achieve desired levels of wearing surface stress in either static or fatigue tests.

Tests can be conducted in a constant temperature environment (room temperature, or prescribed extremes of cold or hot temperatures) or in temperature-varying environments to simulate ranges of temperatures that the bridge deck is normally likely to experience while in service. Temperature-controlled tests are conducted by using insulated chambers enclosing the test specimens (Figure 9-9) along with appropriate control programs to heat and/or cool the test chamber.

Figure 9-9 Simultaneous Testing of Multiple Replicate Specimens to Simulate Load (Stress) Control (Rigdon Et Al., 1990)

Figure 9-10 shows results from a constant temperature test (cold temperature of -18°C [-0°F]) where the relative specimen stiffness (ratio of current stiffness to the stiffness of the virgin specimen) is plotted versus number of fatigue cycles (at 5 Hz) for a composite polymer concrete-steel-plate composite specimen (graph on the left). The upper limit load in this test was increased by 150 lbs (0.67 kN) after every 70,000 fatigue cycles until cracking was observed in the wearing surface. The sudden drop in relative stiffness during this cold-temperature fatigue test can be readily observed, highlighting the critical combination of upper limit fatigue load and test temperature that is likely to cause cracking in the wearing surface. Typically, specimens are not aged since it is practically unrealistic given the numerous parameters that need to be studied (temperature fatigue, freeze-thaw, etc.) in tests of the wearing surfaces.

Results from an alternate test are shown in the right graph of Figure 9-10 where the upper limit load (1,450 lbs [6.45 kN]) and lower limit load (150 lbs [0.67 kN]– enough to ensure no slack on unloading) are held constant during the fatigue test at 5 Hz. The test chamber temperature is varied to simulate real-time variations in summer and winter deck temperatures. Again, the plot shows variation of relative stiffness (ratio of current stiffness to initial stiffness) with number of fatigue cycles. First detection of cracking occurs in this test when the specimen is subjected to the prescribed fatigue loading and the coldest temperature (-18°C [0°F]) of the test.

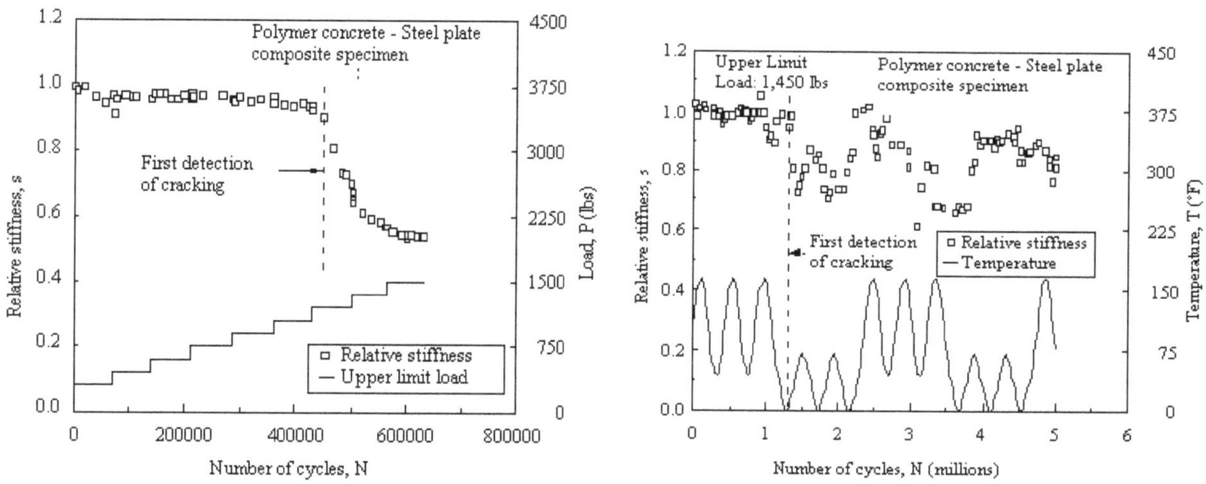

Figure 9-10 Cold Temperature Fatigue Tests on Composite Specimens with Incrementally Increasing Upper Limit Load (Left) and Temperature Varying Fatigue Test at Fixed Upper and Lower Limit Loads (Gopalaratnam Et Al., 1993)

9.6.2. Tensile Bond Testing

Bond test using a tensile pull-out loading configuration (Figure 9-11) has been used for acceptance testing of polymer concrete wearing surfaces on OSDs (Gopalaratnam et al. 1993b). It is used more as a quality control test to ensure good surface preparation of the steel deck and proper placement of the wearing surface to assure minimum specified tensile bond strength. The test originally developed for polymer concrete overlays on concrete decks has been adapted for use on steel decks, generally following the guidelines described in Appendix A of ACI 503R-93 (ACI, 2008).

A 50 mm (2 inch) core (nominal internal diameter of the core drill) is to be drilled at the desired test location through the thickness of the wearing surface (until the surface of the steel deck plate is reached, see Figure 9-11. Average wearing surface thickness at the core location is to be measured after the surface of the core is cleaned for gluing the pipe cap. After adequate curing of the glue (typically between 1 1/2 and 2 hours) the core is subjected to direct tensile pull-out loads (no twisting or torsional loading allowed) using a loading frame specially designed to be in compliance with ACI 503R-93. A manually-controlled tensile loading rate of approximately 89 N/s (20 lb/s) is recommended in ACI 503R. The loading frame can be equipped with a strain-gage based load cell or other spring-based mechanical load cells to measure peak tensile pull-out load. Peak pull-out load is used to compute average bond strength of the wearing surface-steel plate interface, assuming adhesive failure at this interface (failure type 5 in Figure 9-11). In case the failure is due to (1) adhesive failure at the pipe cap, (2) cohesive failure in the glue, (3) adhesive failure at the wearing surface, or (4) cohesive failure in the wearing surface, the test can only be used to establish the minimum tensile bond strength of the wearing surface steel plate interface. Details of the failure type and bond strength are normally reported. Acceptance specification can require minimum tensile bond strength. Bond tests may be required at a select number of locations for a prescribed area of wearing surface placed.

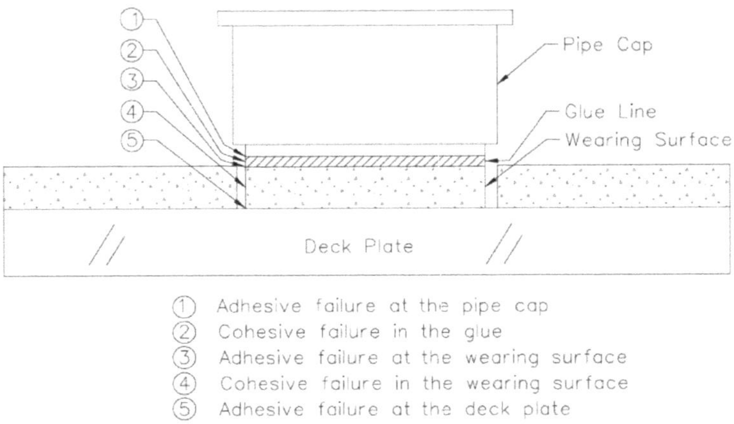

Figure 9-11 Bond Test of the Wearing Surface Deck Plate Interface in Direct Tension (Gopalaratnam Et Al., 1993)

9.6.3. Resistivity Testing

Resistivity tests on the wearing surface-steel plate composites are performed in accordance with ASTM D3633-98 (Reapproved 2006). These tests are used to determine the severity of wearing surface cracking (Figure 9-12). The test can be adapted for use in small laboratory flexural test specimens, while at the same time serving as a good field test of service performance of the actual wearing surface. This type of test can detect incipient cracks that are not visible to the naked eye.

In small laboratory specimens (like the specimen used for the flexural test in Figure 9-8) the test consists of constructing a dam on top of the wearing surface with non-conducting material, such as PVC sheet or plywood along the rectangular edges. Silicone rubber caulking is used around the edges to prevent water from reaching the steel plate. Once the caulk has set up, a soap and water mixture is placed on the surface of the specimen. Soap is added to minimize the surface tension, thus allowing the water to penetrate fine cracks (often invisible to the naked eye), which might be present in the specimen. A thin copper plate is attached to a sponge, which is allowed to soak in the soap water mixture. This "probe" is attached to one lead of an ohm meter; the other lead is attached to the steel plate of the composite specimen. When a reading is desired, the probe is taken from the soap water and placed on top of the specimen and the ohm meter is read immediately. Readings are typically taken at 1 minute, 10 minutes, 30 minutes, 1 hour, and 2 hours after the specimen had been saturated with the soap water mixture. With time, the measured resistance usually decreases. Results from this test provide a qualitative estimate of the severity of cracking in the wearing surface material. Low resistance (less than 10,000 ohms) indicates the wearing surface had a crack through the thickness of the wearing surface. A minimum resistance of 750,000 ohms is specified as the lowest acceptable value for this resistance. The resistance represents a crack-free wearing surface. The same test has been used successfully to detect cracking in the polymer concrete wearing surface on the Poplar Street Bridge (Gopalaratnam et al., 1993-1995).

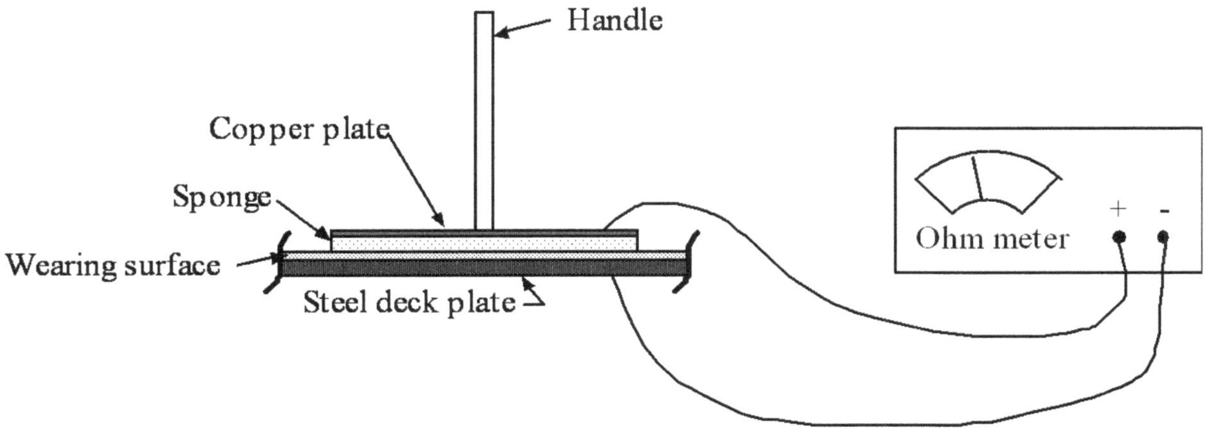

Figure 9-12 Resistivity Test Used on the Bridge Deck to Monitor Potential Cracking in the Wearing Surface (Gopalaratnam Et Al., 1993)

9.7. CONSTRUCTION

Different construction techniques have been used for placement of wearing surfaces on OSDs. These techniques depend on the type of wearing surface materials used. Popular types of wearing surface materials, as described earlier in 9.1, include bituminous, polymeric and cement-based concretes.

Regardless of the types of wearing surface materials used, it is important that the steel deck plate be cleaned and shot-blasted to a SSPC SP 10/NACE 2 (standards for protective coatings and corrosion mitigation, Steel Structures Painting Council, now called Society for Protective Coatings, 2007, National Association of Corrosion Engineers) near-white metal finish and free of all visible oil, grease, dirt, dust, mill scale, rust, paint, oxides, corrosion products, and other foreign matter. Often a zinc-based primer coat (1 mm to 1.25 mm in thickness) is put down immediately following the shot-blasting operation to protect the steel deck from corrosion. Beyond these basic deck treatments, there are different approaches used to put down different wearing surface materials. Details of the various layers placed in some of the more commonly used wearing surface systems are illustrated in Figure 9-1

Bituminous

The construction sequence described here for bituminous materials pertains to an epoxy asphalt wearing surface, commonly used on many OSD bridges in California. The placement techniques for other types of asphalt materials (Mastic asphalt, Gussasphalt, and other similar systems, popularly used in Europe) may differ from the process described here. Typically a bond coat, approximately 0.68 mm thick, of neat epoxy asphalt (without aggregates or fillers) is sprayed on top of the inorganic zinc primer coat. This is followed by the placement of a 25 mm leveling course of epoxy asphalt concrete, using conventional asphalt paving machines. A combination of pneumatic tire and heavy steel rollers provide the desired compaction of the leveling course. An application of a 0.45 mm thick neat epoxy asphalt bond coat precedes the placement of a 25 mm wearing course of epoxy asphalt concrete (Figure 9-1b). The wearing course is again compacted using pneumatic tire and heavy steel rollers.

Polymer

Polymer concrete wearing surfaces are placed in one of two methods that are distinctly different and often dictated by the manufacturers of such systems: the multi-coat overlay method (also sometimes called broom-and-seed method); and the slurry method.

The multi-coat overlay method is a buildup of polymer resin and wearing coarse aggregate, with the specified number of coats required to obtain the specified overlay thickness, usually 10 mm to 12 mm (3/8 inch to 15/32 inch). This method of application has been successfully used for thin overlays on concrete bridge decks for over 20 years. The same method has been used in several recent OSDs. Typically the polymer resin is applied using rollers/squeegees (for calibrated thicknesses) at prescribed application rates. Coarse aggregates are broadcast after the resin coat is partially cured so the aggregates can embed in the resin without complete penetration of the resin layer. Controlled dosage rates of applied aggregate ensure complete coverage of the resin layer with dry aggregates surface. After the resin is cured, excess loose aggregates are removed from surface. A new layer of polymer resin is applied followed by broadcast of additional coarse aggregates. This process is repeated several times until the desired wearing surface thickness is achieved (typically 3 layers are used to build up the 10 mm to 12 mm [3/8 inch to 15/32 inch] wearing surface system). The surface is often sealed with a polymeric resin before the surface is opened to traffic. The placement of the multi-coat overlay polymer concrete wearing surface can be manual or automated as desired.

The slurry overlay method is placed in a three-step operation. First, a neat layer of polymer resin is placed on the steel base plate to serve as the tack coat (applied at a rate of 35 ft2/gallon, providing approximately a 1 mm [0.04 inch] thick layer). This is followed by placement of a 10 mm (3/8 inch) thick layer of polymer concrete slurry. In addition to the neat polymer as used for the tack coat, the polymer slurry contains fine silica sand filler (manufacturer recommended mix proportions used for the slurry). Coarse aggregates for the wearing surface are broadcast at the rate of 1-3 lbs/ft2, after the slurry layer was cured enough to support the aggregate weight, while at the same time being able to embed them. Excess aggregates are brushed off the surface after the polymer concrete cures adequately. The surface is sealed with a polymeric resin before it is opened to traffic. The placement of slurry polymer concrete wearing surface can be manual or automated as desired.

Concrete

Reinforced concrete and fiber reinforced concrete wearing surface have been successfully placed by combining several conventional methods of construction, both in laboratory (Cao, 1998) as well as full-scale field placements (Buitelaar and Braam, 2008 and Kodama et al., 2008). These include use of slipform pavers following placement of mesh made from steel reinforcing bars (Buitelaar and Braam, 2008). Since the thickness of the reinforced concrete overlay used is 65 mm (2 ½ inch), clear cover for the reinforcing bars is significantly smaller (6 mm [1/4 inch]) than is conventionally used. Deck irregularities are also likely to add to difficulties in placement of the wearing surface with tight tolerances. Shear transfer and bonding of the concrete matrix wearing surfaces to the steel deck plate have been accomplished by one of two methods. A bond coat of epoxy material in which angular aggregates are broadcast (prior to full cure) provides mechanical shear transfer to steel fiber reinforced concrete wearing surface, which is placed on top of the bond coat. Alternately, Kodama et al., 2008, also reports using welded shear studs (40 mm [1 ½ inch] tall) to provide shear transfer for an 80 mm (3 1/8 inch) steel fiber reinforced wearing surface that also includes a CFRP (carbon fiber reinforced polymer) reinforcing mesh at the mid-surface. Placement of the SFRC wearing surface was accomplished using conventional concrete pavers. Conventional floating of the concrete surface and providing a broom finish allows good traction for vehicles.

For many new bridges or re-decking projects involving OSDs, prefabricated steel panels already incorporating the wearing surface have been used. The wearing surface in these cases allow for a partial placement under quality controlled shop conditions, while allowing cast-in-place final layers to improve ride quality, as well as facilitate field welding of the prefabricated panels.

9.8. MAINTENANCE AND REPAIR TECHNIQUES

There is little systematic published information available on the maintenance and repair techniques for cracked, delaminated, or spalled wearing surfaces. Information on the effectiveness and service lives of such repair techniques are also not available. Based on the limited information available, it appears that there are primarily five repair techniques that have been used with differing levels of effectiveness and durability.

1. Cracks over stiffeners have been sealed using a low viscosity sealer that is either bituminous (cold pouring type asphalt materials) or polymeric (methacrylates). The sealing mainly provides protection against water infiltration and corrosion of the deck plate (and is not intended to heal the cracked surface). The service life of such sealing will depend on the crack width and the viscosity of the sealants used. It is important that the wearing surface be clean (and preferably dry, although some sealers can also be applied when the surface is wet) and free of debris. The best sealing is achieved when the crack widths are at the largest (for example, when the deck temperatures are at their minimum), as this facilitates good penetration depth.
2. Small delaminations, if detected using an acoustic hammer, can be temporarily repaired by injecting polymers/epoxies under pressure. To verify that the entire void under the wearing surface is filled, it may be necessary to identify the total area of delamination using conventional techniques (such as acoustic hammer or chain drag) and use several outlet vents to drive out the air trapped in the delamination.
3. Larger delaminations or spall areas are repaired by sawing out the damaged wearing surface and replacing the patch with the repair materials (often the original wearing material if practical or alternate polymer/epoxy concretes that may facilitate small scale repairs without using heavy equipment).
4. Placement of overlays or layered buildup of overlays over the damaged wearing surface. This repair technique requires that good bond be provided to the damaged overlay and cracks be sealed adequately. Reflective cracking may result if placement of the repair material is undertaken without sealing existing cracks.
5. Repair techniques involving full-depth patch replacement (described in 3 above) or cast-in-place overlay (described in 4 above) can also use a precast plate of the wearing surface. Often. since it is difficult to match the surface profile of an irregular deck (for full-depth replacement), or damaged wearing surface (for overlays), the bonding of precast plates is

primarily suitable for rapid temporary repairs so traffic can safely use the bridge until more permanent repairs are completed.

Limited laboratory studies of the effectiveness of two of the above repair techniques using post repair flexural fatigue tests have been reported by Cao (1998). Figure 9-13 shows the flexural fatigue specimen used for studying the effectiveness of the direct crack sealing technique. Figure 9-14 illustrates use of a precast bonded overlay made from a special type of steel fiber reinforced concrete called SIMCON, a cement slurry infiltrated steel fiber mat concrete. While both techniques provided good laboratory performance, field trials with real traffic, direct tire pressure loading, and simultaneous exposure to adverse environmental conditions are probably necessary to provide more objective measures of effectiveness and durability of these repair techniques.

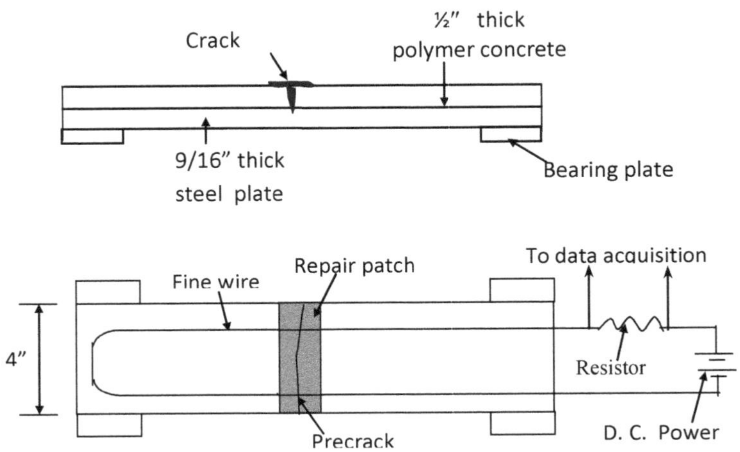

Figure 9-13 Side Elevation (Top) and Plan View (Bottom) of Fatigue Testing of Repair Techniques Used on Specimens with Pre-cracked Wearing Surface (Cao, 1998)

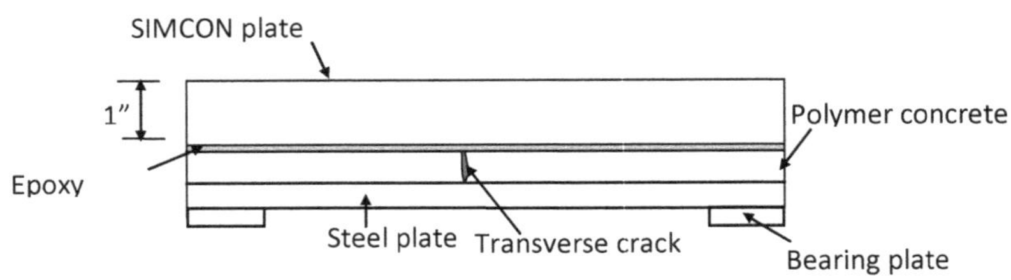

Figure 9-14 Bonding of a Rigid Very Ductile Overlay (Slurry Infiltrated Steel Fiber Mat Concrete – SIMCON) as an Alternate Rapid Repair Technique (Cao, 1998)

10. TESTING

10.1. EXPERIMENTAL TESTING OF DECKS

The experimental testing of bridge components has traditionally been the business of researchers and not practicing engineers. The usual progression is that experimental research leads to design rules that are incorporated into the governing design specification, which is then applied by the engineering practitioner. However, in the development of orthotropic bridge designs, simple comprehensive design rules cannot be developed due to the level of complexity in the behavior. In this case, testing can be used as an integral part of the design verification process, with application by the engineer. That is, test data can be used in place of conventional engineering analysis by structural modeling. This chapter covers the application of experimental testing in the development and verification of orthotropic bridge design. This should not be confused with production testing employed during the construction phase to ensure quality in materials or workmanship. These other types of testing are covered in Chapter 7.

As described in Chapter 5, the approach to design of the orthotropic bridge can vary depending on the level of experimental test data available to the designer. This may include tests conducted as part of ongoing work or by application of trusted test data found in the literature. When appropriate laboratory tests have been conducted for previous projects on specimens similar in design and details to those proposed for a new project, the previous tests may, at the engineer's discretion, be used as the basis for design on the current project per Level 1 design. Testing may include fatigue resistance tests on steel details or performance testing on the wearing surface systems.

The precedent for such an approach to design is found in the PTI Recommendations for Stay Cable Design, Testing, and Installation (PTI, 2000) for acceptance of stay cable systems. This is considered a rational approach since stay cable details are highly repetitive, conventional structural analysis has limited applicability, and this eliminates the significant expense and time devoted to new acceptance tests for every project.

One potential problem with application of test data in design lies in the delay of project development. Tests can be time consuming, and if a problem is discovered, this may delay the project further. Another potential problem is in the interpretation and application of existing test data found in literature. The design engineer must be confident that the test fixture and loading are representative of the conditions that will be in place for the current application (or at least more severe). This is often difficult to ascertain when sufficient details on the test protocols are not divulged. The ideal solution to this dilemma would be for a nationally funded research program or a commercial fabricator to validate a standard steel panel design through testing, and this could be considered as a safe design under certain limitations. Wearing surface systems will likely always remain as proprietary items with performance specifications.

Although performance tests for components and materials used in highway bridges have been available for many years, there are no standardized acceptance testing procedures available that focus on bridge decks and certainly none specifically for orthotropic steel decks (OSD). Consequently, currently neither the AASHTO Standard Specification nor the LRFD Bridge

Design Specifications provide guidelines for deck testing. The only known formal specification for testing of bridge decks was ASTM Specification D6275-98, which provided some guidance for grid decks, although much of the information could be used for other systems. Nevertheless, this specification was subsequently withdrawn by ASTM and not replaced. Interestingly, that specification did not fully address the Fatigue limit state, which typically controls metal deck systems. At the time of this writing (2011), NCHRP Project 10-72 "Bridge Deck Design Criteria and Testing Procedures" is in its second phase, and the results are not currently available.

The primary objectives of NCHRP Project 10-72 are: 1) Determine critical performance factors that affect the durability, strength, and design of deck systems; 2) Develop testing protocols that can be used to objectively evaluate critical performance factors for deck systems; and 3) Develop rational design criteria for various deck systems to address durability, strength, fatigue, deflection, and other performance factors. Once completed, the findings of the research are to be reduced into recommended revisions to the existing AASHTO LRFD Bridge Design Specifications for Highway Bridges. The research is intended to be applicable to all types of deck systems.

Possibly one of the most important factors to consider when developing testing protocols for acceptance testing is that all deck systems should demonstrate a uniform level of design performance, regardless of the material type or deck system. It has become clear that various types of bridge components (e.g. deck systems verses expansion joints) are often held to different standards of design, testing, and performance. Hence, the tests must accurately represent in-situ conditions and the performance criteria must be developed with careful consideration. This includes accurate simulation of specimen size/scale, boundary conditions, load magnitude and position, dynamic impact, and numbers of cycles. Obviously, some of these factors are easier to simulate than others. The following sections provide suggested guidance on how to include these factors in performance testing.

Lastly, when one considers the cost of fabrication, installation, and user costs associated with the construction of a new bridge or the redecking of a major bridge with an OSD, the costs for performance testing will likely amount to less than 1 percent of the total cost of the bridge. Hence, assuming the testing results in improved long-term performance, the benefits will generally far outweigh the initial costs of the testing.

10.2. FATIGUE TESTING OF ORTHOTROPIC STEEL CONNECTION DETAILS

A comprehensive summary of the fatigue resistance testing on orthotropic steel details conducted across the world can be found in the work by Kolstein (2007). Tests have included small specimens of rib/deck to test the rib-to-deck weld (RD), rib/FB sub-assemblies for testing out-of-plane bending, and FBs with sections of ribs for in-plane bending tests. Figure 10 1 (a) and (b) show two tests conducted by Kolstein.

(a) (b)

Figure 10-1 Photos of Fatigue Test Setups Conducted by Kolstein (a) Out-of-Plane Bending of Rib to Floorbeam Subassembly (b) Combined In-Plane and Out-of-Plane Bending of Floorbeam Subassembly with Ribs

The following recommendations are made with respect to fatigue acceptance testing for OSD connection details. The authors believe they are issues that, at a minimum, should be considered in any acceptance testing program. They are listed in no particular order.

- The applied load scheme must accurately simulate the actual in-service stress range cycle. Hence, stress reversals and proportions of stress (e.g., in-plane and out-of-plane stresses) must be simulated. It is acceptable to test at load ranges higher than observed in the field to assess fatigue damage. However, modes of failure should not be altered by increasing the load range (i.e., change the crack orientation and type).
- The specimen must be fabricated using the same procedures as to be used in the field. Caution must be used if the fabricator who makes the test specimen is different than the fabricator who will make the panels for the actual bridge. If this is the case, it must be demonstrated that the performance of the panels is not affected.
- Instrumentation should be installed at locations that would ensure the measured data are comparable to some defined standard. For locations where the nominal stress approach to fatigue evaluation is not applicable, researchers should consider using methods recommended by IIW or other documented methods.
- The boundary conditions must ensure that the behavior of the specimen when loaded is consistent with the behavior anticipated in the field. Modes of failure should not be altered by the boundary conditions used to support the specimen.
- A statistically sufficient number of data should be collected from each detail of interest. Data analysis must be conducted in a manner to provide level of safety consistent with the

AASHTO LRFD Specifications. That is, nominal design resistance must provide 97.5 percent confidence of survival for finite life.

- Measured strains and displacements should be compared to the results of FE analysis. Any significant discrepancies (e.g., greater than 10 percent) should be examined and the reasons either corrected or justified.
- Prototype specimens are also useful, as they present full-scale versions that can be evaluated for future in-service inspectability, ease of construction, and reveal unanticipated problems.

10.3. TESTING OF WEARING SURFACES

Based on the common types of failures in the surfacing systems discussed in Chapter 9, different types of tests have been devised to establish surfacing properties, as well as performance as a composite system. Several acceptable tests have emerged, even if none of them has been standardized. Tests include: (a) static flexural tests of the wearing surface-deck plate composite specimen at different test temperatures, (b) flexural fatigue tests of wearing surface-deck plate composite specimens conducted at several test temperatures to reflect temperature extremes, (c) direct pull-off bond tests, (d) resistivity tests, and (e) rutting tests.

These tests allow investigation of composite action and deflection ratios at the various test temperatures. The fatigue tests will allow investigation of cumulative damage in the surfacing material. The bond pull-out test is typically used to establish the tensile bond characteristics for acceptable surface preparation and quality control. Resistivity tests allow quantification of the extensive nature of tensile cracking in the surfacing. The rutting tests are typically used for asphalt materials to determine propensity to deform permanently under heavy wheel loads due to local contact shear stresses. In addition to laboratory testing, a small section on field testing will also be included (if this is an option available to the designer of the wearing surface system).

10.4. FULL SCALE PROTOTYPES

For the design and construction of major bridge projects, it is often prudent to perform full-scale prototype testing to verify the design prior to mass production of orthotropic steel components. Prototype tests verify the performance of the design, from engineering to construction, and reduce risk of unforeseen circumstances compromising the service life or causing delays. But ultimately, this is a decision to be made by the owner with consideration of cost, schedule, and risk.

The size, aspect ratio, and overall detailing of prototype specimens should be of sufficient proportions to ensure that they can be used to accurately predict the behavior of the in-situ deck system. This statement may suggest to the reader that full-scale testing of multi-span deck systems is always required. However, this is not necessarily true. If properly designed, an experimental program that utilizes smaller subassembly specimens can produce valuable and accurate data. This is an attractive option, as full-scale testing is expensive and can really only be completed by a limited number of laboratories in the United States.

However, this leads to the question of what constitutes a well-designed program for testing subassembly specimens. First, however, it must be recognized that testing is not only intended to gain more information on the known problem areas. For example, it must be determined which welded details are recognized as being a "problem" detail for fatigue resistance. Another objective, and possibly a more important one, is to reveal those problems that may not be initially known.

Thus, the first challenge is to ensure that all issues that are of concern (i.e., failure modes) are well defined and understood. The loading, specimen, and boundary conditions must all allow for proper testing of the details of interest. For example, it is well known that the rib-to-floorbeam (RF) weld is a critical detail that is subjected to in-plane and out-of-plane stresses. If one accurately simulates such stress fields on an individual rib/FB specimen, or multi-rib specimen, by applying the appropriate proportions of equivalent forces, a full-scale multi-span specimen would not be required and the subassembly test is sufficient. However, if improper boundary conditions are applied (i.e., they are different than those in the field), the actual fatigue resistance of a joint may be artificially increased, possibly leading to premature cracking in the field.

As stated, the failure modes that are not always so obvious, nor anticipated by the engineer, are also revealed in the testing. The details we may not recognize as problem areas could be entirely excluded from a subassembly test unless they are somehow accounted for somewhere in the design and analysis process. For example, during the testing of the first prototype OSD panels of the Williamsburg Bridge, the initial details used to attached the deck to the FB and those used to splice the FB plates of the individual deck panels were found to create severe discontinuities in the stress flow in these regions (Connor, 2002). The discontinuities, which were the result of the flexibility of the bolted splices, were not captured in the initial FE studies and therefore not expected. Hence, these splices were not initially believed to have any significant influence on the behavior of the deck. Only after testing of the full-scale multi-span deck were the discontinuities created by these joints realized. Had only a subassembly test been conducted with the applied load effects determined by FE analysis, a critical failure mode would likely have been completely overlooked.

The above example illustrates the importance of ensuring that the prototype test accurately reflects all aspects of the real in-situ conditions to best extent possible. Laboratory tests provide for maximum control of loading and instrumentation and allow time for changes to design, but field testing provides the data under the as-built conditions for final verification without impacts to schedule (assuming success). Unfortunately, there are no established rules for determining the size, loading, and boundary conditions that should be used for laboratory testing. One example is shown in figure 10-2.

Figure 10-2 Test Setup for the Prototype of the Bronx Whitestone Bridge Redecking Tested at Lehigh University showing the Deck, Loading Truck, and Loading Guide

A number of full-scale prototype tests have been conducted in the United States in recent years. Relevant projects are summarized in table 10-1. These tests have led to a continual advancement in the understanding of behavior and fatigue resistance at critical joints. Unfortunately, most of these tests are based on special designs conducted for redecking applications on existing bridges, requiring use of internal rib stiffening measures and special cut-out geometries necessitated by limited FB clearance. Thus, these are not considered optimum solutions for future application or widespread standardization.

Although NCHRP Project 10-72 is not complete and no formal requirements for such testing have been codified, owners would be wise to consider the possibility of testing either complete prototypes, or at a minimum, subassemblies of non-proven details. An argument can be made that there would not be a need to conduct full-scale testing of multi-span prototype OSD systems that are similar to others which: (1) have been tested; or (2) have been demonstrated to be successful designs through successful long-term field performance. However, it is emphasized that minor changes to details may seem like small changes, but in actuality, can have drastic effects on the actual performance of the OSD.

Table 10-1 Summary of Recent Full-Scale Orthotropic Deck Tests Conducted in the United States

Project	Testing Location	Date
Verrazano Narrows Bridge Redecking	Lehigh University	ongoing
California Dept. of Transportation	University of California San Diego	2007
Bronx Whitestone Bridge Redecking	Lehigh University	2001
Williamsburg Bridge Redecking	Lehigh University	1995, 1998

11. DESIGN EXAMPLES

This chapter demonstrates the implementation of the Manual contents in two design examples: a medium-span continuous girder bridge and a long-span cable-stayed bridge. It does not contain any new required information relating to the design of orthotropic steel decks (OSD), but rather it exemplifies the application of the comprehensive information provided in the preceding chapters. These examples include checks on strength, serviceability, and fatigue, with a majority of the focus on fatigue. A commercial finite element program is used to perform all modeling and analysis.

Note that the design examples are not complete designs, nor are the designs optimized. They are intended to only be a demonstration of the design concepts. The focus is to demonstrate to the practitioner the application of Level 3 design (per Chapter 5) by refined analysis. Additionally, important design considerations such as the floorbeam (FB) splice, the FB to girder connection, and other routine design checks are outside the scope of these examples.

11.1. EXAMPLE 1 – MULTIPLE GIRDER CONTINUOUS BRIDGE

This example focuses on the design of the ribs, the deck, and the FB of a medium-span continuous girder bridge. Extra attention is paid to the fatigue design of the rib-to-floorbeam (RF) connection. The following guidelines/assumptions are made:

- ASTM A709 Grade 50 Steel is used for all components.
- A thick bituminous wearing surface (non-structural) is to be used.
- The bridge is required to carry two traffic lanes with full lane width shoulders.
- The bridge is three-span continuous, spanning 30.48 m – 42.67 m– 30.48 m (100 ft – 140ft – 100 ft).
- Welded rib splices are used.
- The region of interest is the FB directly over an interior support.
- No relieving cut-out is used at the RF connection.
- The following checks are required at the region of interest:
    - Strength: Rib moment capacity.
    - Service: Differential deflection between adjacent ribs.
    - Fatigue:
        - Deck portion of the rib-to-deck weld (RD).
        - Rib portion of the RD.
        - Rib splice.
        - Deck plate splice.
        - Rib portion of the RF.
        - FB portion of the RF.
- The fatigue checks are made using the finite element method and local stress analysis as required (Level 3 Design).
- Shell elements are utilized in the region of interest, with weld geometries not modeled.

11.1.1. Description of the Bridge

The cross-section for Example 1 consists of four welded plate girders spanning 30.48 m – 42.67 m – 30.48 m (100 ft – 140 ft – 100 ft). The framing plan consists of an OSD stiffened with closed trough stiffeners (also called closed longitudinal ribs) spanning 3048 mm (10 ft) longitudinally between FBs, as shown in Figure 11-1.

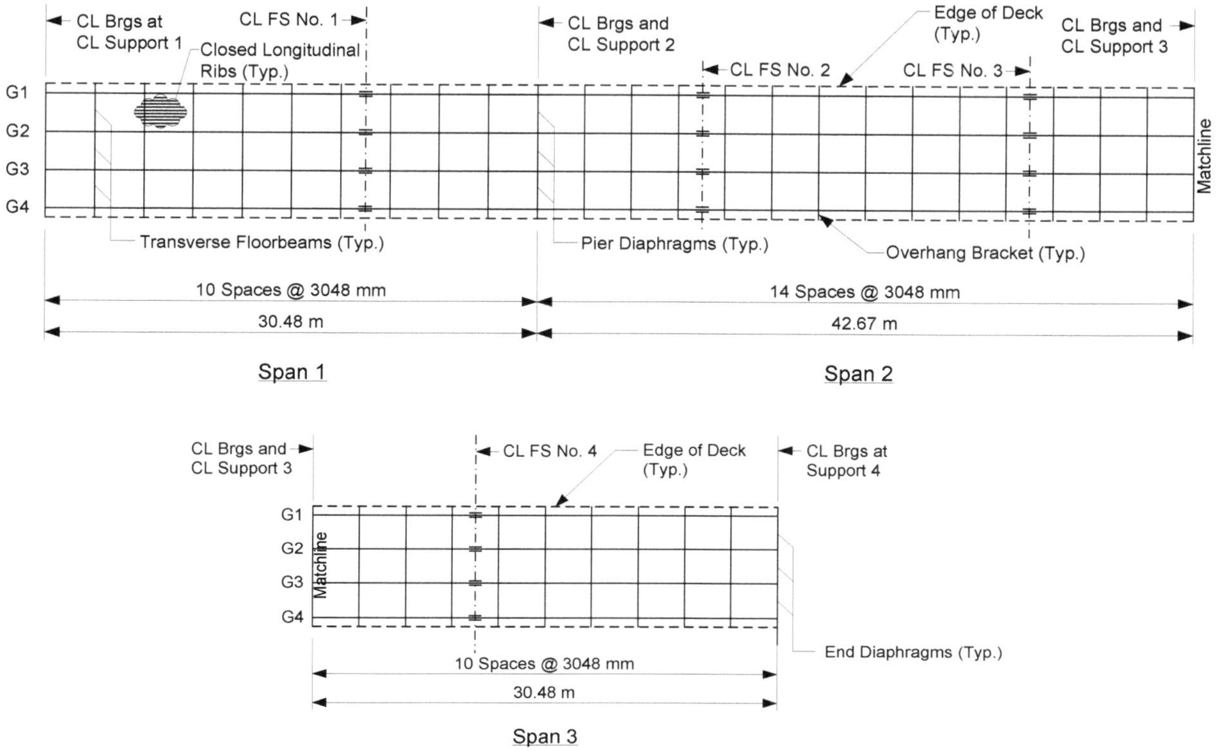

Figure 11-1 Three-Span Multiple Girder Example Framing Plan showing Overall Dimensions, Girder Spacing and Floorbeam Spacing

The deck plate is integral with the FBs and the primary girders, which are spaced at 3,360 mm (11 ft), with 1,067 mm (3.5 ft) overhangs. The RF for the longitudinal ribs do not include a relieving cut-out below the rib. See Figure 11-2 for a graphical representation of the example bridge cross-section and various detailing. Detail A in Figure 11-2 shows the rib dimensions.

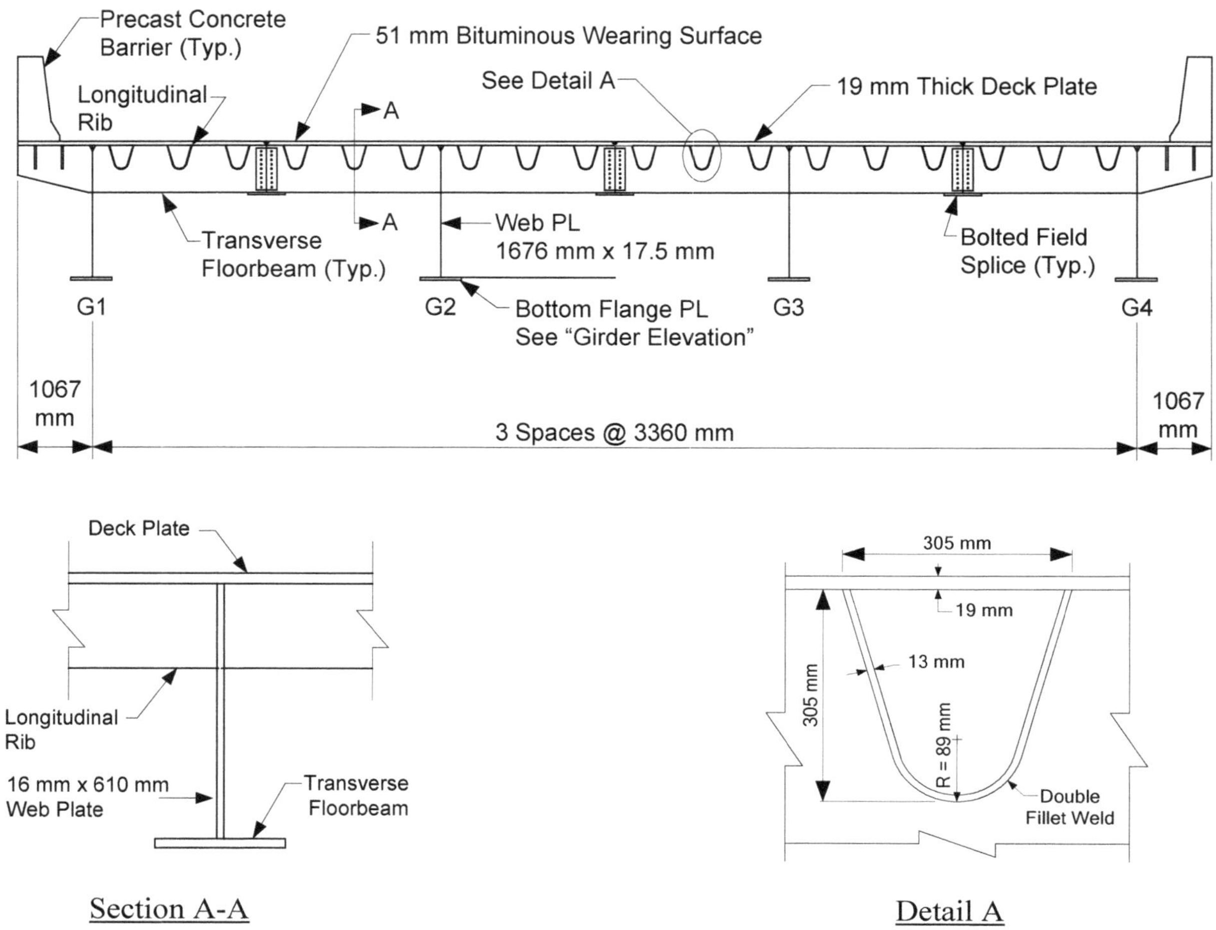

Figure 11-2 Multiple Girder Example Cross-Section, Floorbeam Section (Section A-A), and Rib Section (Detail A) showing Dimensions for each Component

The system is analyzed by the finite element method, in accordance with the guidelines established in previous chapters. A multi-scale modeling approach is employed, with a fine mesh using higher-order elements in the region of interest, while coarsely meshed elements and lower-order elements are used elsewhere. In refined mesh areas, shell elements are used to model the deck plate, rib, and FB. Away from this region, a more coarse mesh of beam and shell elements is used. Mesh sizes are defined in accordance with the provisions of the manual to assess the stresses at the concentrations, but otherwise are kept relatively coarse to keep the size of the model reasonable.

The erection scheme of this OSD system involves the off-site fabrication of the primary girder, FB, deck plate, and longitudinal rib components for a width of the deck plate equal to the girder spacing. The individual field sections are then transported to the site where the FBs are connected to the adjacent field sections through bolted splices. A longitudinal weld at each splice location connects the field section deck plates. For a schematic of the splices, see Figure 11-3.

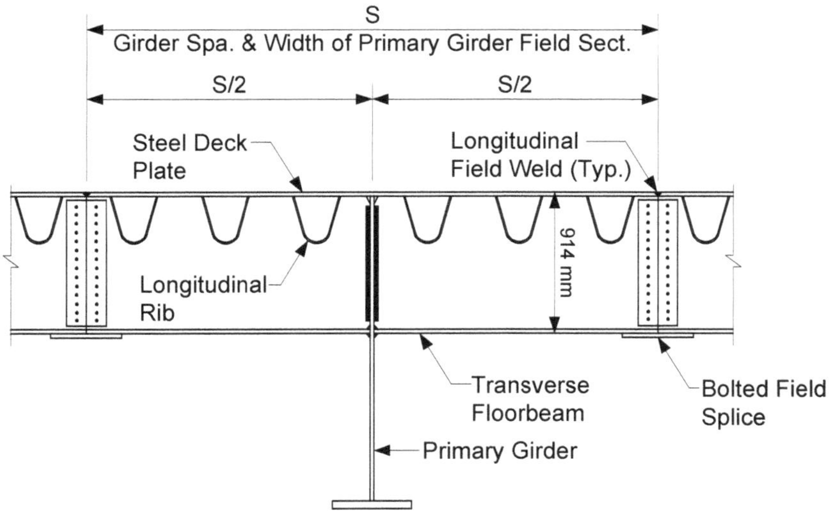

Figure 11-3 Erection Scheme Field Section showing the Interior Girder and Location of Field connections Relative to the Girder

11.1.2. Development of the Finite Element Method

Following the recommendations in this manual, a finite element model is created of the bridge in Example 1. A detailed mesh is used for the section of the bridge directly above the first interior pier, while the remainder of the bridge is modeled using coarse methods. Figure 11-4 shows the full bridge model, with the detailed section at the pier apparent. Figure 11-5 shows a close-up view of the detailed section, and Figure 11-6 shows the region of interest with the deck removed, exposing the ribs, FBs, and girders.

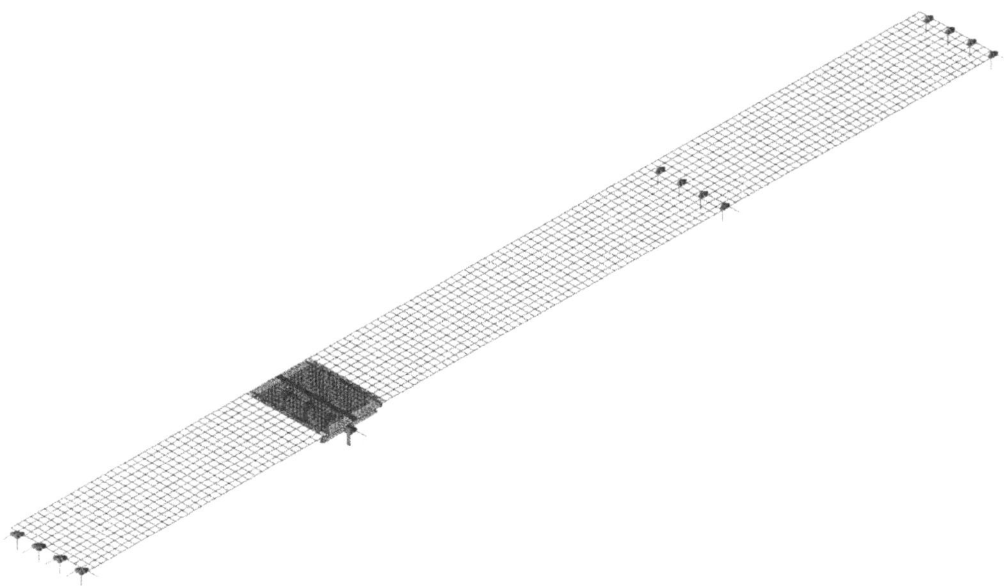

Figure 11-4 Three-span Finite Element Model showing Overall view of Beam Elements and Support Conditions

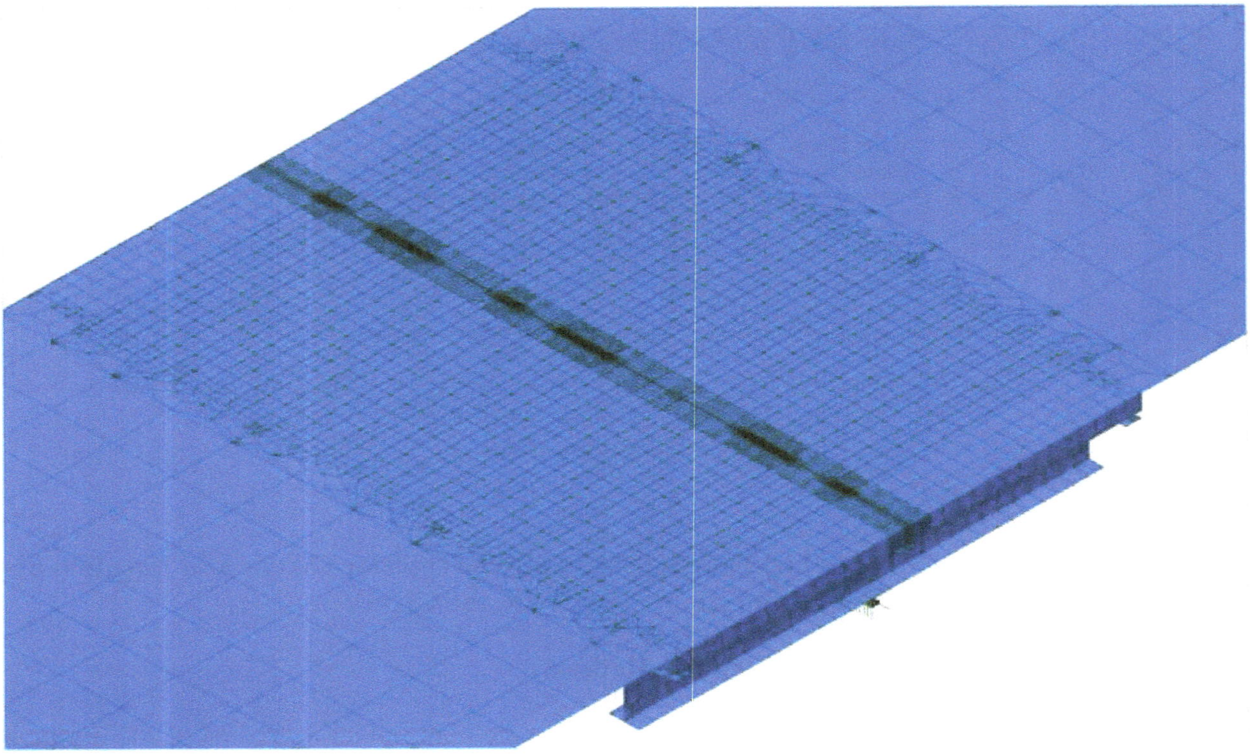

Figure 11-5 Close-up of Detailed Section that is Modeled Using the More Complex Configuration to Capture the Localized Behavior

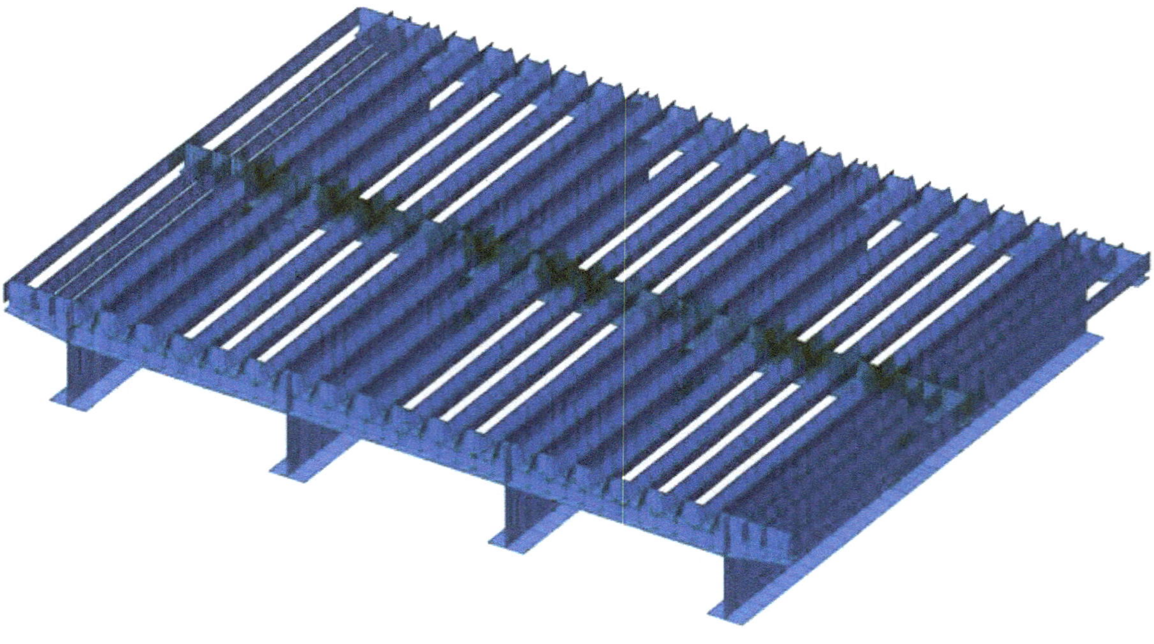

Figure 11-6 Close-up of Detailed Section with Deck Removed to show how the individual Girder, Floorbeam, and Rib Element have been Modeled

In the detailed section, the girders, FBs, ribs, and deck are all modeled with shell elements. The finest mesh size is used at the areas of interest (from a fatigue viewpoint) and the mesh transitions to larger sizes further away. The mesh consists of eight-noded thick shell elements in the refined areas in order to allow for the stress extrapolation techniques used in fatigue evaluation. Thin shell formulation will provide suitable results, but thick formulation should be used if available as it includes deformations in the through-thickness direction due to shear.

The meshing of the FB around the rib is shown in Figure 11-7. The size of the elements directly adjacent to the rib is equal to the thickness of the FB, which is 16 mm (5/8 inch). This mesh size is maintained for two elements away from the rib, to provide a constant element size for use in local stress calculations involving extrapolation. Away from the rib, the mesh is transitioned to much larger element sizes in order to limit the total number of degrees of freedom of the model. Similarly, not all RF regions are modeled with the finest mesh size. Instead, only a few representative connections that are likely to have the highest stresses are modeled with enough detail to allow the stress extrapolation technique to be used. A fine mesh is also used adjacent to the rib, as shown in Figure 11-7, as this is the location of the FB splice, and fatigue stresses will need to be evaluated at this location as well.

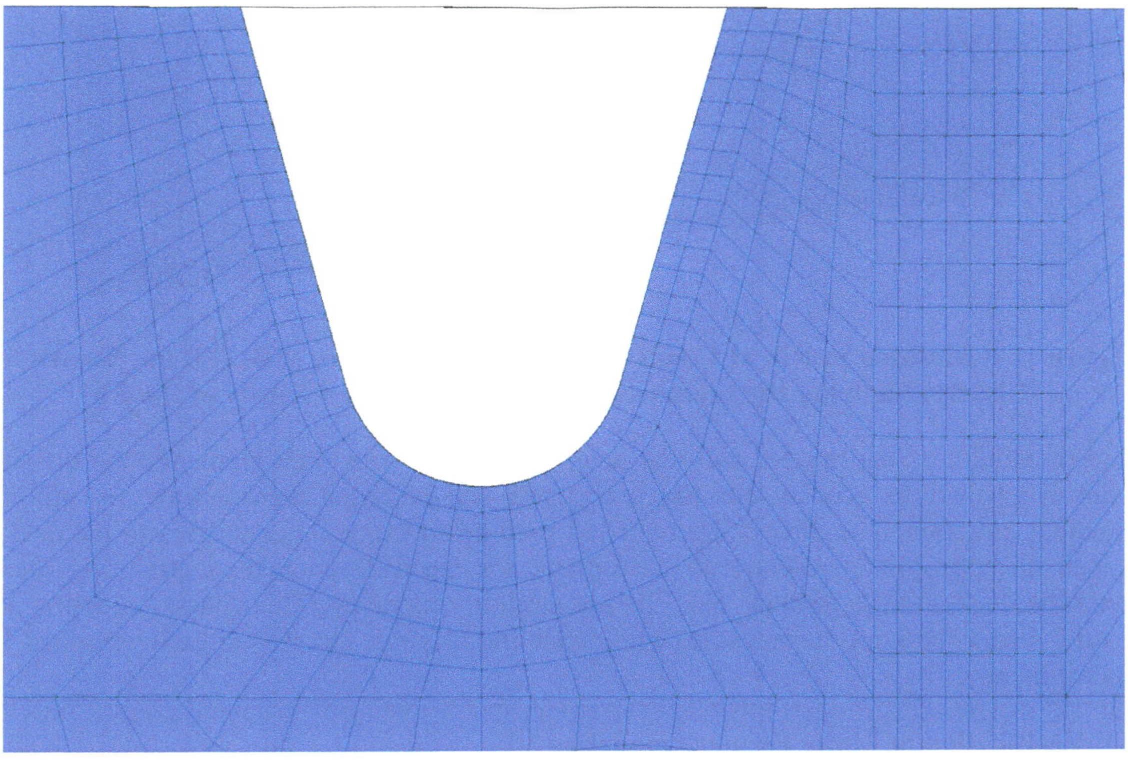

Figure 11-7 Meshing Detail of Floorbeam around Rib

Figure 11-8 shows the rib mesh at the junction with the FB. For clarity, the FB is not shown. Similar to the FB, the mesh size is smallest at the connection area, and is transitioned to a larger size away from this area. Because the rib thickness is less than the FB thickness (13 mm [½ inch] versus 16 mm [5/8 inch]), the mesh size guidelines result in a conflict at this location. The width of the elements is set by the size used in the FB, but the length of the elements in the longitudinal direction of the bridge is 13 mm (1/2 inch) based on the rib thickness. This aspect ratio (≈1.25) does not affect the accuracy of the stresses calculated.

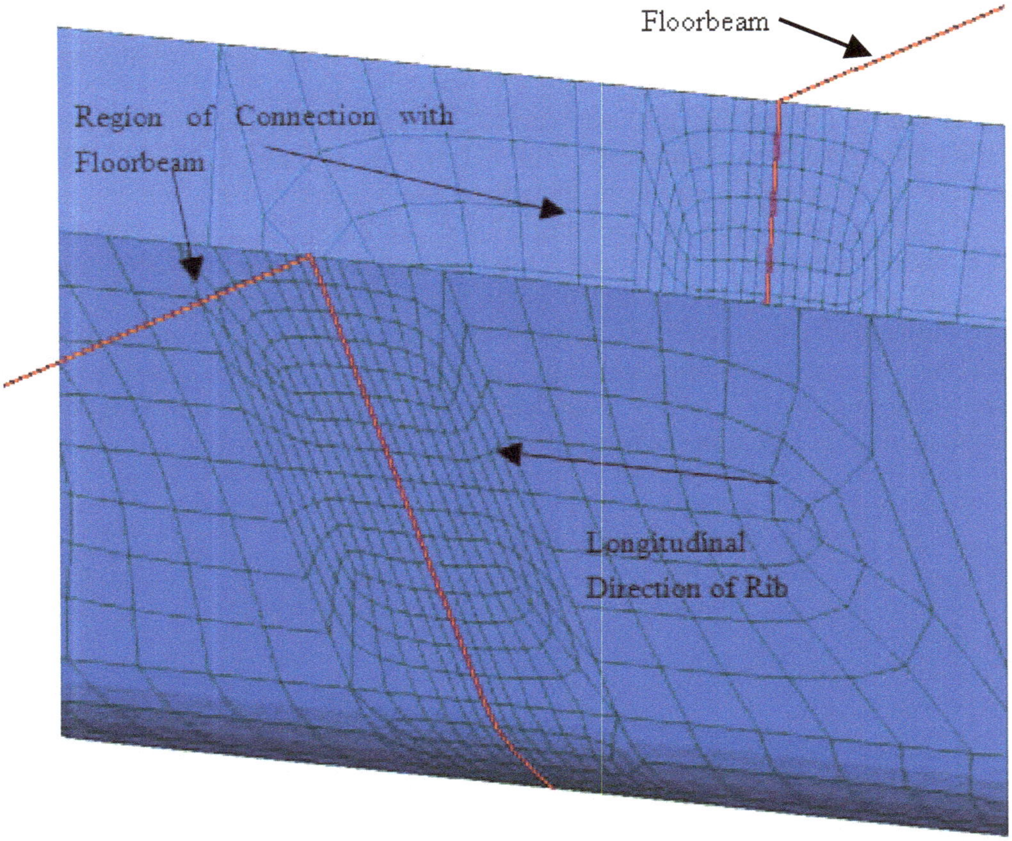

Figure 11-8 Meshing in Rib at Connection Rib-to-Floorbeam (FB) (FB not Shown for Clarity) Showing the Increase Discretization in the Vicinity of the FB

For the remainder of the bridge, only the deck is modeled with shell elements. An orthotropic material is applied to account for the extra area and stiffness provided by the rib in the longitudinal direction. The girders are modeled with beam elements connected to the deck, and are given an appropriate eccentricity to account for the difference in neutral axis locations between the deck and the girders. The FBs are similarly modeled. This coarse modeling allows the model size to be kept manageable, while enabling accurate stresses to be determined in the region of interest.

11.1.3. Verification of Finite Element modeling

11.1.3.1. Floorbeam In-Plane Flexure

The FEA results for the in-plane flexural behavior of the FB are verified using a two-dimensional Vierendeel model. The Vierendeel model is developed using a two-dimensional analysis software. A schematic of the Vierendeel model is shown in Figure 11-9.

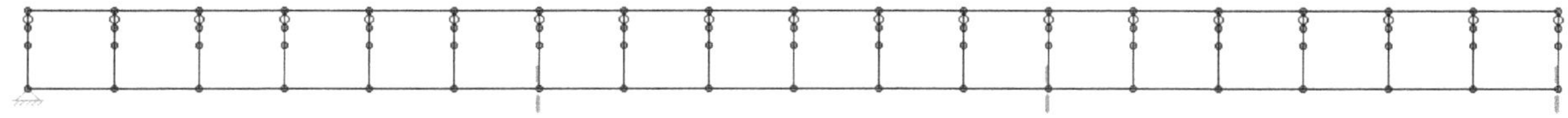

Figure 11-9 2D Vierendeel Model of FB and Deck Plate illustrating the discretization of the Deck Plate and Floorbeam

The horizontal members at the top of the Vierendeel model (the top chords) represent the deck plate. The horizontal members at the bottom (the bottom chords) represent the bottom flange of the FB and the non-disturbed web region of the FB (i.e. the portion of the FB web that is below the rib cut-outs). The bottom chords and top chords are connected by verticals that are fixed to the bottom chords and pinned to the top chords (see Figure 11-10). The upper portion of each vertical represents the portion of FB web between the bottom of the rib cut-out and the deck plate (i.e. the disturbed portion of the FB web), as well as an effective width of rib that extends away from the FB on each side of the web. The lower portion of each vertical represents the undisturbed region of the FB web (between the lower chord of the model and the bottom of the rib cut-out). The supports for the FB are found at the location of the four girders. The left-most support is a pin support, while the other three supports are roller supports.

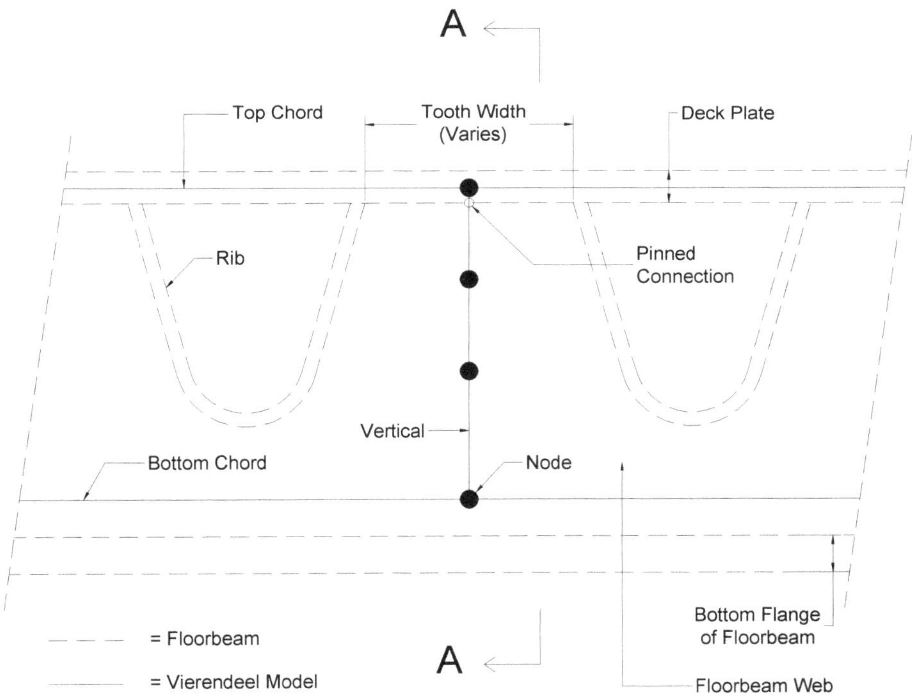

Figure 11-10 Partial Vierendeel Model and Floorbeam Elevation with the Vierendeel Model Superimposed on the Floorbeam to Illustrate the Model

As can be seen in Figure 11-10 the width of the FB "tooth" varies along the depth of the web. This varying width is approximated in the Vierendeel model by using two prismatic members to represent the upper portion of each vertical. The upper prismatic member has a width equal to the width of the tooth at the deck. The lower prismatic member has a width equal to the width at mid-depth of the tooth. The effective width of deck plate and longitudinal rib used in the Vierendeel model is illustrated in Figure 11-11.

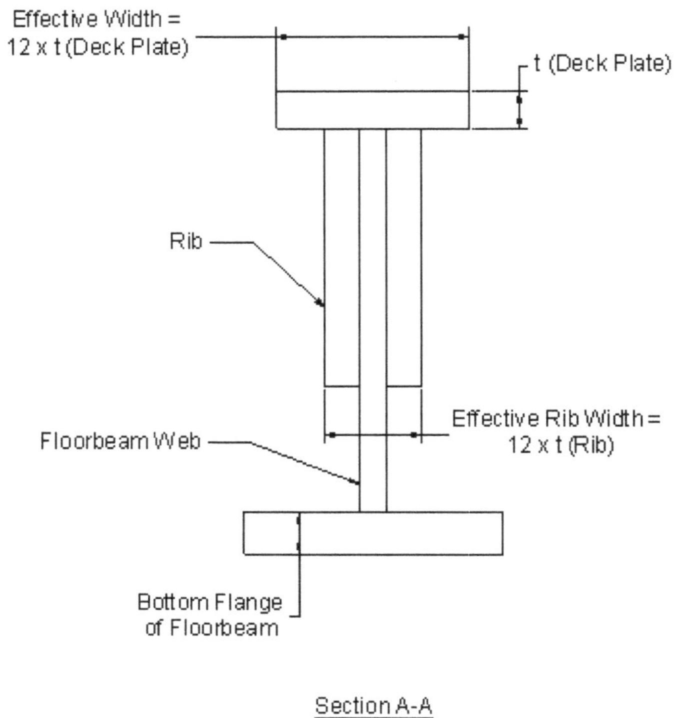

Figure 11-11 Illustration of Effective Width of Deck Plate and Rib Used in the Vierendeel Model

To verify the FEA results, a 445 kN (100 kips) vertical point load is applied directly above the eighth tooth in the FB (which corresponds to 0.33L of span 2, where L = span length). This is shown schematically in Figure 11-12 and Figure 11-13. First, horizontal moments acting on two of the most highly-loaded teeth are checked. The moment on a given tooth is checked at a section of the tooth right before the rib begins to "radius" (as illustrated in Figure 11-12). The FEA moment in a tooth is determined by creating a horizontal slice through a tooth and integrating the stresses over the tooth sectional area to arrive at the net moment on the section. A comparison of the Vierendeel and FEA teeth moment is shown in Table 11-1. As can be seen from the table, the ratios of Vierendeel-to-FEA moments are 0.90 and 0.94 in Tooth 7 and Tooth 10, respectively.

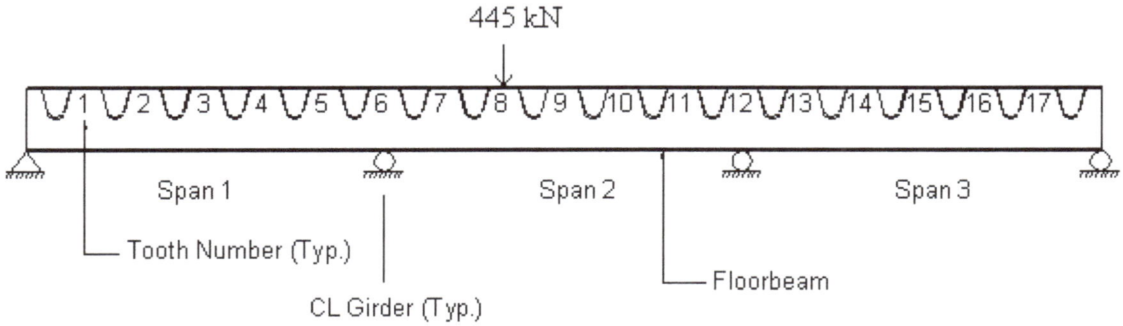

Figure 11-12 Floorbeam Elevation Showing the Verification Load, Tooth Numbering, and the Girders Represented as Supports for the Model

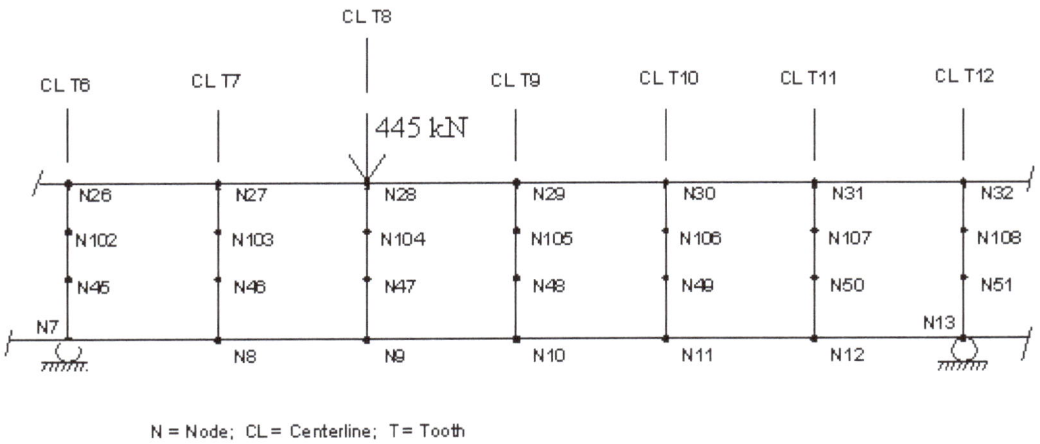

Figure 11-13 Span 2 of the Vierendeel Model showing Verification Load, Node Numbers, and Tooth Numbers for the Model

Table 11-1 Comparison of Vierendeel and FEA Tooth Moments

Vierendeel Node No.	Floorbeam Tooth No.	A = Vierendeel Moment (kN-m)	B = FEA moment (kN-m)	A ÷ B	Location Along Floorbeam (L = Span Length)
N46	7	-48.0	-53.3	0.90	0.17L of Span 2
N49	10	28.2	30.1	0.94	0.67L of Span 2

Next, vertical stresses acting on a horizontal section through the teeth is checked at the same section where the horizontal teeth moments were checked (shown in Figure 11-14). A comparison of the Vierendeel and FEA tooth stress (in two teeth) is shown in Table 11-2. On the left side of Tooth 7, the Vierendeel stress is 18 percent higher compared to the FEA stress. On the right side of Tooth 7, and on the left and right sides of Tooth 10, the Vierendeel and FEA stresses are within 5 percent of each other. A schematic of the FEA vertical stress contours in the FB is shown in Figure 11-15. It shows that the maximum tensile and compressive vertical stresses, excluding the stress directly below the applied concentrated load, occur at the left and right edges of the teeth, near the base of the teeth where the ribs begin to "radius".

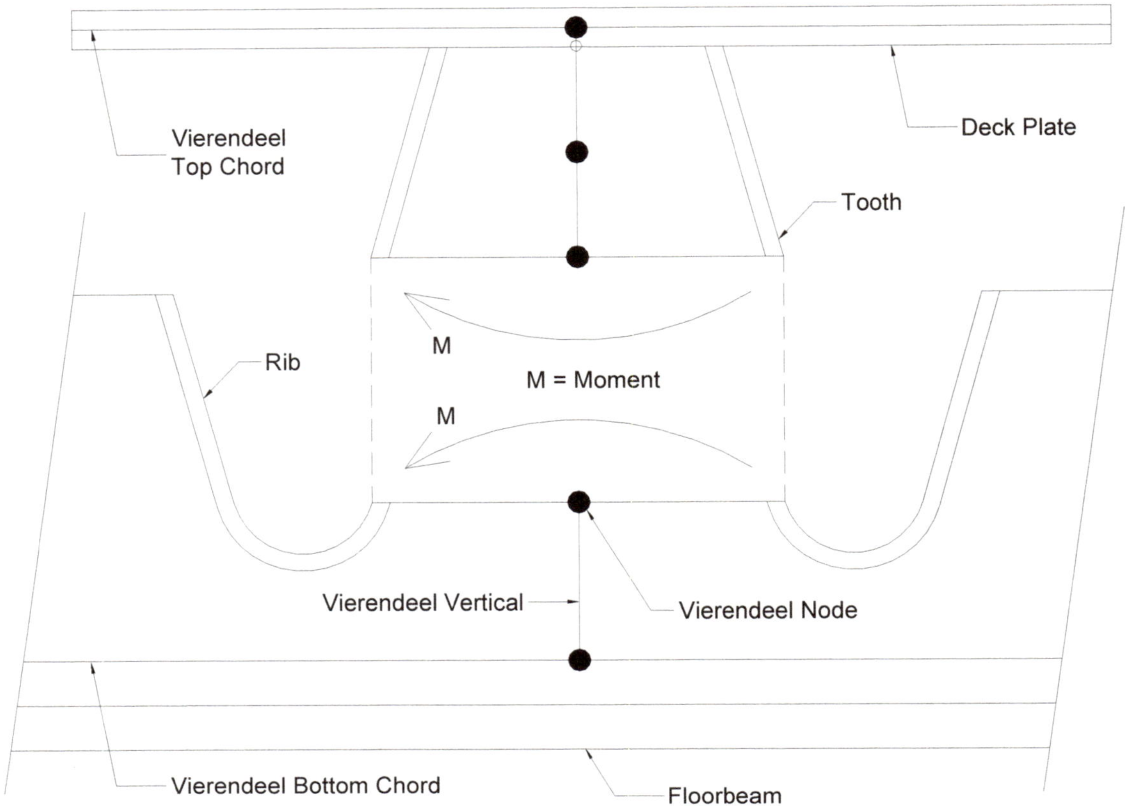

Figure 11-14 Illustration of Horizontal Moment Acting at the Base of a Tooth and at a Vierendeel Node Superimposed on the Floorbeam

Table 11-2 Comparison of Vierendeel and FEA Tooth Stresses

Vierendeel Node No.	Floorbeam Tooth No.	A = Vierendeel Min. Stress (MPa)	B = FEA Min. Stress (MPa)	A ÷ B	C = Vierendeel Max. Stress (MPa)	D = FEA Max. Stress (MPa)	C ÷ D
N46 (bottom)	7	-86.81	-73.64	1.18	86.81	91.63	0.95
N49 (bottom)	10	-50.95	-53.09	0.96	50.95	50.06	1.02
Negative Stress (-) is Compressive.							

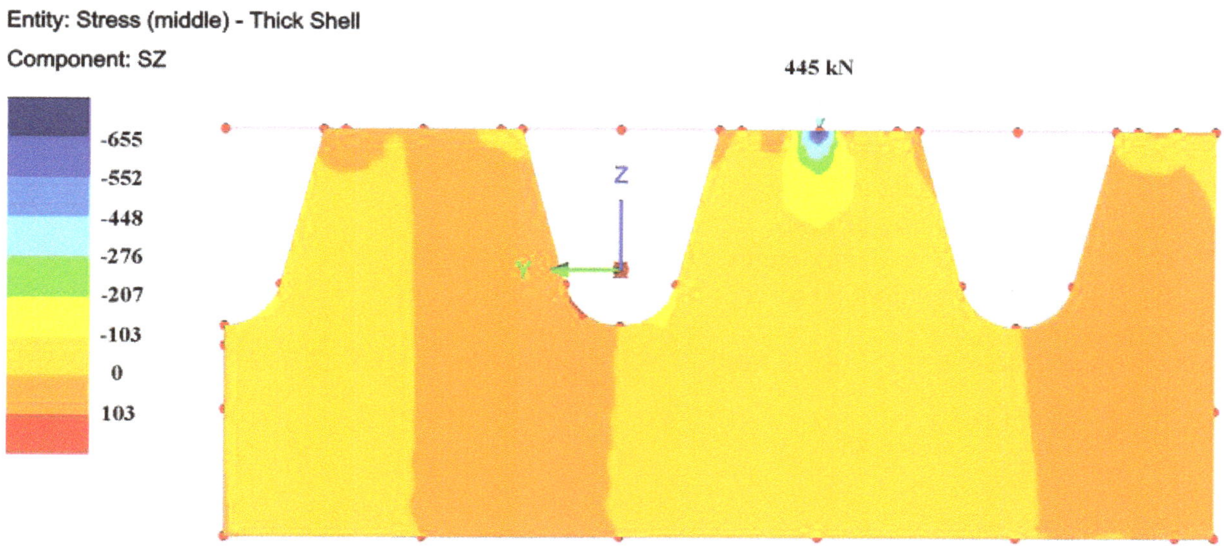

Figure 11-15 Results of FEA Model of the Floorbeam showing the Vertical Stress Contours in the Web and Tooth

The absolute maximum FEA tensile stress in the tooth is 99.63 MPa, (22.4 kips) and occurs on the right side of Tooth 7 along the curved portion at the base of the tooth. This stress is 15 percent higher than the absolute maximum Vierendeel tensile stress in the teeth (86.81 MPa [19.5 kips]). The absolute maximum compressive stress in the tooth is 81.50 MPa (18.3 kips), and occurs on the left side of Tooth 7 along the curved portion at the base of the tooth. This stress is 6 percent lower than the absolute maximum Vierendeel compressive stress in the teeth (86.81 MPa [19.5 kips]).

11.1.3.2. Bending in Rib at the Floorbeam

In the preliminary design of this example, the FB spacing was set at 4,572 mm (15 ft) before being revised to 3,048 mm (10 ft). This section contains a check that was performed on a model with the original 4,572 mm (15 ft) FB spacing, but the conclusions reached are just as valid for the final design.

An estimate of the bending moment in a rib at a FB can be made by assuming a distribution factor and a rigid support at the FB. A distribution factor of 0.5 for closed ribs has been shown to often result in conservative moments.

A single wheel load of 71.2 kN (16 kips) is distributed to a single rib using a distribution factor of 0.5, resulting in an applied load of 35.6 kN (8 kips). This load is spread along a patch of 250 mm (10 inches) in length. The equations and methods outlined in Design Manual for Orthotropic Steel Plate Deck Bridges (AISC, 1963) are used to determine the resulting moment in a rib. The equation given below is used to calculate the moment in a rib at a support when a distributed load is present 0.3804 times the FB spacing away.

$$\frac{\mathbf{M}}{\mathbf{SP}} = -0.085 + 0.1494\left(\frac{\mathbf{C}}{\mathbf{S}}\right)^2 \qquad (11\text{-}1)$$

where:

M	=	moment in rib
S	=	Spacing of FBs
C	=	½ Patch load length
P	=	Total applied force

For the present case, **S** = 15' (4572 mm), **C** = 5" (127 mm) = 0.4165' (126.95 mm)

$$\frac{\mathbf{C}}{\mathbf{S}} = \frac{0.4165'}{15'} = 0.0278$$

$$\frac{\mathbf{M}}{\mathbf{SP}} = -0.085 + 0.1494(0.0278)^2 = -0.0849$$

$$\mathbf{M} = 15' \times 8\ \mathbf{kips} \times -0.0849 = -10.2\ \mathbf{kip \cdot ft} = -13.8\ \mathbf{kN \cdot m}$$

The same loading is applied in the Finite Element model, with the patch load centered on one of the two center ribs in a panel. The loading is applied at 0.3804 times the panel length away from the FB. The moment in the rib is found by integrating the stresses in the shell elements comprising the rib and a portion of the deck extending half-way to the adjacent rib. The integrated moment is found to be -14.8 kN-m (-10.9 kip-ft). This is in very good agreement with the value calculated by the method given in the AISC publication.

11.1.4. Service Limit States Checks

As specified in Section 5.2.2 of this manual, the AASHTO Service I limit state must be satisfied for overall deflection limits of the deck plate, the ribs, and relative deflection between adjacent ribs. The relative deflection limit is set as 2 mm (0.1 in.) and is checked at the mid-span of the ribs between the shell modeled FBs. Dead load and live load are applied to the model, with the live load being an unfactored HL-93 design truck placed at various transverse locations with the rear axle centered at the mid-span between the FBs.

Enveloping the load cases considered, the maximum relative deflection is found to be roughly 0.5 mm (0.02"), which is well below the allowable limit.

11.1.5. Strength Limit State Checks

The negative moment capacity of a rib located between the girders at an interior support is checked. The loading used to create the maximum negative moment at the RF consists of an AASHTO design truck with a 4300 mm (14 ft) rear axle spacing and the 9.34 kN/m (0.64 klf) notional live loading applied across a lane width of 3048 mm (10 ft). The truck is placed so that the rear set of axles are equidistant from the FB. This loading can be placed at various transverse locations to produce maximum moments at different ribs. As specified in this manual, the 142 kN (32 kip) axles are split into two 71 kN (16 kip) axles spaced 1220 mm (4 ft) apart. A pressure of 275 kPa (40 psi) is used for the two front tires, which are each a 250 mm (10 inch) square patch, and also for the remaining eight rear tires, which are each a 510 mm (20 inch) wide x 250 mm (10 inch) long patch.

The effective width of deck plate to use is found in Section 4.3of this Manual. An OSD with closed ribs is classified as a non-composite box section in AASHTO LRFD, and Eq. 6.12.2.2.2-1 gives the equation for M_n. Since the ribs can not buckle laterally, the equation results in M_n being equal to the moment at first yield.

Examining the longitudinal stresses at the bottom of the rib (the extreme fiber in bending) under appropriate load factors, it is apparent that all of the stresses, even the concentrations due to the connection of the rib to the FB, are well below the yield stress. Alternatively, the moments can be extracted from the ribs by integrating over the effective width of the rib, and these moments can be compared with the yielding moment for the section.

11.1.6. Fatigue Limit State Checks

The finite element model developed for Example 1 is analyzed to find the live load fatigue stress range for details of OSDs with ribs without cut-outs. The first five types of details illustrated in Table 5-1 of Chapter 5 are examined and compared with their constant amplitude threshold allowable stress ranges for the applicable category, which can also be found in Table 5-1. An HL-93 design truck with 9000 mm (30 ft) between centerlines of the rear axles groups is used in accordance with the AASHTO LRFD fatigue specifications and the guidelines found in the previous chapters. AASHTO LRFD specifies that the 142 kN (32 kip) axles be split into two 71 kN (16 kip) axles 1220 mm (4 ft) apart as illustrated in Figure 5-1. This results in a uniform tire pressure of 275 kPa (40 psi), which is distributed over a 250 mm (10 inch) square patch for the two 17.75 kN (4 kip) wheels, and over a 510 mm (20 inch) wide x 250 mm (10 inch) long patch for the eight 35.5 kN (8 kip) wheels. This loading model is altered slightly to account for the distribution of the load through the wearing surface and one half of the deck plate, as allowed in Section 5.5.1.1 of this manual. At the ends of each tire edge, an amount equal to the distance from the top of the wearing surface to the mid-depth of the deck is added, and the pressure is then altered to result in the correct total load for the wheel. An impact factor of 15 percent is applied to all loadings, as well as the infinite life fatigue load factor of 2.25 where applicable (see Section 5.3).

Table 11-3 summarizes the results from the analysis, which will be described in the subsequent sections.

Table 11-3 Summary of Fatigue Checks for Example 1

Detail	Point of interest	S_r (MPa)	$S_{r.allow}$ (MPa)
RD (Deck)	Deck at rib connection directly under load mid-span between FBs.	55.9	69.0
RD (Rib)	Rib at deck connection opposite of loaded side mid-span between FBs.	33.9	69.0
Rib Spice (Max)	Lowest point of rib at 0.3 of the span between FBs.	34.3	48.0
Rib Spice (Min)	Lowest point of rib at 0.3 of the span between FBs.		
Deck Plate Splice	Deck splice midway between ribs at mid-span between FBs.	40.0	69.0
RF (Rib)	Lower half of rib at FB connection	39.5	69.0
RF (FB.)	FB at rib-to-deck connection and at base of rib	28.3	69.0

11.1.6.1. Rib-to-Deck (RD) Weld

To get the maximum stress range in the rib-to-deck weld (RD), a single static loading of one rear wheel patch is applied. The wheel load is centered on one leg of a rib, as illustrated in Figure 11-16, and applied at mid-span of the panel between FBs (note the width of the wheel patch is shown, not the distribution of the load through the wearing surface and one half of the deck plate). A schematic showing the longitudinal position of the loading is shown in Figure 11-17. The locations of interest are the deck plate at the rib connection directly under the load, and the rib at the deck connection in the rib leg opposite the load (see Figure 11-18 and Figure 11-19).

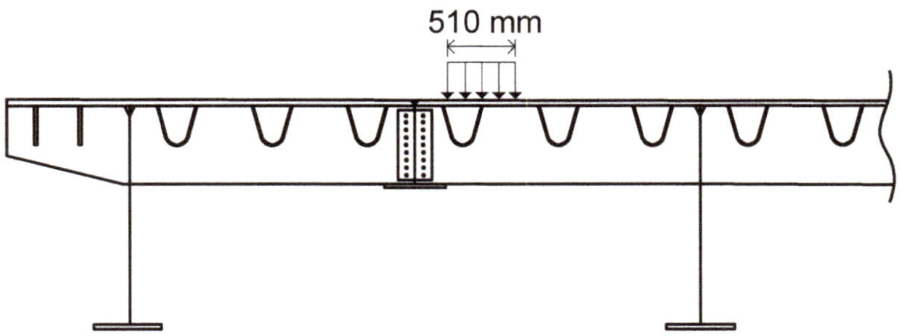

Figure 11-16 Transverse Wheel Location to Produce the Largest Impact on the Rib-to-deck Weld

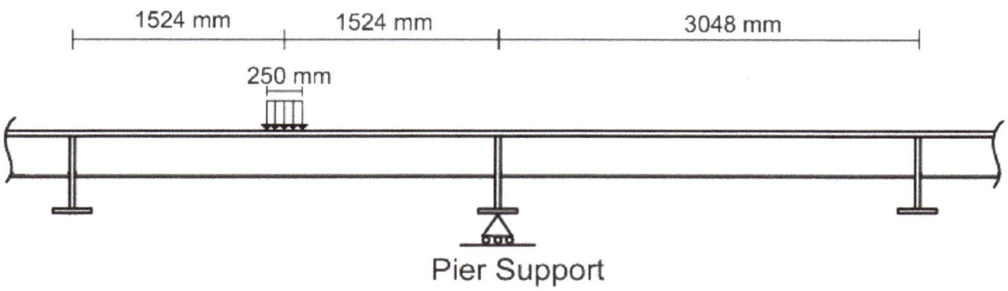

Figure 11-17 Longitudinal Wheel Location to Produce the Largest Impact on the Rib-to-deck Weld

The stress results from the model show a fairly linear relationship up to the points under consideration, so no extrapolation is used for calculating the stress range at this location.

Figure 11-18 and Figure 11-19 show the stress contours plotted on the deck plate and rib portions of the connection, respectively. In Figure 11-18, the stresses are in the element local y-direction, which is the transverse direction of the bridge. In Figure 11-19, the stresses are in the element local x-direction, which is the direction normal to the weld connection with the deck plate.

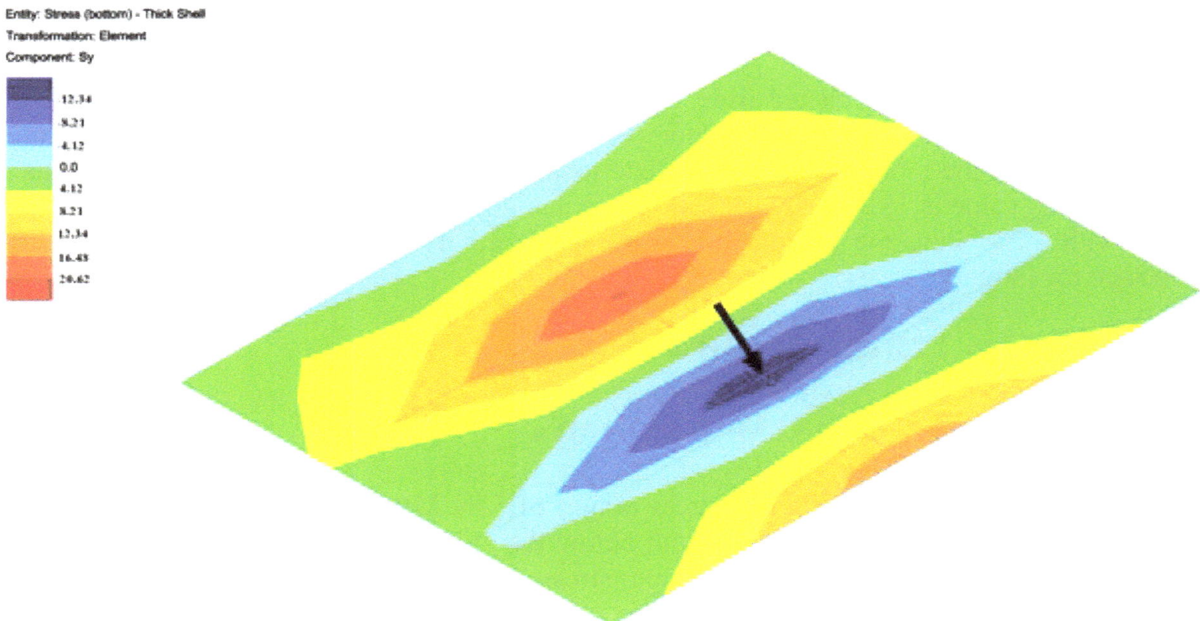

Figure 11-18 Results of FEA Model of the Rib-to-deck Weld showing the Local Stress Contours in the Deck (in MPa)

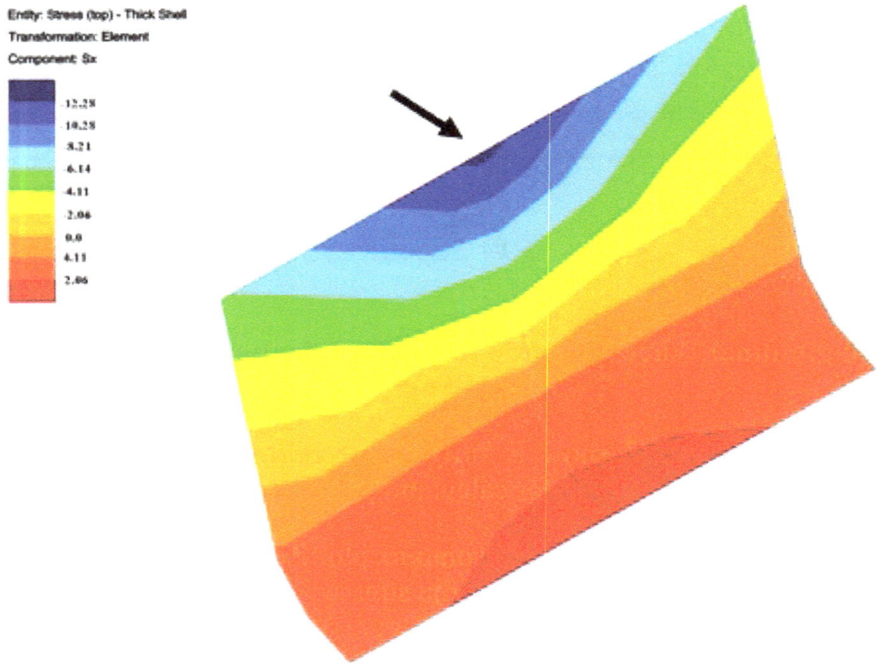

Figure 11-19 Results of FEA Model of the Rib-to-deck Weld showing the Local Stress Contours in the Rib (in MPa)

The nominal stresses are extracted from the model and then multiplied by the appropriate load factors to arrive at the stress range to compare with the allowable stress range. The nominal stress found from the model for the deck plate location is 21.6 MPa (3.1 ksi), while for the rib plate the stress value is 13.1 MPa (1.9 ksi). The calculation of the factored force effect is shown below, with 2.25 being the infinite life load factor, and 1.15 being the 15 percent dynamic allowance for fatigue.

$$\gamma(\Delta\mathbf{f})_{\mathbf{DECK}} = 2.25(1.15 \times 21.6 \ \mathbf{MPa}) = 55.9 \ \mathbf{MPa}$$

$$\gamma(\Delta\mathbf{f})_{\mathbf{RIB}} = 2.25(1.15 \times 13.1 \ \mathbf{MPa}) = 33.9 \ \mathbf{MPa}$$

As seen in Table 11-3, the calculated stress ranges of 55.9 MPa (8.1 ksi) in the deck plate and 33.9 MPa (4.9 ksi) in the rib plate are both below the allowable stress range of 69.0 MPa (10 ksi). Note that for this example, the minimum stresses are 0 MPa. Where minimum stresses are less than zero, they should be included. The total stress range is the difference between the maximum and minimum stress at the detail of concern.

11.1.6.2. Rib splice

The rib splice is assumed to be located at 0.3 times the span length, roughly 1,320 mm (52 inches), from the support. This location is chosen due to its proximity to the dead load contraflexure point. Two static loadings of the entire truck are applied to get the stress range for the rib splice; one for the maximum stress, and the other for the minimum stress. The wheels are centered on the rib as shown in Figure 11-20. A schematic illustrating the longitudinal position of the wheels for the maximum and minimum loadings is shown in Figure 11-21 (a) and (b), respectively. The maximum stress is generated when the front wheels of the rear axle group are directly over the splice. The minimum stress is found when the rear wheels are placed about 0.25 of the span length on the other side of the FB. The base of the rib is the location of the maximum stress.

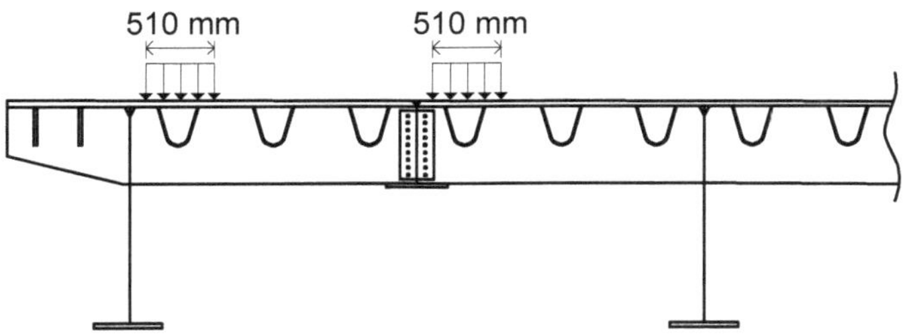

Figure 11-20 Transverse Wheel Location to Produce the Largest Impact on the Rib Splice

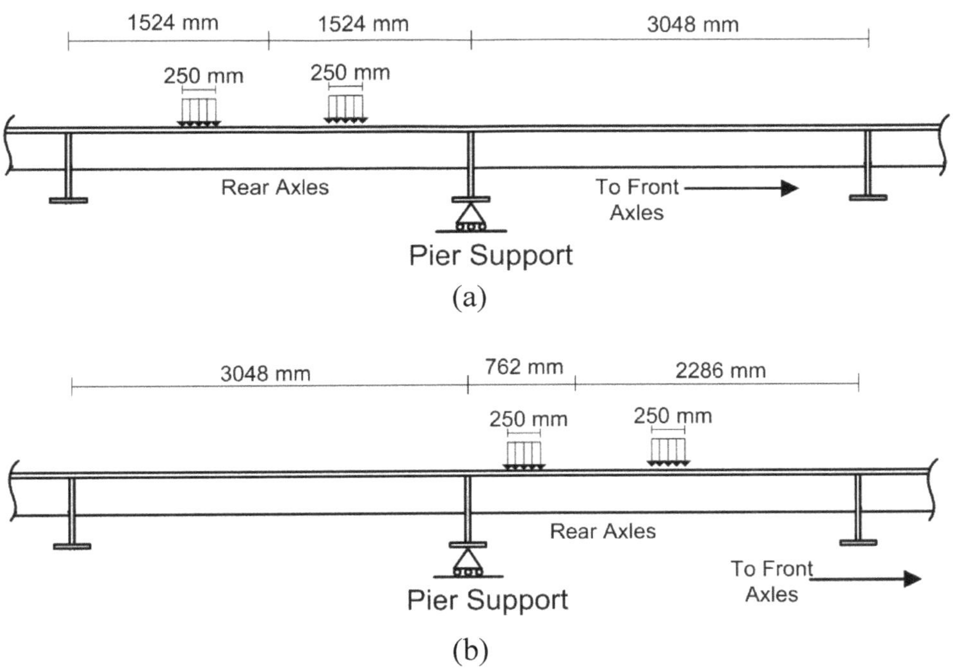

Figure 11-21 Longitudinal Wheel Locations to Produce the Largest Impact on the Rib Splice

There was no discontinuity in the stress field at this location, so the actual stress from the model was used (i.e. no extrapolation was required). Unfactored force effects of 12.97 MPa (1.88 ksi) and -4.7 MPa (0.68 ksi) are extracted from the model and are used to calculate the stress range to compare with the infinite life fatigue threshold stress, as shown below.

$$\gamma(\Delta \mathbf{f})_{\mathbf{RIB\text{-}SPLICE}} = 1.5\{1.15 \times [12.97\ \mathbf{MPa} - (-4.7\ \mathbf{MPa})]\} = 34.3\mathbf{MPa}$$

As seen in Table 11-3, the calculated stress range of 45.7 MPa (6.6 ksi) in the rib plate is less than the 48 MPa (7.0 ksi) allowable stress for a welded splice. Figure 11-22 shows the stress field in the vicinity of the rib splice for the loading causing the maximum (tensile) portion of the stress range. The stresses are in the element local y-direction, which is the longitudinal direction of the rib.

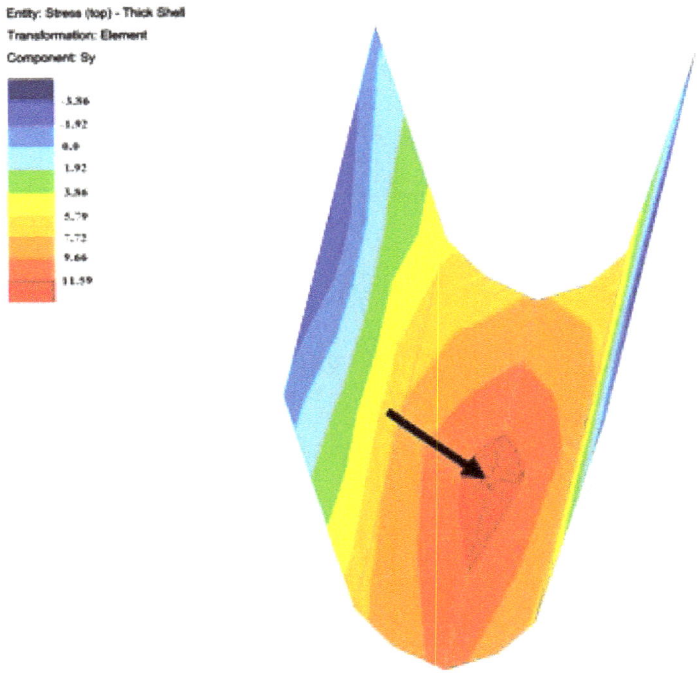

Figure 11-22 Results of FEA Model of the Rib Splice showing the Stress Contours (Maximum Stress Portion of Stress Range, in MPa)

11.1.6.3. Deck Place Splice

To get the maximum stress range in the longitudinal deck plate splice, a single static loading of one front wheel patch is applied. The wheel load is applied at mid-span of the panel between FBs and centered between ribs, as illustrated in Figure 11-23. A schematic illustrating the longitudinal position of the loading is shown in Figure 11-24. The location of interest is the deck plate directly under the load.

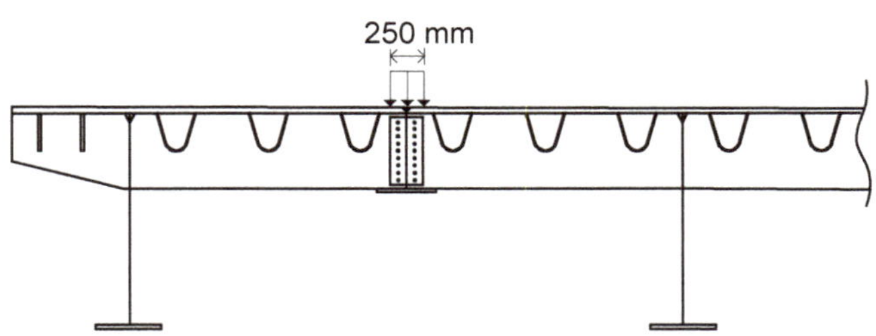

Figure 11-23 Transverse Wheel Location to Produce the Largest Impact on the Deck Plate Splice

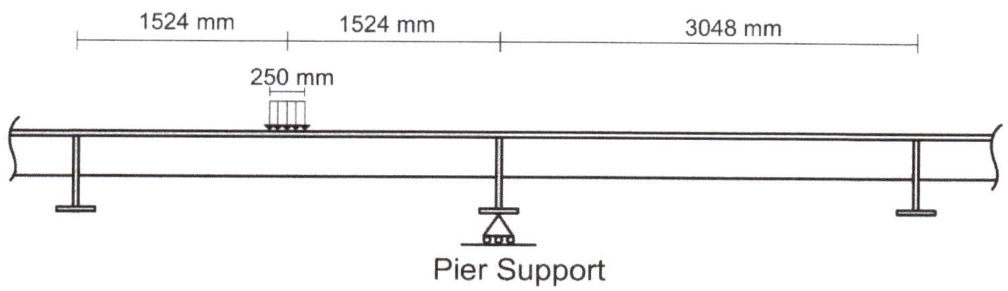

Figure 11-24 Longitudinal Wheel Location to Produce the Largest Impact on the Deck Plate Splice

Again, there is no discontinuity in the stress field at this location, so no extrapolation is required and the actual stress from the model is used to calculate the factored force effect, as shown below.

$$\gamma(\Delta \mathbf{f})_{\text{DECK-SPLICE}} = 2.25(1.15 \times 15.4\mathbf{MPa}) = 40.0\mathbf{MPa}$$

As seen in Table 11-3, the calculated stress range of 40 MPa (5.8 ksi) in the plate is less than the 69 MPa (10.0 ksi) allowable stress for a welded splice. Figure 11-25 shows the stress contours in the region of interest. The stresses are in the element local y-direction, which is the transverse direction of the bridge.

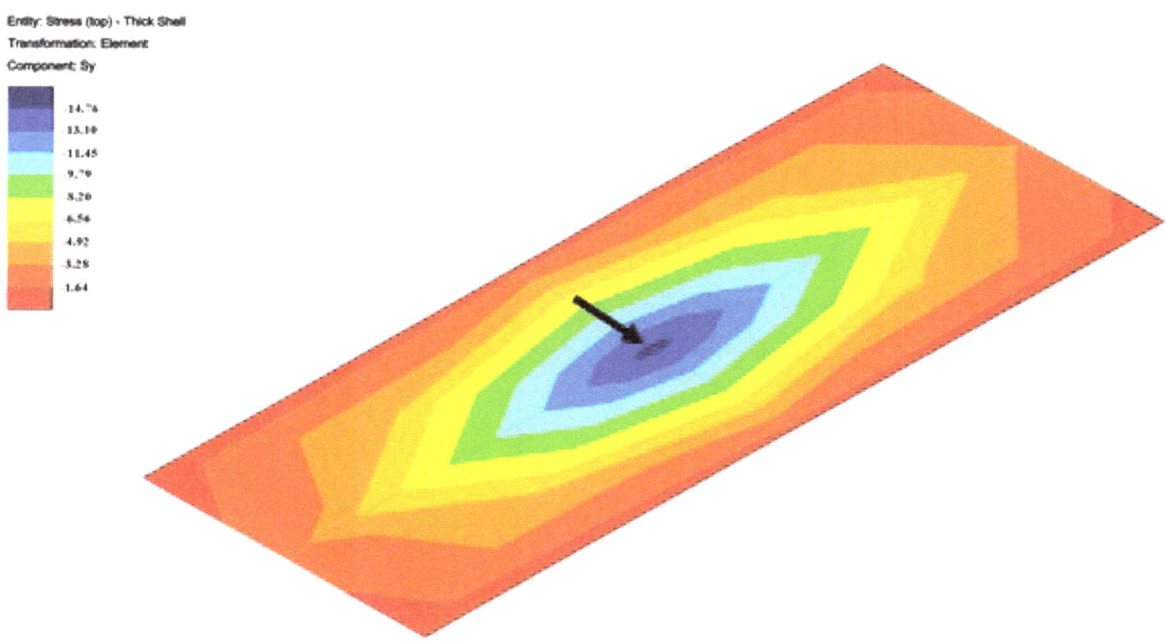

Figure 11-25 Results of FEA Model of the Deck Plate Splice showing the Stress Contours (in MPa)

11.1.6.4. Rib-to-Floorbeam (RF) Weld – Rib detail

The location of the loading for the rib portion of the connection between the rib and FB is based on the continuous beam influence line for bending at a support. The wheels are centered on the rib as shown in Figure 11-26. A schematic illustrating the longitudinal position of the wheels for the loading is shown in Figure 11-27. The maximum stress occurs when the second set of axles straddle the FB, and the location of the maximum stress is near the bottom of the rib.

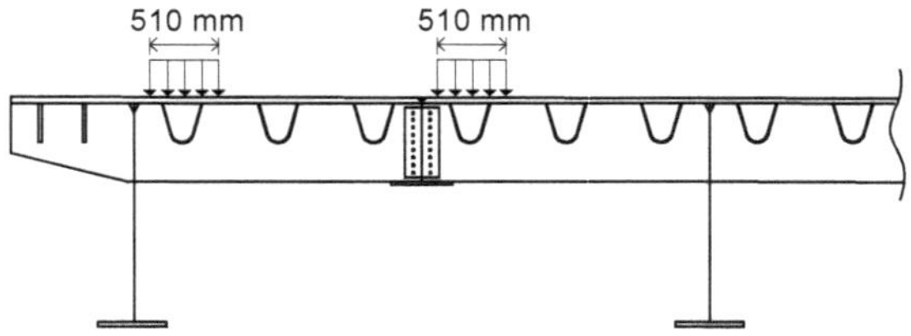

Figure 11-26 Transverse Wheel Location to Produce the Largest Impact on the Rib-to-Floorbeam (RF) Weld (at the Rib)

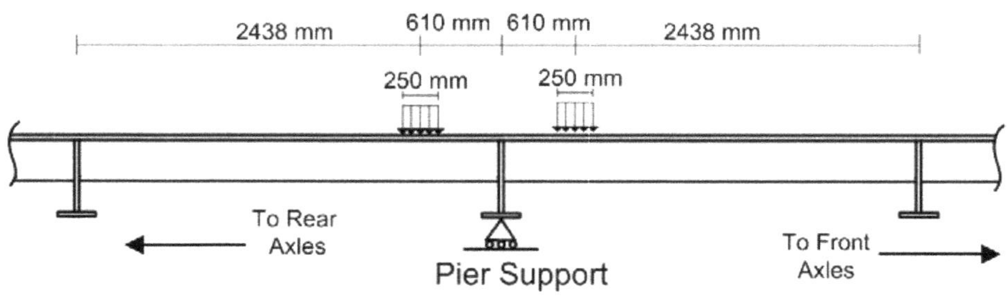

Figure 11-27 Longitudinal Wheel Location to Produce the Largest Impact on the Rib-to-Floorbeam (RF) Weld (at the Rib)

Due to the discontinuity from the FB connection at this location, a stress concentration is present. Stresses in the rib are extrapolated to the FB intersection, according to this Manual's guidelines in Chapter 5 (Eq. 5-5), using the following equation:

$$\sigma = 1.5\sigma_1 - 0.5\sigma_2 \qquad (5\text{-}5)$$

Where σ_1 and σ_2 are the stresses at the midpoint node locations (0.5 t and 1.5 t, respectively) identified in Figure 11-28.

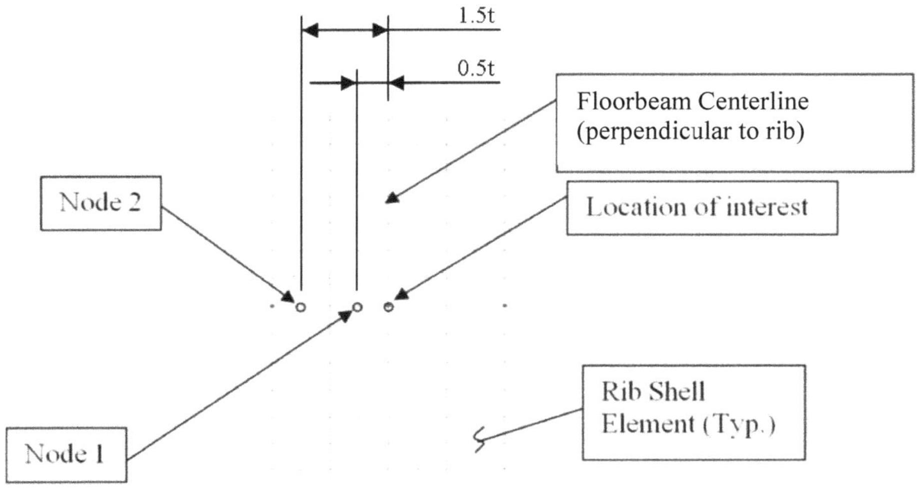

Figure 11-28 Shell Elements near the Rib-to-Floorbeam (RF) Connection showing the Node Locations used for Extrapolation of the Local Structural Stress

This extrapolation is performed around the entire perimeter of the rib at the connection with the FB. At the location of maximum stress, the Node 1 location shown in Figure 11-28 has a value of 19.1 MPa (2.8 ksi), while Node 2 has a value of 11.59 MPa (1.7 ksi). The calculation of the unfactored force effect by extrapolation at the location of maximum stress, as well as the calculation of the factored force effect to compare with the infinite life fatigue threshold stress, is shown below.

$$\sigma_{\mathbf{r}} = 1.5(19.1\ \mathbf{MPa}) - 0.5(11.59\ \mathbf{MPa}) = 22.86\ \mathbf{MPa}$$

$$\gamma(\Delta\mathbf{f})_{\mathbf{RIB@\ DIAPHRAGM}} = 1.5(1.15 \times 22.86\ \mathbf{MPa}) = 39.5\ \mathbf{MPa}$$

Table 11-3 shows that the calculated stress range of 39.5 MPa (5.7 ksi) in the rib plate is less than the 69.0 MPa (10 ksi) allowable stress for a welded connection. Figure 11-29 shows the stress contours in the rib for the maximum component of the stress range. The stresses are in the element local y-direction, which is the longitudinal direction of the rib.

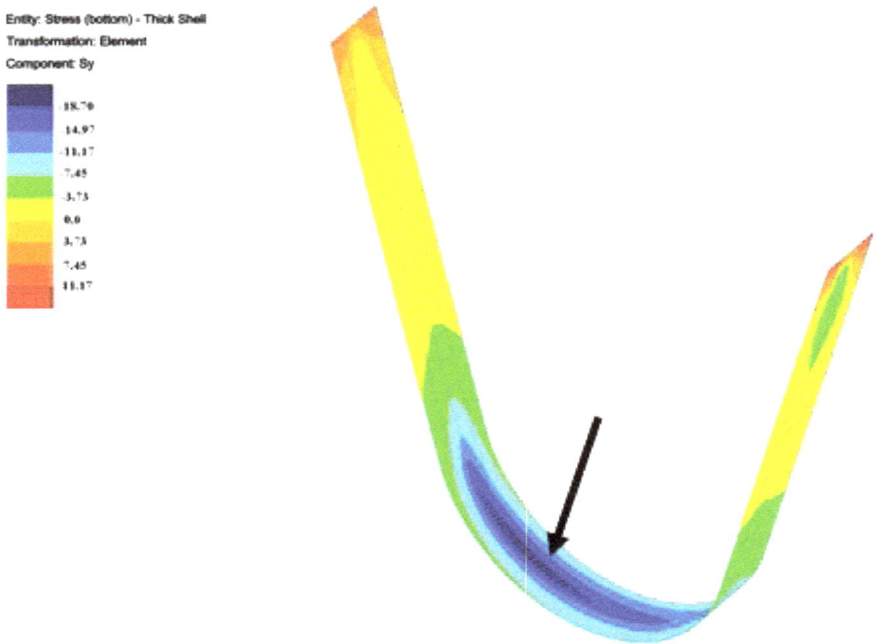

Figure 11-29 Results of FEA Model of the Rib to Floorbeam Weld showing the Rib portion of the Model (Maximum Component of Stress Range) (Rib, in MPa)

11.1.6.5. Rib-to-Floorbeam (RF) Weld – Floorbeam Detail

In order to obtain the stress range in the FB at the rib to FB weld, a series of truck loads are placed on the model at 50 mm (2 inch) increments to capture the location of the maximum and the minimum stresses. For the transverse position of the truck, the wheels are centered on the rib wall as shown in Figure 11-30. A schematic illustrating the longitudinal position of the wheels for the maximum and minimum loadings is shown in Figure 11-31. The maximum stress is generated when the rear wheel loads are about 0.5 times the span past the FB. The minimum stress occurs when the middle axles are placed about 0.5 times the span away from the FB, with the front axle in the span on the other side of the FB. The location of the maximum stress is near the bottom of the rib.

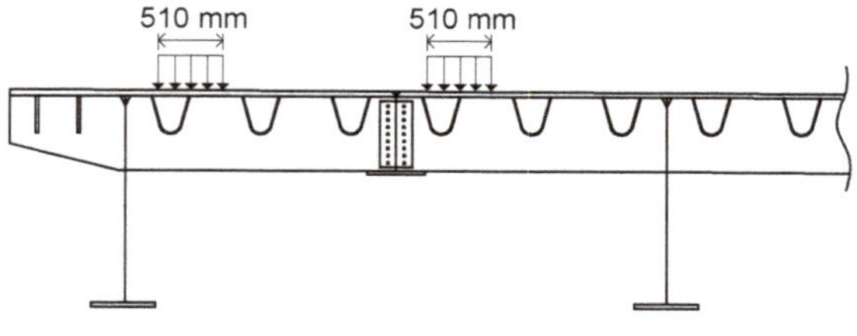

Figure 11-30 Transverse Wheel Location to Produce the Largest Impact on the Rib-to-Floorbeam (RF) Weld (at the Floorbeam)

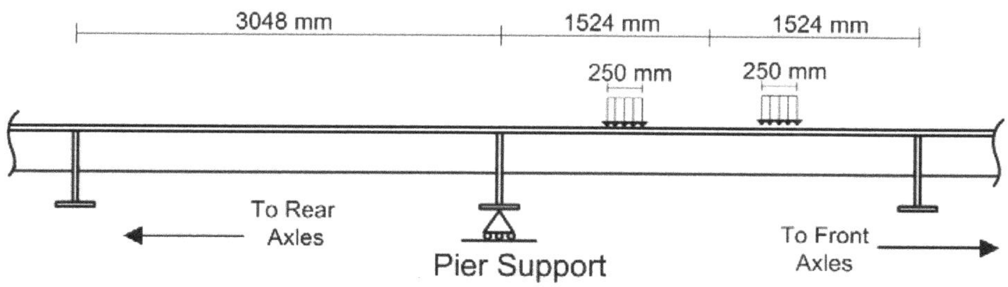

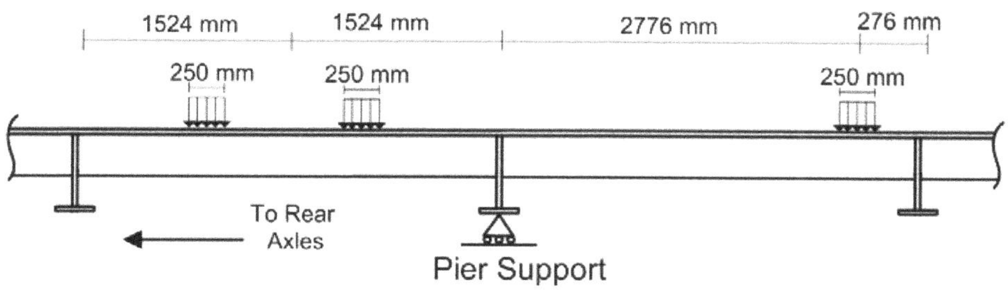

Figure 11-31 Longitudinal Wheel Location to Produce the Largest Impact on the Rib-to-Floorbeam (RF) Weld (at the Floorbeam)

Due to the discontinuity in the FB connection at this location, stresses in the FB plate are extrapolated to the rib intersection according to this manual's guidelines. The calculation of the unfactored force effect by extrapolation, as well as the calculation of the factored force effect to compare with the infinite life fatigue threshold stress, is shown below.

$$\sigma_{max} = 1.5(5.82 \text{ MPa}) - 0.5(4.02 \text{ MPa}) = 6.72 \text{ MPa}$$

$$\sigma_{min} = 1.5(-8.97 \text{ MPa}) - 0.5(-7.52 \text{ MPa}) = -9.7 \text{MPa}$$

$$\gamma(\Delta \mathbf{f})_{\mathbf{DIAPHRAGMWELD}} = 1.5\{1.15 \times [6.72\text{MPa} - (-9.7 \text{ MPa})]\} = 28.3\text{MPa}$$

As seen in Table 11-3, the calculated stress range of 28.3 MPa (4.1 ksi) in the rib plate is less than the 69 MPa (10 ksi) allowable stress for welded connections. Figure 11-32 shows the stress contours in the FB for the minimum component of the stress range. The stresses are in the element local y-direction, which is the direction normal to the weld connection made with the rib.

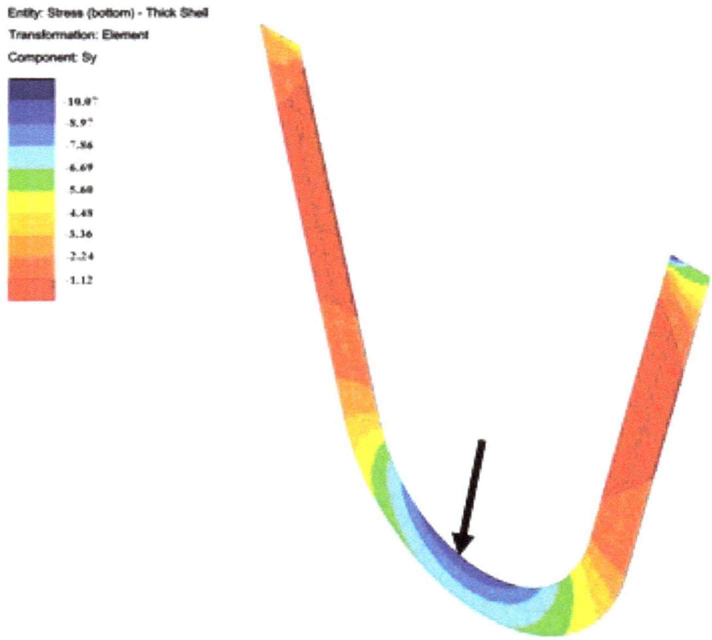

Figure 11-32 Results of FEA Model of the Rib to Floorbeam Weld showing the Rib to Floorbeam (RF) Weld for the Floorbeam Portion of the Connection (Minimum Component of Stress Range, in MPa)

11.1.7. Optimizing the Design

The design shown in this example contains dimensions that can potentially be improved upon to obtain an optimized design. The rib thickness of 13 mm (1/2 inch) and the FB spacing of 3,048 mm (10 ft) are required to satisfy the fatigue checks at the rib splice and the rib-to-floorbeam connection. Alternatively, the geometry of the rib could be altered. By using a deeper rib, the thickness of the rib could likely be made smaller and the rib span length made longer. This decision may result in a lighter and more economical design.

11.2. EXAMPLE 2 – CABLE-STAYED BRIDGE

This example focuses on the design of the ribs, deck plate, and floorbeam. Extra attention is paid to the fatigue design of the RF connection.

The following guidelines/assumptions are made:

- ASTM A709 Grade 345 (Grade 50) Steel is used for all components.
- A thin polymer wearing surface (non-structural) is to be used.
- The bridge is required to carry four traffic lanes.
- The bridge is a cable stay bridge, with a main span of 460 m (1500 ft).
- The region of interest is a four-panel section of the bridge in the main span.
- A relieving cut-out is to be used at the RF connection.
- The bridge is to be designed for finite life, with an $(ADTT)_{SL}$ equal to 485.

- The maximum dead and live load axial forces in the deck plate (per rib strut) due to the cable supports are 1,025 kN (230 kip) and 380 kN (85 kip), respectively.
- The following checks are required at the region of interest.
    - Strength: Rib axial force capacity.
    - Service: Overall rib deflection.
    - Fatigue.
    - Rib portion of the RF.
    - FB portion of the RF.
    - Rib wall at the termination of the weld to the FB.
    - Free edge of the FB cut-out.
- The fatigue checks are to be made using the finite element method and concentration stress analysis as required.

Shell elements are to be utilized in the region of interest, with weld geometries not modeled.

11.2.1. Description of Bridge

The overall structural system consists of two edge girders supported by cables along each side with an OSD stiffened with closed ribs spanning between FBs, as seen in Figure 11-33 and Figure 11-34. The FBs are spaced at 4,572 mm (15 ft).

In this design, the FB cut-outs have additional relieving cut-outs below the longitudinal ribs, since the rib spans cause moderate rotations at the FBs. The cut-out is comprised of several curved cuts of varying radii and provides a minimum gap of 13 mm (0.5 inch) between the FB and longitudinal rib. Detail A in Figure 11-34 shows the dimensions of the rib and the relieving cut-out.

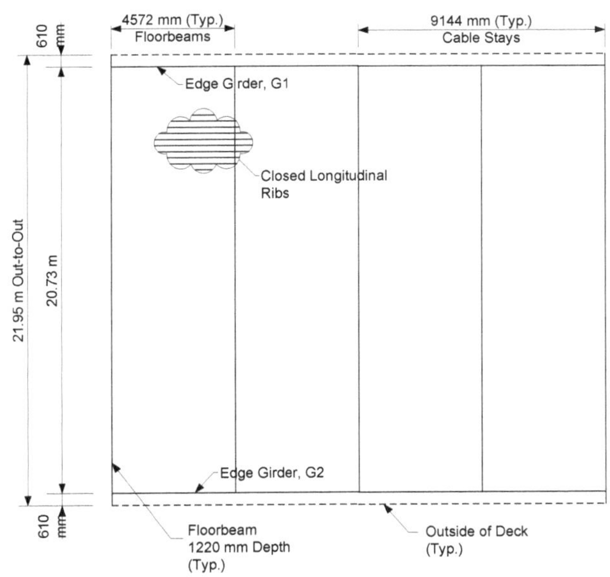

Figure 11-33 Cable-stayed Example Framing Plan Excerpt showing Overall OSD Panel Geometry

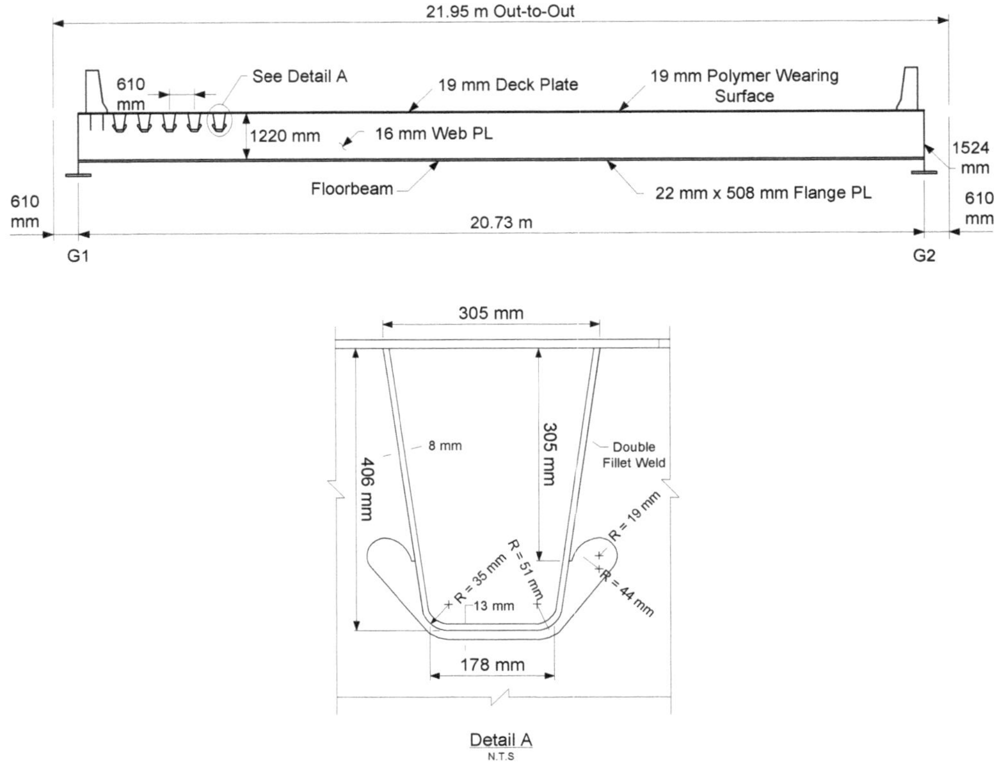

Figure 11-34 Cable-stayed Example Cross-Section and Detailing showing the Floorbeam, Trapezoidal Rib, and Cut-out Dimensions

11.2.2. Development of the finite Element Method

Following the recommendations in this Manual, a finite element model of a portion of the cable-stayed bridge described above is created. The model is comprised of four 4,572 mm (15 ft) spans for a total of 18.3 m (60 ft), with five FBs being modeled. Spring supports modeling cable stay direction and stiffness are placed at every other FB (six total). Additional supports are added to react to the longitudinal forces and to keep the bridge from moving transversely. Figure 11-35 shows an overall view of the model.

The entire model utilizes eight-noded thick shell elements. The mesh size is made smallest in the regions of interest and transitioned to larger sizes further away. This keeps the degrees of freedom, and consequently the model run time, as small as practical while still obtaining valid results in the regions of interest. The eight-noded element is required in the refined areas to perform the stress extrapolation techniques used in the fatigue evaluation. For simplicity, these elements are used throughout the model, so as to avoid transition difficulties.

A detailed mesh is used at two locations; the intersection of two interior ribs with a cable-supported FB, and the intersection of two ribs near one edge girder with a cable supported FB. These two finely meshed areas can be seen in Figure 11-36, which shows a portion of the model with the deck plate removed.

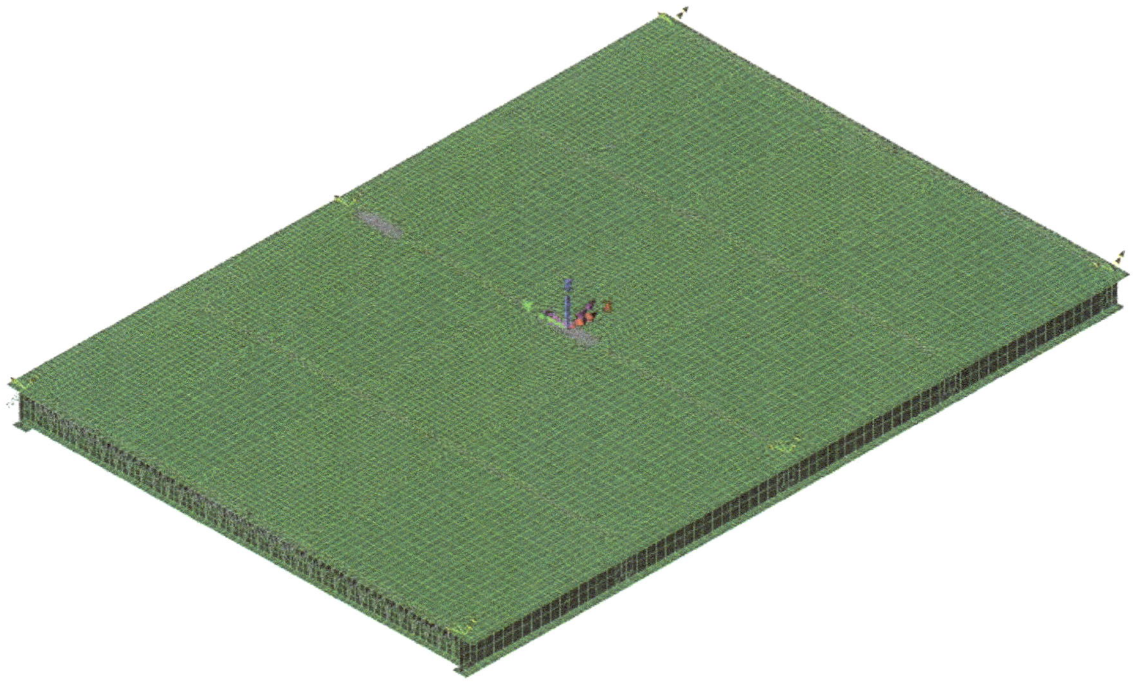

Figure 11-35 Cable-stayed Bridge Section Model showing Overall view of Deck, Floorbeam and Girder Meshing

Figure 11-36 Close-up of Detailed Section with Deck Removed to show how the individual Girder, Floorbeam, and Rib Element have been Modeled. Gray Areas Indicate High Mesh Density

Figure 11-37 shows a typical refined mesh at the rib-deck-Floorbeam (RDF) intersection. As in Example 1, the size of the elements in the FB web directly adjacent to the rib is made equal to the thickness of the web, which is 16 mm (5/8 inch). This mesh size is maintained for two elements away from the rib, to provide a constant element size for use in the stress calculations involving extrapolation. Away from the rib, the mesh is transitioned to much larger element sizes to limit the total number of degrees of freedom of the model.

Figure 11-37 also shows the rib mesh at the junction with the FB web. Similar to the web, the mesh size is smallest at the connection area, and is transitioned to a larger size away from this area. Because the rib thickness is less than the web thickness (8 mm [5/16 inch] versus 16 mm [5/8 inch]), the mesh size guidelines result in a conflict at this location. The width of the elements is set by the size used in the web, but the length in the longitudinal direction of the bridge, is set at 8 mm (5/16 inch), based on the rib thickness. The resulting aspect ratio of roughly 2 is well within the limits of the element types used.

The deck plate mesh at the junction with the FB web is also seen in Figure 11-37. Similarly to the web, the mesh size is smallest at the connection area, and is transitioned to a larger size away from this area. Because the deck plate intersects with the ribs and the FB web, mesh size conflicts occur at this location. In this example, the thickness of the web and the deck plate are similar (16 mm [5/8 inch] and 19 mm [3/4 inch], therefore the aspect ratio is roughly 1.2), so no conflict results in that case. At the rib intersection, the width of the deck plate elements is set by the size used in the deck, but the length in the longitudinal direction of the bridge is set based on the rib thickness. This results in an aspect ratio of 2.4.

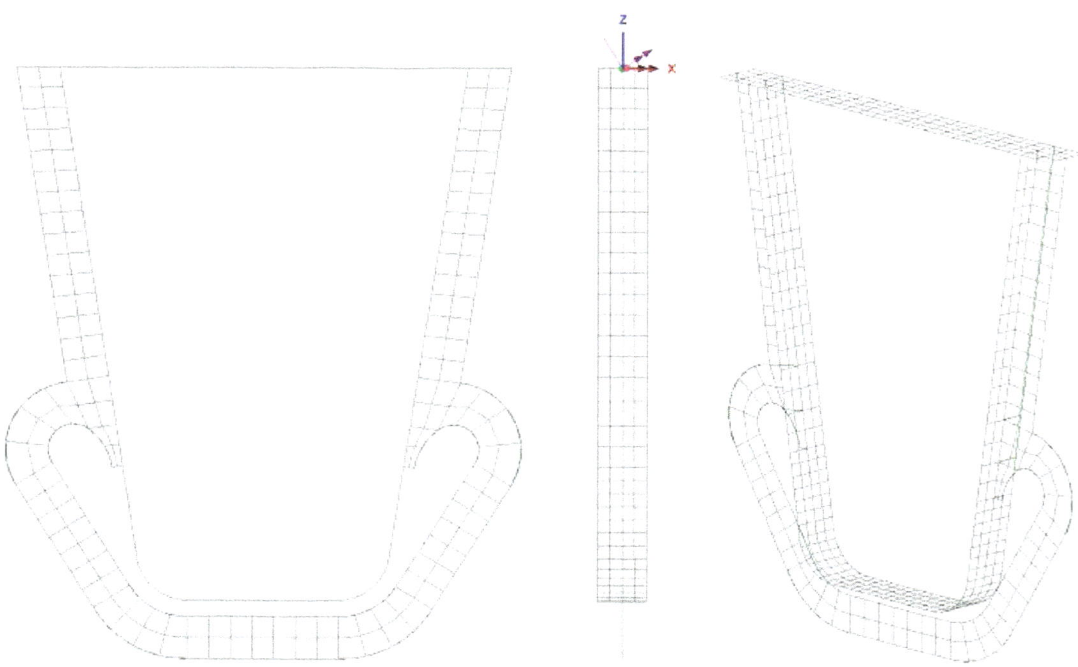

Figure 11-37 Section of Model Showing Refined Mesh at Rib-to-Floorbeam Intersection including the Adaptation of the Cut-out Smooth Transitions

11.2.3. Service Limit State Checks

Considering only the modeled section of the bridge, the live load deflection of the deck is checked against the limit, L/1000, where L is the total bridge span. The worst case loading in the model occurs from five lanes placed symmetrically about the bridge centerline. Superimposed with each of the lane loads is HL-93 truck loadings. The resulting maximum deflection value is calculated to be roughly 200 mm (7 3/4 inch), which is well below the L/1000 value of 460 mm (18 inch). Similar deflection checks for the rib and FB are also satisfied.

11.2.4. Strength Limit State Checks

The presence of large compression forces in the deck plate due to the stay cables can cause local buckling (of the rib or deck plate) or panel buckling. In this analysis the capacity of a single rib strut under compression is checked.

The width to thickness ratios for each component of the rib strut must be checked for local buckling.

When the inequality of the Equation 11-2 is true, local buckling must be considered.

$$\frac{\mathbf{b}}{\mathbf{t}} \geq 1.4\sqrt{\frac{\mathbf{E}}{\mathbf{F_y}}} \qquad (11\text{-}2)$$

$$\frac{\mathbf{b}}{\mathbf{t}} \geq 1.4\sqrt{\frac{200000\,\mathbf{MPa}}{345\,\mathbf{MPa}}} = 33.7$$

For the deck plate between rib walls and between adjacent ribs:

$$\frac{\mathbf{b}}{\mathbf{t}} = \frac{305\,\mathbf{mm}}{19\,\mathbf{mm}} = 16 \qquad \text{OK, No Local Buckling}$$

For the rib flange:

$$\frac{\mathbf{b}}{\mathbf{t}} = \frac{178\,\mathbf{m}}{8\,\mathbf{mm}} = 22.25 \qquad \text{OK, No Local Buckling}$$

For the rib webs:

$$\frac{\mathbf{b}}{\mathbf{t}} = \frac{412\,\mathbf{mm}}{8\mathbf{mm}} = 51.5 \qquad \text{NG, Slenderness Reduction Factor Required}$$

The effective width of a component deemed slender by Equation 11-2 is calculated by using the following Equation 11-3.

$$\mathbf{b_e} = 1.92\mathbf{t}\sqrt{\frac{\mathbf{E}}{\mathbf{f}}}\left[1 - \frac{0.38}{{}^{\mathbf{b}}\!/_{\mathbf{t}}}\sqrt{\frac{\mathbf{E}}{\mathbf{f}}}\right] \leq \mathbf{b} \qquad (11\text{-}3)$$

For rib webs, conservatively taking $\mathbf{f} = \mathbf{F_y}$:

$$\mathbf{b_e} = 1.92(8\,\mathbf{mm})\sqrt{\frac{200000\,\mathbf{MPa}}{345\,\mathbf{MPa}}}\left[1 - \frac{0.38}{412\,\mathbf{mm}/8\,\mathbf{mm}}\sqrt{\frac{200000\,\mathbf{MPa}}{345\,\mathbf{MPa}}}\right] \leq 412\,\mathbf{mm}$$

$$\mathbf{b_e} = 304\ \mathbf{mm}$$

The reduction factor for slender elements, **Q**, is then found by determining the ratio between the effective cross-sectional area and the total cross-sectional area of a strut, as seen in Equation 11-4.

$$\mathbf{Q} = \frac{\mathbf{A_{eff}}}{\mathbf{A}} \qquad (11\text{-}4)$$

$$\mathbf{Q} = \frac{2(304\,\mathbf{mm} \times 8\,\mathbf{mm}) + 610\,\mathbf{mm} \times 19\,\mathbf{mm} + 178\,\mathbf{mm} \times 8\,\mathbf{mm}}{2(412\,\mathbf{mm} \times 8\,\mathbf{mm}) + 610\,\mathbf{mm} \times 19\,\mathbf{mm} + 178\,\mathbf{mm} \times 8\,\mathbf{mm}} = 0.91$$

Next, the section properties of the strut are found in order to calculate the slenderness ratio of the strut and to determine whether to use Equation (11-5) or Equation (11-6).

$$\mathbf{F_{cr}} = \mathbf{Q}\left[0.685^{\frac{\mathbf{QF_y}}{\mathbf{F_e}}}\right]\mathbf{F_y} \qquad (11\text{-}5)$$

$$\mathbf{F_{cr}} = 0.877\mathbf{F_e} \qquad (11\text{-}6)$$

Calculate the slenderness ratio:

$$\mathbf{A} = 195.6\,\mathbf{cm}^2$$

$$\mathbf{I} = 43450\,\mathbf{cm}^4$$

$$\mathbf{r} = \sqrt{\frac{\mathbf{I}}{\mathbf{A}}} = 14.9\,\mathbf{cm}$$

With L = 457.2 cm and K assumed to equal 1.0:

$$\frac{\mathbf{KL}}{\mathbf{r}} = \frac{1.0 \times 457.2\,\mathbf{cm}}{14.9\,\mathbf{cm}} = 30.68 \leq 4.71\sqrt{\frac{\mathbf{E}}{\mathbf{QF_y}}} = 4.71\sqrt{\frac{200000\,\mathbf{MPa}}{0.91 \times 345\,\mathbf{MPa}}} = 118.9$$

Therefore, Equation (11-5) is used to calculate the $\mathbf{F_{cr}}$ value:

$$\mathbf{F_e} = \frac{\pi^2\mathbf{E}}{\left(\frac{\mathbf{KL}}{\mathbf{r}}\right)^2} = \frac{\pi^2 200000\,\mathbf{MPa}}{(30.68)^2} = 2100\,\mathbf{MPa}$$

$$\mathbf{F_{cr}} = 0.91\left[0.658^{\frac{0.91\times345\,\mathbf{MPa}}{2100\,\mathbf{MPa}}}\right]345\,\mathbf{MPa} = 295\ \mathbf{MPa}$$

$$\phi\mathbf{P_n} = \phi(\mathbf{A}\times\mathbf{F_{cr}}) = 0.9(195.6\,\mathbf{cm}^2\times295\,\mathbf{MPa}\times\frac{1\,\mathbf{kN}}{1000\,\mathbf{N}}\times\frac{(10\,\mathbf{mm})^2}{(1\ \mathbf{cm})^2}) = 5190\ \mathbf{kN}$$

The dead and live load axial forces in the deck plate (per rib strut) were given to be 1,025 kN (230 kips) and 380 kN (85 kips), respectively. Using the Strength I load factors, the factored load is 2,000 kN (440 kips) for each strut, which is well below the calculated φP_n value of 5190 kN (1170 kips).

11.2.5. Fatigue Limit State Checks

The finite element model developed for Example 2 is used to find the live load fatigue stress ranges for details in OSDs with relief cut-outs at four locations. The allowable stress ranges for the various details are calculated based on finite life of the details, using the given (ADTT)SL, a 75-year design life, and n (cycles per truck passing) equal to 1. Note that this example is calculating the fatigue life for details not directly connected to the deck plate. Those details connecting to the deck plate experience one cycle per axle, i.e. n = 5. See Chapter 5 for additional information on the number of cycles.

As in Example 1, a design truck with a constant 9,000 mm (30 ft) between rear axles groups is used in accordance with the AASHTO LRFD fatigue specifications and the guidelines in the previous chapters. The 142 kN (32 kip) axles are split into two 71 kN (16 kip) axles 1,220 mm (4 ft) apart, as illustrated in Figure 5-1. This results in a uniform tire pressure of 275 kPa (40 psi), which is distributed over a 250 mm (10 inch) square patch for the two 17.75 kN (4 kip) wheels, and over a 510 mm (20 inch) wide x 250 mm (10 inch) long patch for the eight 35.5 kN (8 kip) wheels. This loading model is altered slightly to account for the distribution of the load through the wearing surface and one half of the deck plate, as allowed in Section 5.5.1.1 of this Manual. At the ends of each tire edge, an amount equal to the distance from the top of the wearing surface to the mid-depth of the deck is added, and the pressure is then altered to result in the correct total load for the wheel.

The truck is moved along the model in 610 mm (2 ft) increments, with the wheel centered over an interior rib to generate an influence line for the stresses. After determining the approximate longitudinal positions of the wheels that resulted in the highest stresses, the truck is shifted transversely so the wheels are centered over the rib, the rib wall, and the tooth of both an interior rib and the edge rib for a limited number of longitudinal positions. From these loadings the position of the truck that resulted in the largest stress for each of the four details is determined. An impact factor of 15 percent is applied to all loadings, as well as the finite life fatigue load factor of 0.75.

Table 11-4 summarizes the results of the analysis, with explanations of the calculations given in the subsequent sections.

Table 11-4 Summary of Fatigue Checks for Example 2

Detail	Point of Interest	S_r (MPa)	$S_{r,allow}$ for $(ADTT)_{SL}$ = 485 (MPa)
RF (Rib)	At start of cut-out	38.6	47.7
Rib Wall @ Cut-out	At cut-out	47.7	47.7
RF (Floorbeam)	At start of cut-out	17.9	47.7
Floorbeam Cut-out	Around free edge of cut-out	65.3	85.2

The calculation of the allowable stress ranges for the two types of details is performed using Section 6.6.1.2.5 of the AASHTO LRFD Bridge Design Specifications (AASHTO 2010). The calculations are shown below.

For the Category C details (RF, rib wall): AASHTO Ref.

$$\mathbf{A} = 44 \times 10^{8}\ \mathbf{ksi}$$ Table 6.6.1.2.5-1

$$\mathbf{N} = 365(75)\,\mathbf{n}(\mathbf{ADTT})_{SL} = 365(75)(1)(485) = 13{,}280{,}000$$ Equation 6.6.1.2.5-3

$$\Delta\mathbf{F} = \left(\frac{\mathbf{A}}{\mathbf{N}}\right)^{1/3} = \left(\frac{44 \times 10^{8}}{13{,}280{,}000}\right)^{\frac{1}{3}} = 6.92\ \mathbf{ksi} = 47.7\ \mathbf{MPa}$$ Equation 6.6.1.2.5-2

For the Category A detail (FB cut-out):

$$\mathbf{A} = 250 \times 10^{8}\ \mathbf{ksi}$$ Table 6.6.1.2.5-1

$$\mathbf{N} = 13{,}280{,}000$$

$$\Delta\mathbf{F} = \left(\frac{\mathbf{A}}{\mathbf{N}}\right)^{1/3} = \left(\frac{250 \times 10^{8}}{13{,}280{,}000}\right)^{\frac{1}{3}} = 12.35\ \mathbf{ksi} = 85.2\ \mathbf{MPa}$$ Equation 6.6.1.2.5-2

11.2.5.1. Rib-to-Floorbeam (RF) Weld – Rib Detail

The maximum longitudinal stress in the rib at the FB web occurs in the edge rib with the truck positioned so the wheel load is located on the tooth (as seen in Figure 11-42a), with the FB positioned halfway between the front axle and the first middle axle, as illustrated in Figure 11-43. The maximum stress range occurs in the rib at the termination of the FB web cut-out, as illustrated in the stress contour of Figure 11-38. The stresses are in the element local y-direction, which is the longitudinal direction of the rib.

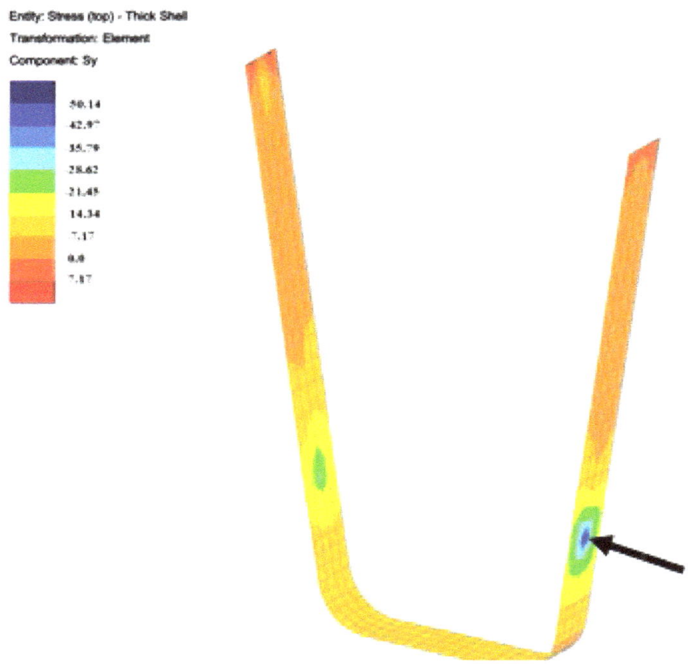

Figure 11-38 Results of FEA Model of the Rib to Floorbeam Weld showing the Rib portion of the Model for the Longitudinal Stress Contours in Rib Wall (in MPa)

Due to the stress concentration seen in Figure 11-38, the stress extrapolation technique is used to find the unfactored force effect. This value is then multiplied by the appropriate factors to determine the factors' live load stress range, as seen below.

$$\sigma_{\mathrm{r}} = 1.5(-39.6\ \mathbf{MPa}) - 0.5(-29.3\ \mathbf{MPa}) = -44.75\ \mathbf{MPa}$$

$$\gamma(\Delta \mathbf{f})_{\mathbf{RIB\text{-}LONGITUDINAL}} = 0.75(1.15 \times 44.75\ \mathbf{MPa}) = 38.6\ \mathbf{MPa}$$

Table 11-4 shows that the calculated stress range of 38.6 MPa for this detail is less than the finite life allowable stress range of 47.7 MPa for the anticipated ADTT.

11.2.5.2. Rib Wall at Cut-out

The maximum vertical stress in the rib at the FB web cut-out also occurs in the edge rib with the truck positioned such that the wheel load is located on the tooth, with the FB positioned halfway between the front axle and the first middle axle. Figure 11-39 shows the stress contours and illustrates the location of the stress. The stresses are in the element local x-direction, which is the direction along the perimeter of the rib cross-section.

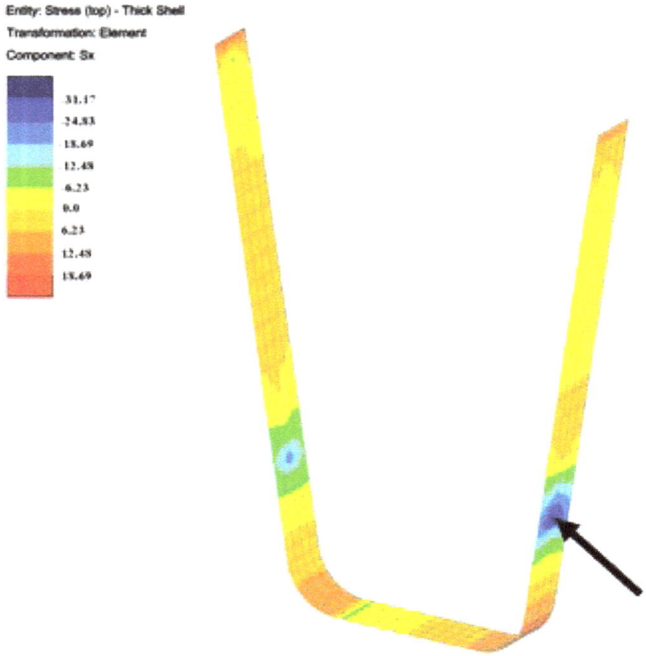

Figure 11-39 Results of FEA Model of the Rib to Floorbeam Weld showing the Rib portion of the Model for the Vertical Stress Contours in Rib Wall (in MPa)

At this location, the stresses are extrapolated to the start of the cut-out to determine the unfactored force effect, which is then multiplied by the appropriate factors to arrive at the factored live load stress range.

$$\sigma_{\mathbf{r}} = 1.5(44.9\ \mathbf{MPa}) - 0.5(24.1\ \mathbf{MPa}) = 55.3\ \mathbf{MPa}$$

$$\gamma(\Delta \mathbf{f})_{\mathbf{RIB\text{-}VERTICAL}} = 0.75\,(1.15 \times 55.3\mathbf{MPa}) = 47.7\ \mathbf{MPa}$$

Table 11-4 shows that the calculated stress range of 47.7 MPa (7.0 ksi) for this detail is equal to the finite life allowable stress range of 47.7 MPa (7.0 ksi).

11.2.5.3. Rib-to-Floorbeam (RF) Weld – Floorbeam Detail

The maximum stress in the FB web normal to the rib weld also occurs at the edge rib with the truck positioned such that the wheel load was located on the tooth, with the FB positioned halfway between the front axle and the first middle axle, as illustrated in Figure 11-42a and Figure 11-43a. Similarly to the previous two cases, the highest stress occurs at the cut-out termination. Figure 11-40 shows the stress contours and illustrates the location of the stress. The stresses are in the element local y-direction, which is the direction normal to the weld and free edge.

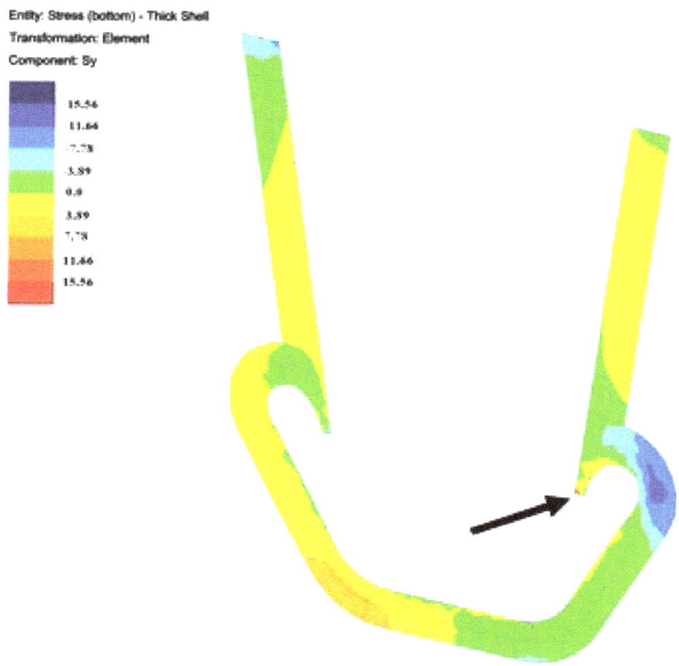

Figure 11-40 Results of FEA Model of the Rib to Floorbeam Weld showing the Floorbeam portion of the Model for the Stress Contours Normal to the Weld (in MPa)

The stresses are extrapolated to the start of the cut-out to determine the unfactored force effect, which is then multiplied by the appropriate factors to arrive at the factored live load stress range.

$$\sigma_r = 1.5(14.4\ \mathbf{MPa}) - 0.5(1.66\ \mathbf{MPa}) = 20.77\ \mathbf{MPa}$$

$$\gamma(\Delta \mathbf{f})_{\mathrm{FLOORBEAM}} = 0.75(1.15 \times 20.77\mathbf{MPa}) = 17.9\ \mathbf{MPa}$$

Table 11-4 shows that the calculated stress range of 17.9 MPa (2.6 ksi) for this detail is less than the allowable stress range of 47.7 MPa (7.0 ksi).

11.2.5.4. Floorbeam Cut-out

The maximum stress in the FB web around the edge of the cut-out occurs in the edge rib with the truck positioned such that the wheel load was located on the web of the rib, as seen in Figure 11-42b, with the FB positioned halfway between the two middle axles, as illustrated in Figure 11-43b. Figure 11-41 shows the stress contours and illustrates the location of the largest stress range. The stresses are in the element local x-direction, which is the direction along the free edge of the cut-out.

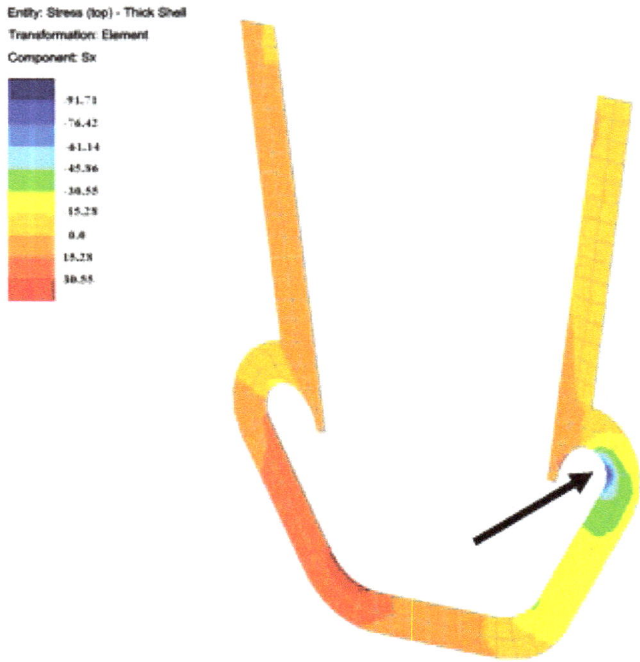

Figure 11-41 Results of FEA Model of the Rib to Floorbeam Weld showing the Floorbeam portion of the Model for the Stress Contours Tangential to Cut-out Stress (in MPa)

The stresses are extrapolated to the start of the cut-out to determine the unfactored force effect, which is then multiplied by the appropriate factors, to arrive at the factored live load stress range.

$$\sigma_r = 1.5(-60.2\ \mathbf{MPa}) - 0.5(-29.2\ \mathbf{MPa}) = 75.7\ \mathbf{MPa}$$

$$\gamma(\Delta \mathbf{f})_{\mathrm{FLOORBEAM\text{-}CUTOUT}} = 0.75(1.15 \times 75.7\mathbf{MPa}) = 65.3\ \mathbf{MPa}$$

Table 11-4 shows that the calculated stress range of 65.3 MPa (9.5 ksi) for this detail is less than the finite life allowable stress range of 85.2 MPa (12.0 ksi).

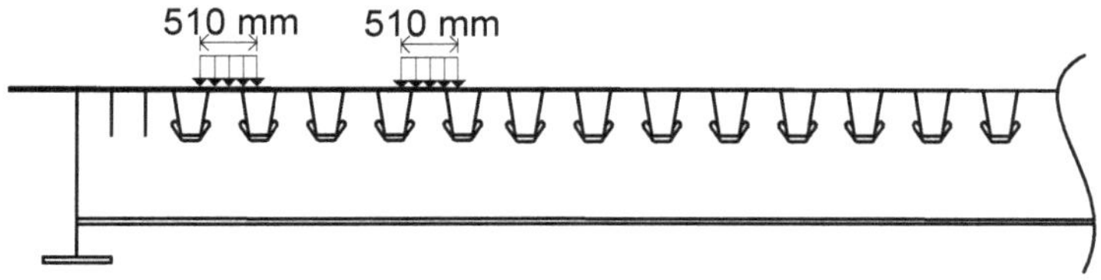

(a) Centered on tooth

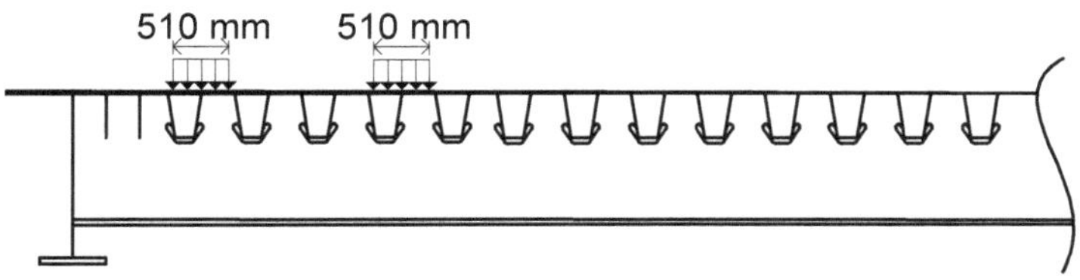

(b) Centered over Rib Web

Figure 11-42 Transverse Wheel Location to Produce the Largest Impact on the Cut-out

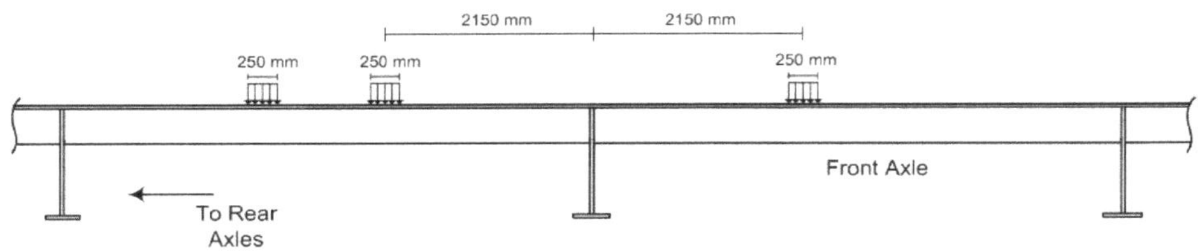

(a)

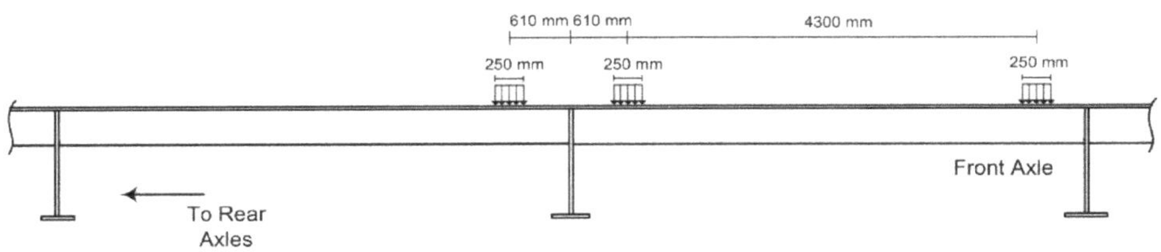

(b)

Figure 11-43 Longitudinal Wheel Locations to Produce the Largest Impact on the Cut-out

11.2.6. Optimizing the Design

The calculated stress ranges for the various details examined are all less than the finite life allowable stress limits, which are found based on an (ADTT) SL of 485. Given a larger daily volume of trucks, changes would be required to ensure that the design is adequate. Since the strength and serviceability of the OSD system as a whole are sufficient, this implies that the geometry of the rib and cut-out should be altered to limit the fatigue stresses. Possible geometry changes include increasing the radii of the relieving cut-out and grinding smooth the connection of the FB to the rib at the cut-out, as discussed in Section 6.5 of this Manual. Also, more accurate (and less conservative) calculation of local stresses could be obtained by employing more detailed modeling techniques at the critical locations, such as solid element modeling, weld profile consideration, and so forth.

12. REFERENCES

AASHTO Highways Subcommittee on Bridges and Structures, (2008), **The Manual for Bridge Evaluation**, First Edition, American Association of State Highway and Transportation Officials, Washington, D.C.

AASHTO Highways Subcommittee on Bridges and Structures, (2010), **LRFD Bridge Construction Specifications**, Third Edition, American Association of State Highway and Transportation Officials, Washington, D.C.

AASHTO LRFD Bridge Design Specifications, 5th Ed., (2010) American Association of State Highway and Transportation Officials, Washington D.C.

AASHTO Subcommittee on Bridges and Structures, (1998 with 2002 Interim Revisions), **AASHTO Guide for Commonly Recognized (CoRe) Structural Elements**, American Association of State Highway and Transportation Officials, Washington, D.C.

Abrahams, M. and Hirota, M. (2004). "Fremont Orthotropic Bridge Deck Longest Steel Arch in Western Hemisphere." Proc. of 2004 Orthotropic Bridge Conference, ASCE, Sacramento, CA, Aug. 25-27.

AISC (1963). **Design Manual for Orthotropic Steel Plate Deck Bridges**, American Institute of Steel Construction. Chicago, IL.

AISC 2005. **Specification for Structural Steel Buildings**. American Institute of Steel Construction.

AISI (2001), **North American Specification for the Design of Cold-Formed Steel Structural Members**, American Iron and Steel Institute, Washington, DC.

Allen, T.M., Nowak, A.S., Bathurst, R.J., (2005). "Calibration to Determine Load and Resistance Factors for Geotechnical and Structural Design," Transportation Research Circular E-C079. Transportation Research Board. Washington D.C.

American Concrete Institute Committee 503 (ACI503R-93 – Reapproved 2008), "Use of Epoxy Compounds with Concrete," (Appendix A).

American Welding Society D1.5. (2008), Bridge Welding Code, AWS, Miami, FL

Anderson, T.L. (2005). Fracture Mechanics Fundamentals and Applications, 3rd Edition, Taylor & Francis, New York.

ASTM A6 / A6M – 11, (2011) Standard Specification for General Requirements for Rolled Structural Steel Bars, Plates, Shapes, and Sheet Piling, American Society for Testing and Materials

ASTM D3633-98 (Reapproved 2006), Standard Test Method for Electrical Resistivity of Membrane-Pavement Systems, American Society for Testing and Materials

Barrett, D. (2003), "Fatigue Performance of Polymer Concrete Wearing Surfaces," M. S. Thesis, University of Missouri-Columbia, May 2003, 90 pp.

Barsom, J.M., and Rolfe, .S.T. (1987). Fracture & Fatigue Control in Structures Applications of Fracture Mechanics, 2nd Edition, Prentice Hall, Englewood Cliffs, NJ.

Battista, R.C., Pfeil, M.S., Carvalho, E.M.L. (2008) "Fatigue Life Estimates for a Slender Orthotropic Steel Deck," J. Constructional Steel Research.

Bethlehem Steel Corporation (unknown date) **Orthotropic Bridge Decks Using Bethlehem Standard** Ribs, Bethlehem, PA

Bjorhovde, R., Brozzetti, J., Alpsten, G.A., and Tall, L. (1972), "Residual Stresses in Thick Welded Plates," **Weld J**, Vol. 51, No. 8.

Boersma, P.D. (2003) "Techniques and Solutions for Rehabilitation of Orthotropic Steel Bridge Decks in the Netherlands", Proceedings, 10th Annual International Conference and Exhibition – structural Faults and Repair Conference, London, England

Braam, R., Kolstein, H., Romeijn, A, and Buitelaar, P. (2008), "Concrete Overlays to Improve the Fatigue Life of Movable Orthotropic Steel Bridges". Proceedings 2008 International Orthotropic bridge Conference, Sacramento, CA

Buckland, P. (2004). "Four Decades of Experience with Orthotropic Decks." Proc. of 2004 Orthotropic Bridge Conference, ASCE, Sacramento, CA, Aug. 25-27.

Buitelaar, P. and Braam, R. (2008), "Application of Reinforced High Performance Concrete Overlays on Orthotropic Steel Bridge Decks – Theory and Practice", Proceedings 2008 International Orthotropic bridge Conference, Sacramento, CA

Burgan, B.A. and Dowling, P.J. (1985). "The Collapse Behavior of Box Girder Compression Flanges – Numerical Modeling of Experimental Results," CESLIC Report BG 83, Imperial College, University of London.

Callaghan, J. (2010) "Connecting the Bay Bridge" AB Connections, American Bridge Company, Coraopolis, PA

Cao, W.M. (1998), "Polymer Concrete Wearing Surface System for Orthotropic Steel Plate Bridges," M. S. Thesis, University of Missouri-Columbia, May, 1998

Chen, S.S., Aref, A.J., Ahn, I.-S., Chiewanichakorn, M., Carpenter, J.A., Nottis, A and Kalpakidis, I. (2005). "Effective Slab Width for Composite Steel Bridge Members," NCHRP Report 543, National Cooperative Highway Research Program, Washington, DC.

Choi, D., Kim, Y., Yoo, H., Seo, J. (2008). "Orthotropic Steel Deck Bridges in Korea." Proc. of 2008 Orthotropic Bridge Conference, ASCE, Sacramento, CA, Aug. 25-27.

Chou, C.C., Uang, C.M., Seible, F. (2006), "Experimental Evaluation of Compressive Behavior of Orthotropic Steel Plates for the New San Francisco-Oakland Bay Bridge," Journal of Bridge Engineering, ASCE, Volume 11, No. 2.

Connor, R.J., Fisher, J.W. (2001). "Results of Field Measurements on the Williamsburg Bridge Orthotropic Deck – Final Report," ATLSS Report No. 01-01, Department of Civil and Environmental Engineering, Lehigh University, Bethlehem PA, January.

Connor, R.J. (2002) "A Comparison of the In-service Response of an Orthotropic Steel Deck with Laboratory Studies and Design Assumptions," Ph.D. dissertation, Department of Civil Engineering, Lehigh University, Bethlehem, PA May 2002

Connor, R.J., Fisher, J.W. (2004). "Results of Field Measurements Made on the Prototype Orthotropic Deck on the Bronx-Whitestone Bridge - Final Report," ATLSS Report No. 04-03, Center for Advanced Technology for Large Structural Systems, Lehigh University, Bethlehem PA.

Connor, R.J., Fisher, J.W. (2006) "A Consistent Approach to Calculating Stresses for Fatigue Design of Welded Rib-to-Web Connections in Steel Orthotropic Bridge Decks", Journal of Bridge Engineering, Vol. 11, No. 5, September/October 2006, pp. 517-525.

Connor, R.J. Urban, M.J., Kaufmann, E.J. (2008) "Heat-Straightening Repair of Damaged Steel Bridge Girders: Fatigue and Fracture Performance" NCHRP Report 604, National Cooperative Highway Research Program, Washington, DC

Cullimore, M.S.G., Flett, I.D., and Smith, J.W. (1983), "Flexure of Steel Bridge Deck Plate with Asphalt Surfacing, IABSE Periodica, Vol. 1.

DeCorte, W. and Van Bogaert, P. (2006) "The Effect of Shear Deformations in Floorbeams on the Moment Distribution in Orthotropic Plated Bridge Decks." Journal of Constructional Steel Research 62. Elsevier.

DeCorte, W., Delesie, C., Van Bogaert, P. (2007), "Examination of Local Stresses in Relation to Fatigue Failure at the Rib to Floorbeam Joint of Orthotropic Plated Bridge Decks," Bridge Structures: Assessment, Design, and Construction, Volume 3, Nos. 3-4. Taylor and Francis.

DeCorte, W., Van Bogaert, P. (2007) "Improvements for the Analysis of Floorbeams with Additional Webs Cutouts for Orthotropic Plated Decks with Closed Continuous Ribs." J. Steel Composite Struct., 2007, 7, 1-18.

De Jong, F.B.P., and Kolstein, H. (2004), "Strengthening a Bridge Deck with High Performance Concrete, Proceedings 2004 International Orthotropic bridge Conference, Sacramento, CA

Dexter, R.J., Kelly, B.A., (1997). Fatigue Performance of Repair Welds for Ship Structures, ATLSS Report No. 97-09, Department of Civil and Environmental Engineering, Lehigh University, Bethlehem PA.

Dong, P., Hong, J.K. (2006) "A Robust Structural Stress Parameter for Evaluation of Multiaxial Fatigue of Weldments" Journal of ASTM International (JAI), West Conshohocken, PA

Dowling, P.J. and Harding, J.E. (1992). "Box Girders," Chapter 2.7, Constructional Steel Design, An International Guide, P.J. Dowling, J.E. Harding and R. Bjorhovde (ed.), Elsevier, Essex, England, 175-196.

Dowling, P.J., Harding, J.E. and Frieze, P.A. (eds.) (1977). Steel Plated Structures, Proceedings, Imperial College, Crosby Lockwood, London.

ECS, European Committee for Standardization (1992) Eurocode 3 Design of Steel Structures

FHWA Office of Engineering, Bridge Division, Bridge Management Branch, (1995), **Recording and Coding Guide for the Structural Inventory and Appraisal of the Nation's Bridge** Report No. FHWA-PD-96-001, Washington, D.C.

Fisher, J.W., Albrecht, P.A., Yen, B.T., (1974). Klingerman, D.J., McNamee, B.M., Fatigue Strength of Steel Beams with Welded Stiffeners and Attachments, NCHRP Report 147, National Cooperative Highway Research Program, Washington, DC.

Fisher J.W., Jin J., Wagner, D.C., Yen, B.T. (1990). Distortion-Induced Fatigue Cracking in Steel Bridges NCHRP Report 236, National Cooperative Highway Research Program, Washington, DC, December.

Fisher, J.W. et al. (1993). Structure Failure Modes for Advanced Double Hull: Fatigue and Fracture Failure Modes, TDL 91-01, Phase I.3 (a), Volume 3a, March.

Freitas, S., Kolstein, M.H., Bijlaard, F.S.K., Souren, W.H.M (2009) "Orthotropic Deck Renovation of the Movable Bridge Scharsterrijn" Proceedings, Nordic Steel Construction Conference, Malmö, Sweden

Gimsing, N. (Editor) (1998). "The Storebaelt Publications: East Bridge." A/S Storebaeltsforbindelsen. Copenhagen, Denmark.

Goldberg, J.E. and Leve, H.L. (1957). "Theory of Prismatic Folded Plate Structures," IABSE, Vol. 16, International Association for Bridge and Structural Engineers, Zurich, Switzerland, 59-86.

Gopalaratnam, V. S., (2009) "Stresses and Composite Action of the Wearing Surface in Orthotropic Steel Bridge Decks", University of Missouri Report,

Gopalaratnam, V. S., Baldwin, Jr., J.W., and Hartnagel, B.A. (1993-1995), Performance of Polymer Concrete Wearing Surface System on the Poplar Street Bridge Inspection Reports I-III, Report 92-5. Submitted to the Missouri Highway and Transportation Department.

Gopalaratnam, V.S., Baldwin Jr., J.W., Hartnagel, B.A., and Rigdon, R.A. (1993a), "Evaluation of Wearing Surface Systems for Orthotropic Steel-Plate-Bridge Decks," Final Report 89-2. Submitted to the Missouri Highway and Transportation Department, 64 pp.

Gopalaratnam, V.S., Baldwin Jr., J.W., and T. R. Krull (1993b) Application of Polymer Concrete Wearing Surface on the Poplar Street Bridge Deck: Construction Report," Final Report 92-5. Submitted to the Missouri Highway and Transportation Department, 30 pp.

Grondin, G.Y., Wang, E., and Elwi, A.E. (2002), "Interaction Buckling of Stiffened Steel Plates," Proc. Structural Stability Research Council Annual Technical Session, Structural Stability Research Council, Washington D.C.

Hague, L, "The Storebaelt Publications: East Bridge", Section 5.2, A/S Storebaeltforbindelsen, Københarn, 1998.

Haibach, E., Plasil, I., (1983) "Untersuchungen zur Betriebsfestigkeit von Stahlleichtfahrbahnen mit Trapezhohlsteifen im Eisenbahnbruckenbau." Der Stahlbau, 9, 269-274

Hartnagel, B.A. (1993), "Long-Term Deck Strain Measurements on an Orthotropic Steel Plate Bridge Deck," M. S. Thesis, University of Missouri-Columbia, May 1993, 64 pp.

Hartnagel, B.A., Gopalaratnam, V.S., and Baldwin Jr., J.W. (1991), "Challenges in Long-Term Deck Strain Measurements on Orthotropic Steel-Plate Decks," Proceedings of the 9th national Conference on Microcomputers in Civil Engineering, University of Central Florida, October 1991, pp. 101-105.

Heins, C.P. and Firmage, D.A., Design of Modern Steel Highway Bridges, John Wiley & Sons, New York, NY, 1979.

Hemeau, G., Puch, C., Ajour, A-M. (1981), **Comportment a la fatigue en flexion sous moment negative, Bulletin de Liaison des Laboratoires des Ponts et Chaussees**, Paris (in French).

Higgins, C. (2003). "LRFD Orthotropic Plate Model for Live Load Moment in Filled Grid Decks." J. Bridge Engineering, 8(1). ASCE.

Higgins, C. (2004). "Orthotropic Plate Model for Estimating Deflections in Filled Grid Decks." J. Bridge Engineering 9(6). ASCE.

Hindi, W.A. (1991). "Behavior and Design of Stiffened Compression Flanges of Steel Box Girder Bridges," Ph.D. Thesis, University of Surrey, January.

Hoopah, W. (2004). "Orthotropic Decks for Small and Medium Span Bridges in France – Evolution and Recent Trends." Proc. of 2004 Orthotropic Bridge Conference, ASCE, Sacramento, CA, Aug. 25-27.

Horne, M.R., and Narayanan, R. (1977), "Design of Axially Loaded Stiffened Plates," ASCE J. Struct. Div., Vol. 103, ST11, pp. 2243-2257.

Huang, C., Mangus, A. (2008). "Redecking Existing Bridges with Orthotropic Steel Deck Panels." Proc. of 2008 Orthotropic Bridge Conference, ASCE, Sacramento, CA, Aug. 25-27.

Huang, C., Mangus, A., Murphy, J., Socha, M.J., (2008). "Twenty Four Icons of Orthotropic Steel Deck Bridge Engineering," ICONS Project – An International Discussion. Professional Engineers in California Government.

Hulsey, J.L., Yang, L., and Raad, L. (1999), "Wearing Surfaces for Orthotropic Steel Bridge Decks," Transportation Research Record 1654, Paper 99-1049, pp.141-150.

International Institute of Welding, (2007). RECOMMENDATIONS FOR FATIGUE DESIGN OF WELDED JOINTS AND COMPONENTS IIW document XIII-2151-07 / XV-1254-07 May.

Jetteur, P. et al. (1984). "Interaction of Shear Lag with Plate Buckling in Longitudinally Stiffened Compression Flanges," Acta Technica CSAV, (3), 376.

Jong, F.B.P. de, Kolstein, M.H. (2004). "Strengthening a Bridge Deck with High Performance Concrete." Proc. of 2004 Orthotropic Bridge Conference, ASCE, Sacramento, CA, Aug. 25-27.

Jong, F.B.P. de, Kolstein, M.H., Bijlaard, F.S.K. (2004). "Research Project TU Delft; Behaviour Conventional Bridge Decks and Development of Renovation Techniques." Proc. of 2004 Orthotropic Bridge Conference, ASCE, Sacramento, CA, Aug. 25-27.

JRA 2002, **Fatigue Design by Structural Details**, Japan Road Association, Tokyo, Japan

Kelly, B., Dexter, R.J. (1997) "Fatigue Performance of Repair Welds", ATLSS Report No. 97-09, ATLSS Engineering Research Center, Lehigh University, Bethlehem, PA

Kodama, T, Kondo, A., Goto, K and Iwashita, Y (2008), "A Study of SFRC Pavement on the Steel Deck Bridge", Proceedings 2008 International Orthotropic bridge Conference, Sacramento, CA, 10 pp.

Kolstein, H., and Wardenier, J. (1997), "Stress Reduction Due to Surfacing on Orthotropic Steel Decks," IABSE Workshop Proceedings, Lausanne, 1997.

Kolstein, M.H. (2007). "Fatigue Classification of Welded Joints in Orthotropic Steel Bridge Decks," Ph.D. Dissertation. Delft University of Technology. The Netherlands. ISBN 978-90-9021933-2.

Kolstein, MH, De Ridder, R (2008) "Enhancing the fatigue strength of trough to deck plate, joint in bridge decks using ultrasonic impact treatment", Proceedings, 2nd International Orthotropic Bridge Conference, pp. 586-598, August 25-30,2008, Sacramento, CA

Kolstein, H. and Sliederecht (2008), "Reduction of traffic induced stresses using reinforced high strength concrete, Proceedings 2008 International Orthotropic bridge Conference, Sacramento, CA, 11 pp.

Korniyiv, M. (2004). "Orthotropic Deck Bridges in Ukraine." Proc. of 2004 Orthotropic Bridge Conference, ASCE, Sacramento, CA, Aug. 25-27.

Kulicki, J.M. and Mertz, D. (2006). "Evolution of Vehicular Live Load Models During the Interstate Design Era and Beyond, 50 Years of Interstate Structures: Past, Present, and Future" Transportation Research Circular, Number E-C104, Transportation Research Board of the National Academies, Washington, DC, September.

Kurrer, K.E. (2011) "On the history of the orthotropic bridge deck" 12th International Conference on Metal Structures, ICMS, Wroclaw, Poland

Lamas, A.R.G. and Dowling, P.J. (1980). "Effect of Shear Lag on the Inelastic Buckling Behavior of Thin-Walled Structures," Thin-Walled Structures, J. Rhodes & A.C. Walker (eds.), Granada, London, p. 100.

Leshko, B.J. (2008), **Access Methods for Bridge Inspections**, STRUCTURE magazine, Copper Creek Publishing, October 2008.

Mangus, Alfred R., and Sun, Shawn (2000). "Orthotropic Deck Bridges," **Bridge Engineering Handbook**, Chapter 14, edited by Chen, Wai-Fah and Duan, Lian. CRC Press LLC, Boca Raton, Florida

Mangus, A. (2001). "Innovative Movable Bridges with Welded Orthotropic Steel Decks." Proc. of the International Bridge Conference, Pittsburgh PA

Mangus, A. and Mistry, V. (2006). "Get In, Get Out, Stay Out," Public Roads Magazine Vol. 70 No. 3.

Moffatt, K.R. and Dowling, P.J. (1975). "Shear Lag in Steel Box Girder Bridges," The Structural Engineer, 53, October, 439-447.

Moffatt, K.R. and Dowling, P.J. (1976). "Shear Lag in Steel Box Girder Bridges," The Structural Engineer, Discussion, 54, August, 285-297.

Moses, F., Schilling, C.G., Raju, K.S. (1987). "Fatigue Evaluation Procedures for Steel Bridges," NCHRP Report 299. Transportation Research Board. Washington D.C.

Nowak, A.S. (1999). "Calibration of LRFD Bridge Design Code." NCHRP Report No. 368, Transportation Research Board, Washington, D.C.

NCHRP (2003). **Mechanistic Empirical Pavement Design Guide (MEPDG).** Publication NCHRP 0.910. Developed for projects NCHRP 1-37a and 1-40d, Applied research Associates, Arizona State University.

Pelikan, W. and Esslinger, M. (1957). "Die Stahlfahrbahn-Berechnung und Konstruktion," M.A.N. Forschungsheft, No. 7.

Pelphrey, J. and Higgins, C. (2006). "Calibration of LRFR Live Load Factors for Oregon State-Owned Bridge Using Weigh in Motion Data." Publication Oregon Department of Transportation and Federal Highway Administration. Department of Civil Engineering, Oregon State University.

Pontis Bridge Inspection Manual (2007), Michigan Department of Transportation.

Post Tensioning Institute, (2000), **PTI Recommendations for Stay Cable Design, Testing and Installation**

Rigdon, R.L. (1990), "Wearing Surface for Orthotropic Steel Plate Bridges," M. S. Thesis, University of Missouri-Columbia, August 1990

Rigdon, R.L., Gopalaratnam, V.S., and Baldwin Jr., J.W. (1991), "Automated Temperature Varying Flexural Fatigue Tests for Steel-Polymer Concrete Composite Specimens," Proceedings of the 9th national Conference on Microcomputers in Civil Engineering, University of Central Florida, October 1991

Ryan, T.W., Hartle, R.A., Mann, J.E., and Danovich, L.J., (2006), **Bridge Inspector's Reference Manual**, Publication No. FHWA NHI 03-001, U.S. Government Printing Office, Washington, D.C.

Sadlacek, Gerhard (1992). "Orthotropic Plate Deck Bridges," **Constructional Steel Design – An International Guide**, Chapter 2.10, ed. by Dowling P., Harding J., and Bjorhovde R. Elsevier Science Publishers LTD. Essex, England.

Seim, C., and Ingham, T. (2004). "Influence of Wearing Surfacing on Performance of Orthotropic Steel Plate Decks," Transportation Research Record No. 1892, pp. 98-106.

Sim, H. and Uang, C. (2007). "Effects of Fabrication Procedures and Weld Melt-Through on Fatigue Resistance of Orthotropic Steel Deck Welds." Report No. SSRP-07/13. University of California, San Diego. August, 2007.

SSPC (2007), "Surface Preparation Specification No. 10.: Near-White Blast Cleaning," Steel Structures Painting Council

SSRC (1998) Structural Stability Research Council, Guide to Stability Design Criteria for Metal Structures 5th Edition. Galambos, T (ed). John Wiley and Sons Inc., New York, NY. ISBN 0-471-12742-6.

Sorensen, O. (2004). "Box Girders, Design, Fabrication, Operation, and Maintenance." Proc. of 2004 Orthotropic Bridge Conference, ASCE, Sacramento, CA, Aug. 25-27.

Touran, A., and Okereke, A. (1991), "Performance of Orthotropic Bridge Decks", ASCE Journal of Performance of Constructed Facilities, 5(2), May 1991
Troitsky, M.S. (1977), Stiffened Plates, Bending, Stability, and Vibrations. Elsevier, New York.
Troitsky, M.S. (1987) **Orthotropic Bridges Theory and Design** 2nd Ed. The James F. Lincoln Arc Welding Foundation. Cleveland OH.
Tsakopoulos, P.A., Fisher, J.W. (1999). Williamsburg Bridge Replacement Orthotropic Deck As-Built Full-Scale Fatigue Test, Final Report on Phase 2B/, ATLSS Report No. 99-02, Department of Civil and Environmental Engineering, Lehigh University, Bethlehem PA, May.
Tsakopoulos, P.A., Fisher, J.W. (2002). Fatigue Resistance Investigation for the Orthotropic Deck on the Bronx Whitestone Bridge – Final Report, ATLSS Report No. 02-05, Department of Civil and Environmental Engineering, Lehigh University, Bethlehem PA, September.
Walter, R., Olesen, J.F., Stang, H., and Vejrum, T. (2007), "Analysis of an Orthotropic Deck Stiffened with a Cement-Based Overlay," ASCE Journal of Bridge Engineering, 12 (3), May-June 2007
Weidlinger Associates, Inc. (2002) "Bronx-Whitestone Bridge Deck Replacement Project," Personal correspondence,
Winter, G. (1947), "Strength of Thin Steel Compression Flanges," Transactions of the ASCE, Vol. 112.
Wolchuk, R. and Mayrbaurl, R.M. (1980). "Proposed Design Specifications for Steel Box Girder Bridges," Final Report, FHWA-TS-80-205, Washington, DC, 288 pp.
Wolchuk, Roman (1999). **Steel Orthotropic Decks- Developments in the 1990s** National Academy Press, Washington, D.C.
Wolchuk, R. (2002), "Structural Behavior of Surfacings on Steel Orthotropic Decks and Considerations for Practical Design," Structural Engineering International, IABSE, 12 (2), May 2002,
Wolchuk, R. (2004). "Orthotropic Decks with Long Rib Spans." Proc. of 2004 Orthotropic Bridge Conference, ASCE, Sacramento, CA, Aug. 25-27.
Wolchuk, R., (2006), "Prefabricating Standard Orthotropic Steel Decks," Modern Steel Construction, December 2006
Wright, W. (2011) **Orthotropic Deck Discussion Group Meeting**, Washington D.C. Jan. 24
Xiao (2008) Effect of Fabrication Procedures and Weld Melt-Through on Fatigue Resistance of Orthotropic Steel Deck Welds, Final Report No. CA08-0607, Department of Structural Engineering University of California, San Diego, CA
Ya, S., Yamada, K., Ishikawa, T. (2011) "Fatigue Evaluation of Rib-to-Deck Welded Joints of Orthotropic Steel Bridge Deck" Journal of Bridge Engineering, ASCE, Volume 16, No. 4, pp.492-499
Yarnold, M.T., Wilson, J.L, Jen, W.C., Yen, B.T. (2007), "Local Buckling Analysis of Trapezoidal Rib Orthotropic Bridge Deck Systems," Bridge Structures: Assessment, Design, and Construction, Volume 3, No. 2. Taylor and Francis.

APPENDIX A

STANDARD ORTHOTROPIC PANEL DETAILS AND SECTION PROPERTIES

The standard shape details in this appendix originated in the out-of-print design aid titled **OrthotropicBridge Decks Using Bethlehem Standard Ribs**published by the Bethlehem Steel Corporation, Bethlehem, PA (unknown date) and have been converted to metric with updated sketches.

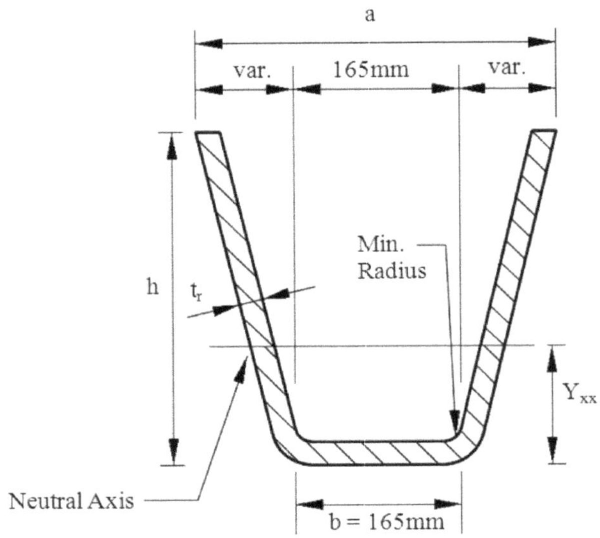

Standard Rib Designation	Depth of Rib, d (mm)	Width at Top, a (mm)	Rib Thickness, t_r (mm)	Weight per meter (kN)	I_{xx} (mm^4)	Y_{xx} (mm)
85	196	282	8	342	1.67×10^7	76
86			9	408	1.97×10^7	76
87			11	473	2.26×10^7	77
95	221	297	8	374	2.30×10^7	87
96			9	447	2.72×10^7	88
97			11	519	3.13×10^7	88
105	245	312	8	407	3.07×10^7	99
106			9	486	3.63×10^7	99
107			11	564	4.15×10^7	100
115	270	328	8	439	3.98×10^7	111
116			9	525	4.72×10^7	111
117			11	609	5.44×10^7	112

Figure A-1 Typical Rib Dimensions for Standardized Trapezoidal Shapes

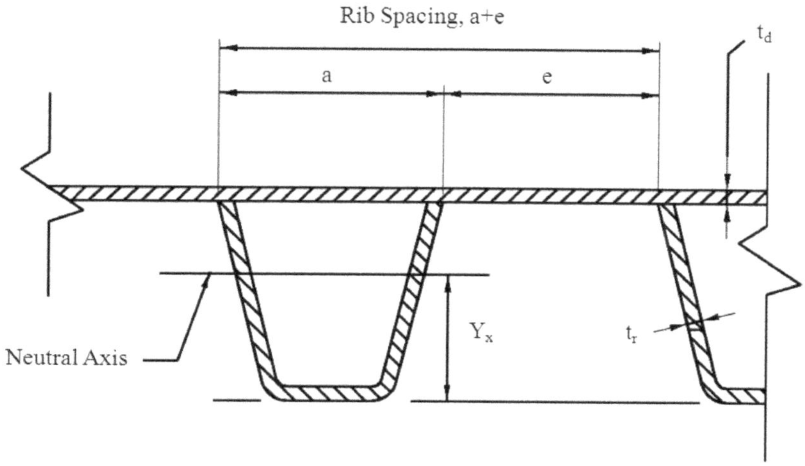

15mm Thickness		Standard Rib Section Designation											
a+e		85	86	87	95	96	97	105	106	107	115	116	117
22	Wt. (kN/m)	559	594	630	576	616	655	594	638	664	611	658	705
	I_X (mm^4)	6.31E+07	7.06E+07	7.85E+07	8.29E+07	9.37E+07	1.03E+08	1.06E+08	1.20E+08	1.32E+08	1.29E+08	1.50E+08	1.65E+08
	Y_x (mm)	161	159	155	181	176	171	199	193	187	216	209	203
24	Wt. (kN/m)	543	576	609	559	595	632	575	614	654	591	635	677
	I_X (mm^4)	6.41E+07	7.28E+07	8.07E+07	8.43E+07	9.58E+07	1.06E+08	1.08E+08	1.23E+08	1.35E+08	1.35E+08	1.53E+08	1.70E+08
	Y_x (mm)	166	161	157	184	179	174	202	196	190	219	210	208
26	Wt. (kN/m)	530	560	591	544	576	611	560	597	632	575	614	654
	I_X (mm^4)	6.52E+07	7.42E+07	8.21E+07	8.61E+07	9.76E+07	1.08E+08	1.10E+08	1.25E+08	1.39E+08	1.38E+08	1.57E+08	1.74E+08
	Y_x (mm)	168	164	160	186	181	177	204	198	193	222	216	210
28	Wt. (kN/m)	518	547	575	533	563	594	546	581	614	560	597	633
	I_X (mm^4)	6.63E+07	7.53E+07	8.36E+07	8.72E+07	9.94E+07	1.10E+08	1.12E+08	1.28E+08	1.41E+08	1.41E+08	1.60E+08	1.77E+08
	Y_x (mm)	170	166	162	188	184	179	207	201	196	225	218	213
30	Wt. (kN/m)	509	536	533	522	550	579	534	566	598	547	582	616
	I_X (mm^4)	6.70E+07	7.67E+07	8.50E+07	8.86E+07	1.01E+08	1.12E+08	1.13E+08	1.29E+08	1.44E+08	1.43E+08	1.62E+08	1.81E+08
	Y_x (mm)	172	143	164	190	185	181	209	203	198	227	221	216

17mm Thickness		Standard Rib Section Designation											
a+e		85	86	87	95	96	97	105	106	107	115	116	117
22	Wt. (kN/m)	595	778	667	613	652	692	630	674	717	649	696	741
	I_X (mm^4)	6.49E+07	7.35E+07	8.14E+07	8.54E+07	9.30E+07	1.07E+08	1.09E+08	1.24E+08	1.37E+08	1.37E+08	1.55E+08	1.71E+08
	Y_x (mm)	167	162	158	185	179	175	202	196	191	220	213	208
24	Wt. (kN/m)	581	613	647	597	633	668	613	652	692	629	657	714
	I_X (mm^4)	6.59E+07	7.49E+07	8.32E+07	8.68E+07	9.87E+07	1.10E+08	1.11E+08	1.26E+08	1.40E+08	1.39E+08	1.59E+08	1.75E+08
	Y_x (mm)	169	165	160	187	182	178	205	199	194	223	217	211
26	Wt. (kN/m)	568	597	628	582	616	648	597	633	670	611	651	690
	I_X (mm^4)	6.70E+07	7.64E+07	8.50E+07	8.83E+07	1.01E+08	1.12E+08	1.13E+08	1.29E+08	1.43E+08	1.42E+08	1.60E+08	1.79E+08
	Y_x (mm)	171	167	160	190	185	180	208	202	197	226	220	214
28	Wt. (kN/m)	556	584	611	569	601	632	584	617	651	597	635	670
	I_X (mm^4)	6.81E+07	7.78E+07	8.65E+07	8.97E+07	1.02E+08	1.14E+08	1.15E+08	1.31E+08	1.46E+08	1.44E+08	1.65E+08	1.83E+08
	Y_x (mm)	173	169	165	192	184	183	210	205	200	229	222	219
30	Wt. (kN/m)	546	572	598	559	588	617	572	604	635	585	619	654
	I_X (mm^4)	6.88E+07	7.89E+07	8.79E+07	9.08E+07	1.04E+08	1.16E+08	1.17E+08	1.33E+08	1.48E+08	1.47E+08	1.67E+08	1.86E+08
	Y_x (mm)	174	170	142	194	189	185	212	207	202	231	225	220

18mm Thickness		Standard Rib Section Designation											
a+e		85	86	87	95	96	97	105	106	107	115	116	117
22	Wt. (kN/m)	633	668	705	651	690	730	668	712	755	686	733	779
	I_X (mm^4)	6.67E+07	7.57E+07	8.39E+07	8.76E+07	9.94E+07	1.10E+08	1.12E+08	1.28E+08	1.41E+08	1.41E+08	1.59E+08	1.44E+08
	Y_x (mm)	170	165	161	188	183	178	206	200	195	224	217	211
24	Wt. (kN/m)	617	651	668	633	670	706	649	689	728	665	709	752
	I_X (mm^4)	6.77E+07	7.71E+07	8.58E+07	8.94E+07	1.02E+08	1.13E+08	1.14E+08	1.30E+08	1.44E+08	1.43E+08	1.63E+08	1.81E+08
	Y_x (mm)	172	167	164	190	185	181	209	203	198	227	221	215
26	Wt. (kN/m)	604	635	665	619	652	686	635	671	706	649	689	728
	I_X (mm^4)	6.88E+07	7.85E+07	8.76E+07	9.08E+07	1.03E+08	1.15E+08	1.16E+08	1.33E+08	1.47E+08	1.46E+08	1.66E+08	1.84E+08
	Y_x (mm)	174	170	166	193	188	184	211	206	201	230	223	218
28	Wt. (kN/m)	591	622	649	607	638	668	620	655	689	635	671	708
	I_X (mm^4)	6.99E+07	7.96E+07	8.90E+07	9.19E+07	1.05E+08	1.17E+08	1.18E+08	1.35E+08	1.50E+08	1.48E+08	1.69E+08	1.88E+08
	Y_x (mm)	175	172	168	195	190	186	214	208	203	232	226	221
30	Wt. (kN/m)	584	610	636	597	625	649	609	641	673	620	657	690
	I_X (mm^4)	7.06E+07	8.07E+07	9.01E+07	9.30E+07	1.06E+08	1.19E+08	1.20E+08	1.37E+08	1.52E+08	1.50E+08	1.71E+08	1.91E+08
	Y_x (mm)	177	173	170	196	192	188	216	210	206	234	229	223

Figure A-2 Typical Rib Panel Dimensions for Standardized Trapezoidal Shapes

APPENDIX B

ADDITIONAL RECOMMENDATIONS FOR FINITE ELEMENT ANALYSIS

The finite element analysis (FEA) method has contributed to revolutionary leaps in design and research in the field of Mechanics to include work on materials in any state or medium (plasma, fluid, solid, heat transfer, two-phase, and many others). The number of software programs applicable to structural mechanics is very large and is a testimony to the market demand for solving problems in broad or specialized fields. Software companies continue to make improvements to render their codes more user-friendly and to enable engineers to solve more complex problems.

For FEA work on orthotropic steel decks (OSD), it is sufficient that a software program be capable of analyses in the elastic domain without considerations to material or geometric nonlinearity. First order displacements are sufficient.

General Characteristics of Structural FEA Models

All 3D structural analysis models (including those used in solving structural problems in a continuum such as OSDs, contain either shell or brick elements) at the engineer's discretion may decide and execute the following general procedures:

- The models organize stiffness or flexibility matrices in which displacements or loads are given as inputs, respectively. The sizes of the matrices depend on the number of finite elements and the degrees of freedom at element nodes.
- All stiffness matrices include material properties (E, G, and ν) and geometric properties (thickness and element length) of the finite elements that are used to make up for the influence coefficients that determine the displacement at any given node. Each node, unless otherwise specified, has six degrees of freedom (three translations and three rotations).
- All element deformations are calculated assuming forces applied at the nodes. Equilibrium is tested and enforced at the nodes. Element stresses are calculated from element deformations.
- Global displacements are integrated along the elements from fixed boundaries.
- Most modern computer models provide stress contours.

Additional characteristics of finite elements are:

- Greater accuracy is obtained with greater mesh refinement.
- Hybrid shell/brick element models are possible with the judicious placement of rigid links at the boundaries where they come together. This will be illustrated in later sections.

General Approach to Modeling of Orthotropic Deck Structures

Current software enables even the smallest engineering firms to produce 3-D models for large bridge projects. There is no reason why 3-D models cannot be made for even the smallest bridges. Three levels of refinement are generally used: gross, intermediate (local), and stress concentration. Different programs offer greater or lesser ease in application of two basic options to address the changes from large-scale meshing to stress concentration meshing: a) substructuring, and b) continuous meshing. In the first, separate models are made and the boundary conditions from the large are forced onto the smaller model, down to the smallest, despite the fact that certain nodes do not coincide. In the second, transition meshing is provided to the more refined meshing. The latter method is more time consuming, but it leaves no discrepancy relative to boundary conditions. Substructuring, however, may be satisfactory if the elements under investigation are far from such boundaries, to the extent that there is nodal mismatching, which should be minimized.

Generally, all OSD bridge models can be designed with shell meshing. Some stress concentrations may be better analyzed with brick elements. Shell elements give thickness properties for calculation of stress on an element that lies in a plane. They assume that the rotations and displacements at each corner of the plate they represent are the same. This is illustrated in Figure B-1.

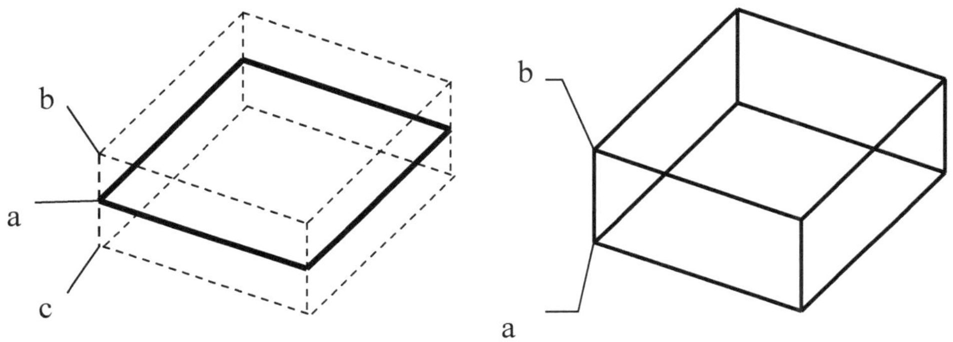

Shell Element
Nodes are at corners of middle plane (continuous lines). All extensions and rotations are assumed to take place at node "a." Rotations at points "b" and "c" are hypothetical and are assumed to be the same as at node "a."

Brick Element
Brick elements have eight nodes. Nodes are at corners of the brick. All nodes have independent displacements, including rotations at nodes such as "a" and "b."

Figure B-1 A Comparison of using Shell Elements verses Brick Elements for Finite Element Modeling

Finite elements that have bulk (three-dimensional) are available in most software programs. The basic unit of these elements is the tetrahedron, with four nodes and four faces, but brick and prismatic elements are also available. Also, constraint equations (also called master-slave relations), are sometimes applied when nodes of adjacent elements do not coincide.

- Gross Meshes

  To reduce the amount of computer time, the finite elements away from the stress concentrations or panels on which greater focus is placed, are of large sizes. For example, if a bay spans 20 feet, elements of 1/5 this size have been used (i.e., 48 inches). This is deemed sufficient to approximate global stiffness. Naturally, mesh size must also be consistent with floor beam web and roadway width. It is left to the judgment of the engineer to select the appropriate mesh dimensions.

- Local Meshes

  These are typically applied to two or three bays that contain the stress concentrations stresses that are being investigated. Meshes are controlled by dimensions such as rib widths or depths, or FB depth. They are generally on the order of 6 to 12 inches and may vary between deck plate and FBs.

- Stress Concentrations

  The stress concentrations are included in a highly local model, which usually contains two ribs and adjacent FB. The ribs' mesh may be on the order of several feet long to meet with the next largest local mesh. In general, the meshes for the deck plate are on the order of the deck plate thickness if shell meshing is used, but there must be a gradual cascading of dimension sizes to the level of concentrations to exclude aspect ratio effects in the intermediate finite elements. Local elements around the concentrations and curved surfaces at cut-outs require high mesh refinement, in which brick elements may not be avoided. Good analysts always try to keep as close to a 1:1 in aspect ratio, but this is not an absolute necessity nor feasible everywhere in the model. Judgment must be exercised such that, at location of worst expected effects, aspect ratios are reasonable.

Suggested Meshes for Stress Concentrations

Several areas of welded joints in OSDs require careful analysis to perform a fatigue life evaluation. These areas are:

- Rib- to-deck plate weld (RD) (away from the diaphragm).
- Rib-to-deck plate at the floorbeam joint (RDF) (at the FB).
- Rib-to-FB (RF) cut-out transition where rib and FB meet.
- Sharpest curvature of the cut-out.
- Top and bottom of the cutout and bulkhead terminations.
- Joint of rib to diaphragm at bottom of round belly rib welded all-around, without a cut-out.

These areas are discussed below, in relation to possible effects and the best ways of evaluating them by FEA.

Rib-to-Deck Plate Weld (RD)

Modeling of this connection is relatively straight forward and follows the guidance set forth in Chapters 4 and 5.

Rib-to-Deck at the Floorbeam Joint (RDF)

In Figure B-2, shell meshing is shown for both partial penetration and for complete penetration welding of the RDF. This joint consists of a system of two intersecting welds and three intersecting planes.

The mesh shows a node at the root or toe, where the fatigue stress range must be checked (red line). The mesh can, and should, be made to model the welds joining the diaphragm as shown in Figure B-3.

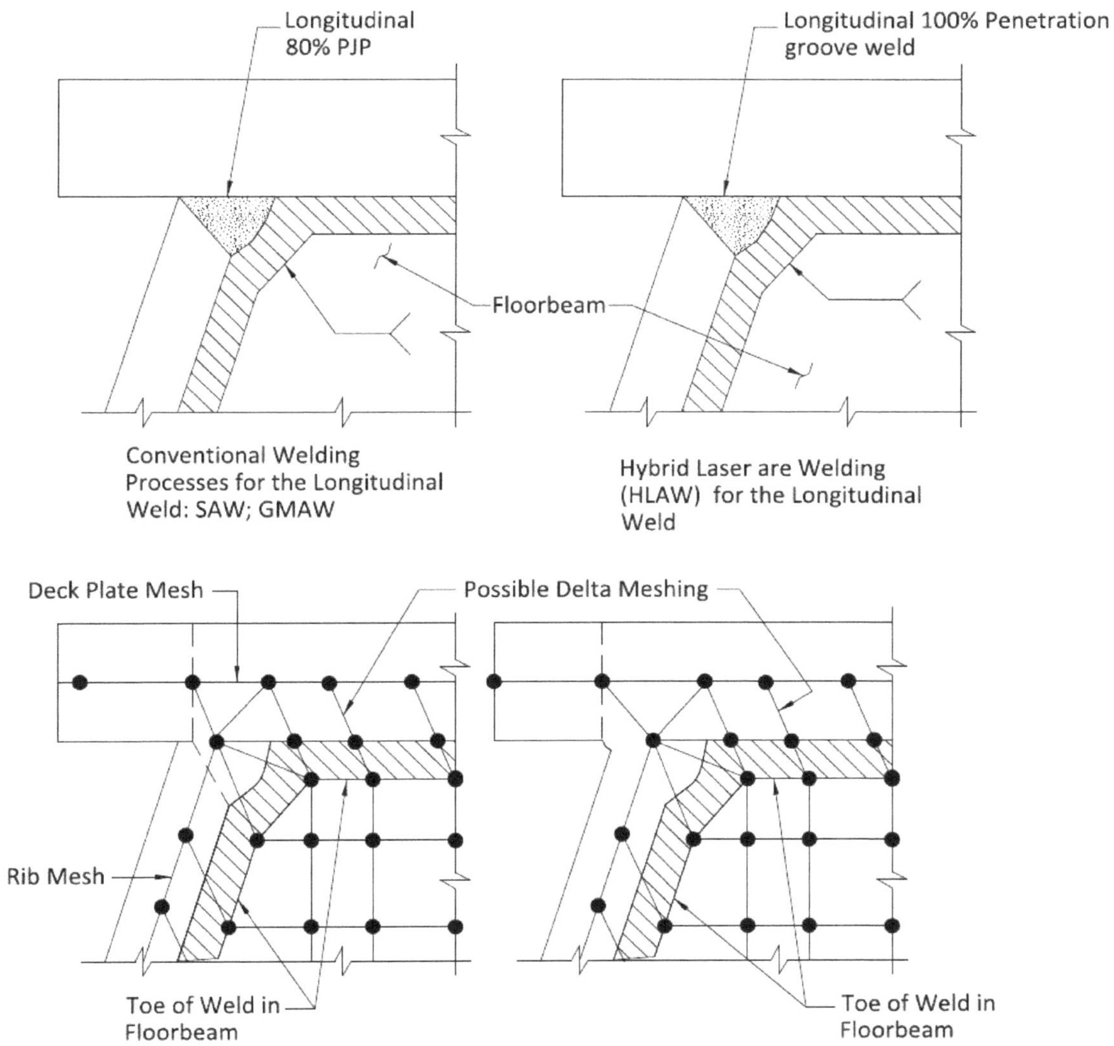

Figure B-2 Modeling Techniques for the rib-to-Deck (RD) Weld showing a comparison of Meshing Techniques

Finite element analyses have shown that a relatively thick FB better distributes the stress concentration where it takes the greatest proportion of force that engages the deck plate (at the leading edge of the tooth). For this distribution to occur in the physical joint, the deck plate should be fully joined to the FB at that point, as illustrated in Figure B-4. This figure shows, in plan, shell nodes aimed at characterizing to some degree of accuracy the joint and associated stress in conformance with a fully welded condition, and at the root of the 80 percent penetration weld where the crack is presumed to start. Brick meshing is thought to be too complicated for such detail and unnecessary. The meshing shown allows for stress calculation by extrapolation techniques.

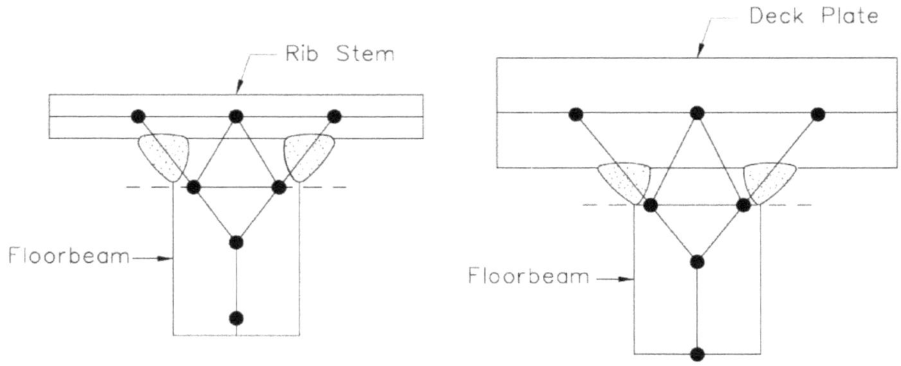

Figure B-3 Modeling Techniques for the Floorbeam to Deck Connection Comparing the Difference between Thick and Thin Deck Plates

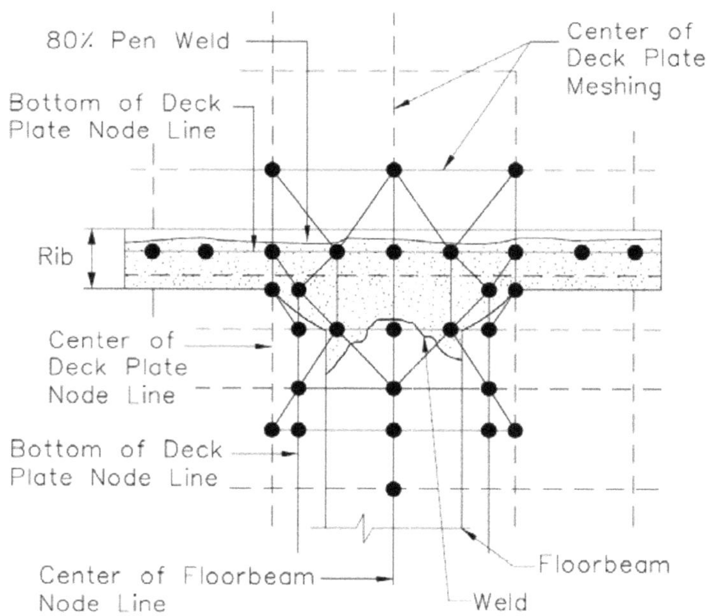

Figure B-4 Shell Nodes Aimed at Characterizing to some Degree of Accuracy the Floorbeam to Deck Connection Superimposed on the Connection Detail (Looking Down "Through" the Deck Plate, Deck Meshing Shown with Dashed Lines)

Rib-to-FB Cut-Out Transition

Figure B-5 shows a hybrid shell-brick meshing of the transition, where the transition proper is idealized with brick elements. The shell elements take over at areas away from the stress concentrations. Although shell elements may be satisfactory, they may be too conservative such that, if used, the design might be invalidated and a more costly or heavier deck plate alternative adopted. The figure illustrates a method for using hybrid meshing only close to stress concentrations where it might be needed. Alternately, it can be readily seen how the entire detail can be accomplished in shell elements, if the engineer chooses.

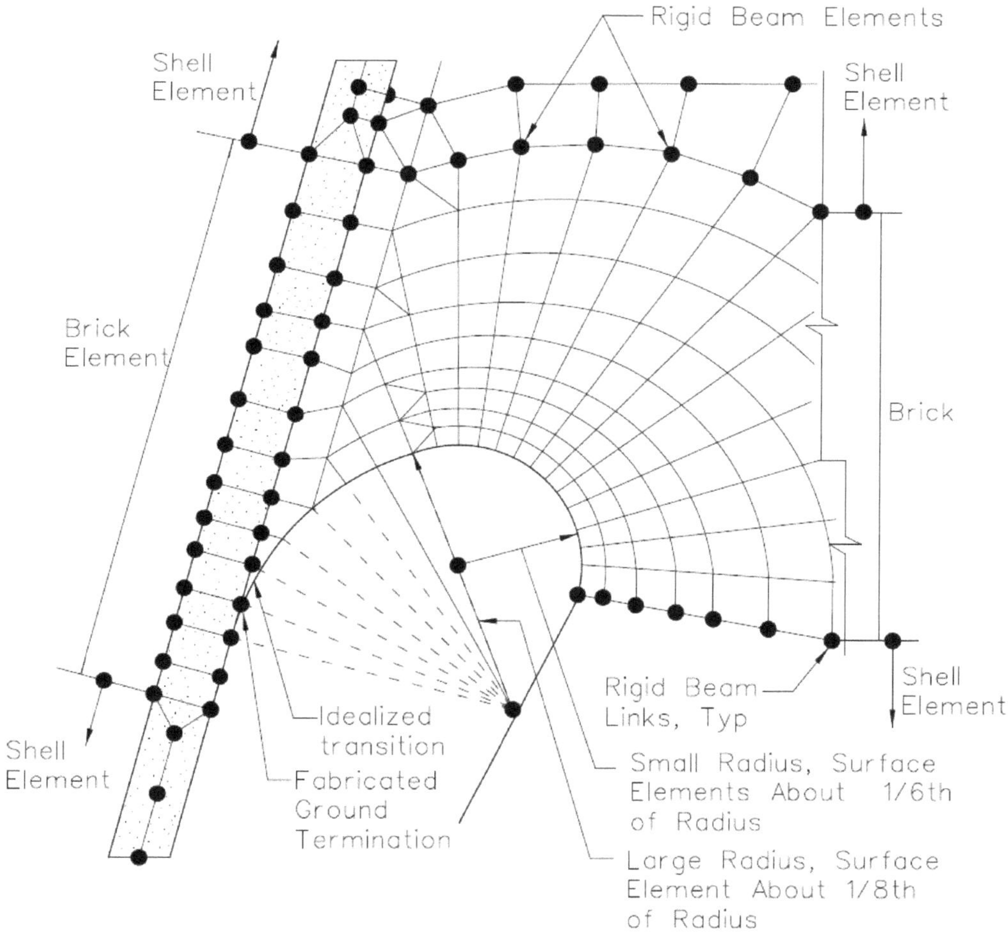

Figure B-5 Cut-out Radius Detail showing the Detailed Meshing Transition along the Curved Radius of the Cut-out using both Shell and Brick Elements

The model contains three areas of evaluation: a) at the curve-to-tangent transition, where rib-to toe effects are largest; b) at the toe of the weld in the FB; and c) at the principal stresses in the sharpest area of the curvature.

Bulkhead Terminations

Two meshes are shown in Figure B-6, one with abrupt and one with gradual transition in the bottom termination. For the detail on the left, if round belly is presumed with an all-around weld detail, an abrupt transition may not be harmful but it will not likely be possible with a cut-out detail shown on the right, for which a gradual transition is presumed. In a round belly all-around welded detail, the rib must be sufficiently rigid to keep stresses low at the bottom of the rib. In that case, a full depth bulkhead without abrupt transitions may also be used.

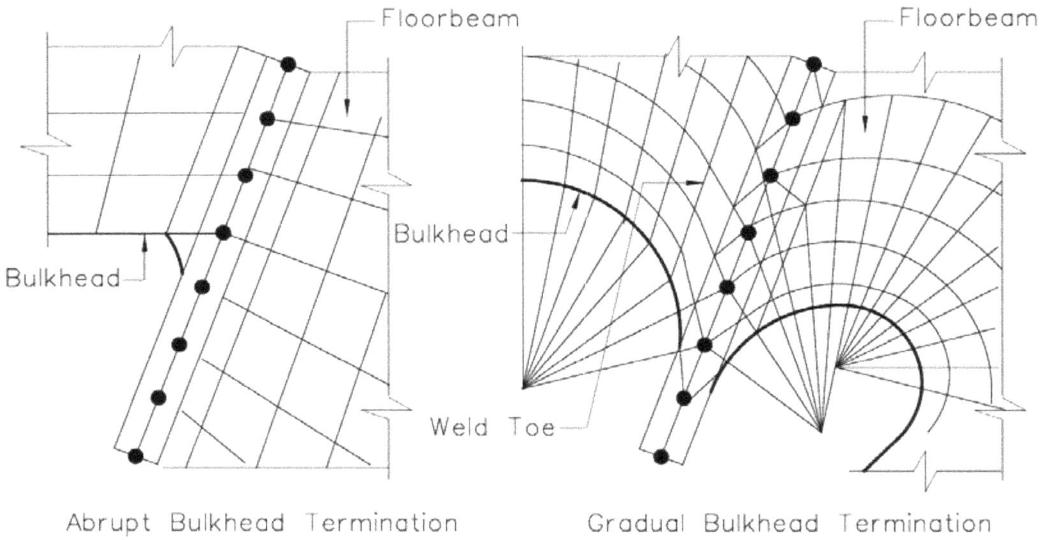

Figure B-6 Meshing Strategies for the Cut-out Radius Detail when a Bulkhead Detail is Required

Designers are reminded that, in a cantilever floor beam, the deck plate and web are in global tension. In a simply supported beam, partial penetration or even fillet welds may be acceptable, depending on stress level. Figure B-7 shows how a shell mesh can be produced to model a FB to bulkhead connection through the rib, assuming perfect alignment.

This simple model does not address possible eccentricity of FB relative to the bulkhead. This is considered satisfactory as the both the Design and Fabrication chapters (Chapters 5 and 7, respectively) indicate that if a bulkhead is used it must be in alignment with the floorbeam. Moreover, current knowledge of such a misaligned detail is incomplete and needs to be expanded.

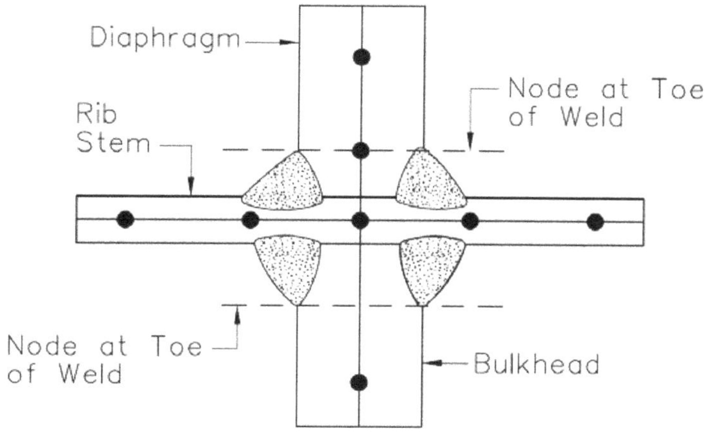

Figure B-7 Node Alignment through the Bulkhead Detail at the Rib

All-Around Weld with Round Belly Rib

Figure B-8 shows the effects of rib rotation and VQ/I on a round belly rib welded all-around. The VQ/I effects are shown for a cantilever floor beam. Several checks must be conducted for this detail:

- Longitudinal effects at the bottom of the rib at the weld toe in the rib.
- Combined effects in the FB at the bottom of the rib. These effects are non-proportional.
- Stress concentrations at the tangent-to-curve areas and in the weld toe in the FB.

In a simply supported beam, the global tension in the diaphragm would become compression. The concentration stresses at the tangent-to-curve would be reversed. The direction of the principal stresses, at these points, is at a sharp angle with the weld line, as graphically indicated.

Fatigue criteria for dealing with this issue are discussed in Chapter 5, as are general criteria for dealing with non-proportionality.

Figure B-9 shows suitable shell meshing of this area. The weld toe line can be used as a delta meshing point with divergent shell meshes going from the center of the diaphragm to opposite meshing nodes in the rib mesh surface.

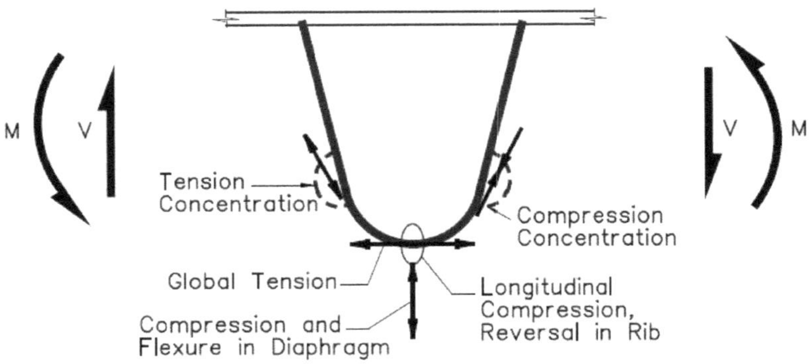

Figure B-8 Detail of the Rib-to-Floorbeam (RF) Connection showing the Effect of Rib Rotation and Floorbeam Shear on a Rounded Rib

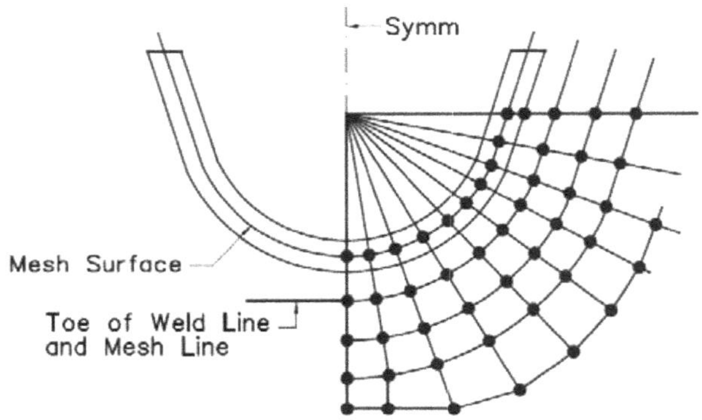

Figure B-9 Suitable Shell Meshing in vicinity of a Rounded Rib

APPENDIX C

LEVEL 1 STANDARD DETAILS

Development of a Standard Panel Design

This Appendix demonstrates the application of Level 1 design of an orthotropic steel deck (OSD) based on available experimental test data from a previous project. The basis of design is the prototype OSD for redecking of the Bronx-Whitestone Bridge, which was tested at Lehigh's ATLSS Research Center in 2002 (ref). This full-scale laboratory test simulated 4.1M cycles of 3 times the AASHTO HS15 (HS15 = 0.75*HS20) Fatigue Truck plus an additional 2M cycles of 4 times the HS15, producing effectively 239M cycles of the HS15 loading. Fatigue cracks were not found in any of the primary connections, which demonstrates outstanding fatigue resistance and verifies the design performance for the given conditions.

The prototype specimen included a conventional OSD with 343mm deep ribs spaced at 660mm and 390/530mm intermediate diaphragms supported by bridge floorbeams with span of 22.5m and spacing of 6020mm. Local response of the deck panel at wheel loads plus the interactions of the panel with the floorbeams were considered in the testing. Global response (from cable deflections) does not impact fatigue performance to large degree and was neglected.

For future redecking projects with floorbeam span and spacing within the limits of those utilized in the prototype test, this can be considered a "standard deck panel" that can be used without need for additional refined engineering analysis. The fabrication must be performed with the same level of quality that was achieved in the prototype specimen, which requires that tolerances and welding satisfy the applicable provisions of AASHTO and AWS. The standard deck details are shown in the following drawings. For implementation, owners should consider adoption of this standard design and incorporation into their standards.

It is noted that this prototype design was influenced by the limited available clearance betwe the top of the existing floorbeams and the finished deck surface, which demanded shall diaphragms, large cutouts, CJP welds, and internal rib stiffeners. This design may not b optimum when additional clearance is available. The engineer should evaluate the potential co savings of design improvements vs. the cost of fabrication and testing of a new pr[illegible]ty

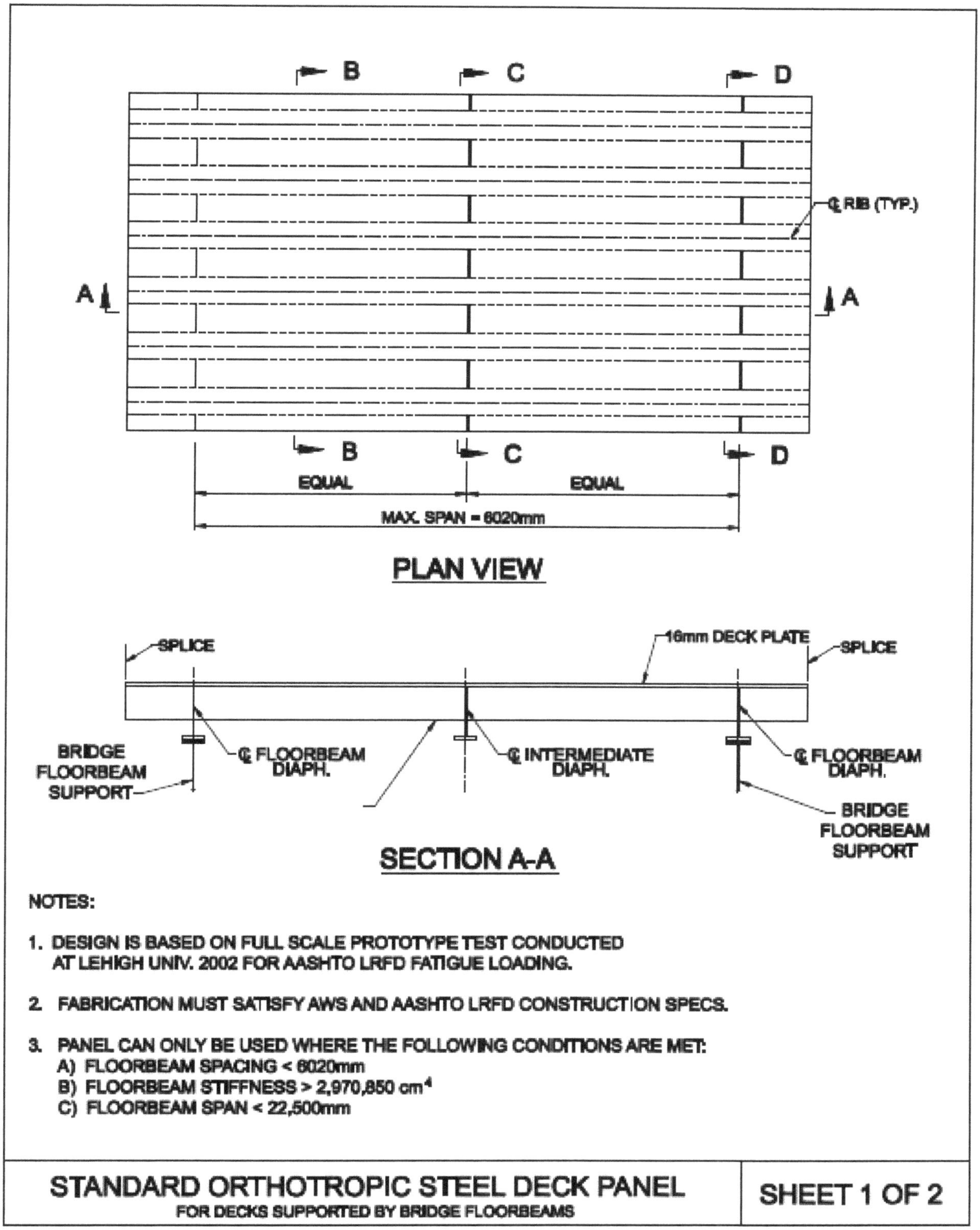

Figure C-1 Plan View and Secion of the prototype OSD for redecking of the Bronx-Whitestone Bridge

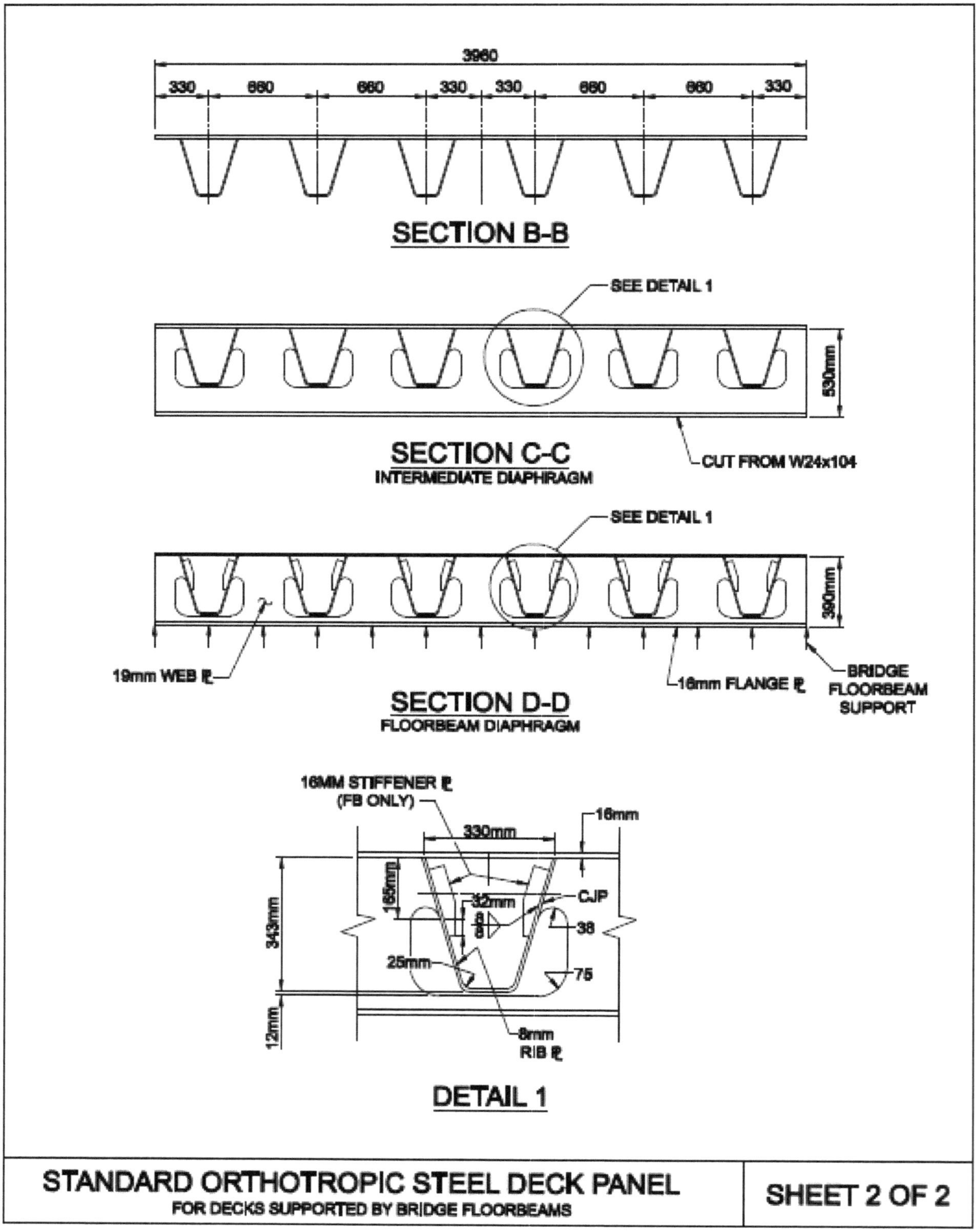

Figure C-2 Various Secions and Rib Detail of the prototype OSD for redecking of the Bronx-Whitestone Bridge

www.ingramcontent.com/pod-product-compliance
Ingram Content Group UK Ltd.
Pitfield, Milton Keynes, MK11 3LW, UK
UKHW051040091026
15624UKWH00011B/598

9 781782 66